Advances in Intelligent Systems and Computing

Volume 466

Series editor

Janusz Kacprzyk, Polish Academy of Sciences, Warsaw, Poland
e-mail: kacprzyk@ibspan.waw.pl

About this Series

The series "Advances in Intelligent Systems and Computing" contains publications on theory, applications, and design methods of Intelligent Systems and Intelligent Computing. Virtually all disciplines such as engineering, natural sciences, computer and information science, ICT, economics, business, e-commerce, environment, healthcare, life science are covered. The list of topics spans all the areas of modern intelligent systems and computing.

The publications within "Advances in Intelligent Systems and Computing" are primarily textbooks and proceedings of important conferences, symposia and congresses. They cover significant recent developments in the field, both of a foundational and applicable character. An important characteristic feature of the series is the short publication time and world-wide distribution. This permits a rapid and broad dissemination of research results.

Advisory Board

Chairman

Nikhil R. Pal, Indian Statistical Institute, Kolkata, India
e-mail: nikhil@isical.ac.in

Members

Rafael Bello, Universidad Central "Marta Abreu" de Las Villas, Santa Clara, Cuba
e-mail: rbellop@uclv.edu.cu

Emilio S. Corchado, University of Salamanca, Salamanca, Spain
e-mail: escorchado@usal.es

Hani Hagras, University of Essex, Colchester, UK
e-mail: hani@essex.ac.uk

László T. Kóczy, Széchenyi István University, Győr, Hungary
e-mail: koczy@sze.hu

Vladik Kreinovich, University of Texas at El Paso, El Paso, USA
e-mail: vladik@utep.edu

Chin-Teng Lin, National Chiao Tung University, Hsinchu, Taiwan
e-mail: ctlin@mail.nctu.edu.tw

Jie Lu, University of Technology, Sydney, Australia
e-mail: Jie.Lu@uts.edu.au

Patricia Melin, Tijuana Institute of Technology, Tijuana, Mexico
e-mail: epmelin@hafsamx.org

Nadia Nedjah, State University of Rio de Janeiro, Rio de Janeiro, Brazil
e-mail: nadia@eng.uerj.br

Ngoc Thanh Nguyen, Wroclaw University of Technology, Wroclaw, Poland
e-mail: Ngoc-Thanh.Nguyen@pwr.edu.pl

Jun Wang, The Chinese University of Hong Kong, Shatin, Hong Kong
e-mail: jwang@mae.cuhk.edu.hk

More information about this series at http://www.springer.com/series/11156

Radek Silhavy · Roman Senkerik
Zuzana Kominkova Oplatkova
Petr Silhavy · Zdenka Prokopova
Editors

Automation Control Theory Perspectives in Intelligent Systems

Proceedings of the 5th Computer Science On-line Conference 2016 (CSOC2016), Vol 3

Editors
Radek Silhavy
Faculty of Applied Informatics
Tomas Bata University in Zlín
Zlín
Czech Republic

Petr Silhavy
Faculty of Applied Informatics
Tomas Bata University in Zlín
Zlín
Czech Republic

Roman Senkerik
Faculty of Applied Informatics
Tomas Bata University in Zlín
Zlín
Czech Republic

Zdenka Prokopova
Faculty of Applied Informatics
Tomas Bata University in Zlín
Zlín
Czech Republic

Zuzana Kominkova Oplatkova
Faculty of Applied Informatics
Tomas Bata University in Zlín
Zlín
Czech Republic

ISSN 2194-5357   ISSN 2194-5365   (electronic)
Advances in Intelligent Systems and Computing
ISBN 978-3-319-33387-8   ISBN 978-3-319-33389-2   (eBook)
DOI 10.1007/978-3-319-33389-2

Library of Congress Control Number: 2016937381

Preface

This book constitutes the refereed proceedings of the Automation Control Theory Perspectives in Intelligent Systems Section and of the Intelligent Information Technology, System Monitoring and Proactive Management of Complex Objects Section of the 5th Computer Science On-line Conference 2016 (CSOC 2016), held in April 2016.

The volume Automation Control Theory Perspectives in Intelligent Systems brings 47 of the accepted papers. Each of them presents new approaches and methods to real-world problems and exploratory research that describes novel approaches in the field of cybernetics, automation control theory and proactive management of complex objects.

CSOC 2016 has received (all sections) 254 submissions, 136 of them were accepted for publication. More than 60 % of all accepted submissions were received from Europe, 20 % from Asia, 16 % from America and 4 % from Africa. Researchers from 32 countries participated in CSOC 2016.

CSOC 2016 intends to provide an international forum for the discussion of the latest high-quality research results in all areas related to computer science. The addressed topics are theoretical aspects and applications of computer science, artificial intelligence, cybernetics, automation control theory and software engineering.

Computer Science On-line Conference is held online and broad usage of modern communication technology improves the traditional concept of scientific conferences. It brings equal opportunity to participate to all researchers around the world.

The editors believe that readers will find the proceedings interesting and useful for their own research work.

March 2016
Radek Silhavy
Roman Senkerik
Zuzana Kominkova Oplatkova
Petr Silhavy
Zdenka Prokopova

Program Committee

Program Committee Chairs

Zdenka Prokopova, Ph.D., Associate Professor, Tomas Bata University in Zlín, Faculty of Applied Informatics, email: prokopova@fai.utb.cz

Zuzana Kominkova Oplatkova, Ph.D., Associate Professor, Tomas Bata University in Zlín, Faculty of Applied Informatics, email: kominkovaoplatkova@fai.utb.cz

Roman Senkerik, Ph.D., Associate Professor, Tomas Bata University in Zlín, Faculty of Applied Informatics, email: senkerik@fai.utb.cz

Petr Silhavy, Ph.D., Senior Lecturer, Tomas Bata University in Zlín, Faculty of Applied Informatics, email: psilhavy@fai.utb.cz

Radek Silhavy, Ph.D., Senior Lecturer, Tomas Bata University in Zlín, Faculty of Applied Informatics, email: rsilhavy@fai.utb.cz

Roman Prokop, Ph.D., Professor, Tomas Bata University in Zlín, Faculty of Applied Informatics, email: prokop@fai.utb.cz

Program Committee Chairs for Special Sections

Intelligent Information Technology, System Monitoring and Proactive Management of Complex Objects

Prof. Viacheslav Zelentsov, Doctor of Engineering Sciences, Chief Researcher of St. Petersburg Institute for Informatics and Automation of Russian Academy of Sciences (SPIIRAS)

Program Committee Members

Boguslaw Cyganek, Ph.D., D.Sc., Department of Computer Science, University of Science and Technology, Krakow, Poland

Krzysztof Okarma, Ph.D., D.Sc., Faculty of Electrical Engineering, West Pomeranian University of Technology, Szczecin, Poland

Monika Bakosova, Ph.D., Associate Professor, Institute of Information Engineering, Automation and Mathematics, Slovak University of Technology, Bratislava, Slovak Republic

Pavel Vaclavek, Ph.D., Associate Professor, Faculty of Electrical Engineering and Communication, Brno University of Technology, Brno, Czech Republic

Miroslaw Ochodek, Ph.D., Faculty of Computing, Poznań University of Technology, Poznań, Poland

Olga Brovkina, Ph.D., Global Change Research Centre Academy of Science of the Czech Republic, Brno, Czech Republic

Elarbi Badidi, Ph.D., College of Information Technology, United Arab Emirates University, Al Ain, United Arab Emirates

Luis Alberto Morales Rosales, Head of the Master Program in Computer Science, Superior Technological Institute of Misantla, Mexico

Mariana Lobato Baes, M.Sc., Research-Professor, Superior Technological of Libres, Mexico

Abdessattar Chaâri, Professor, Laboratory of Sciences and Techniques of Automatic Control and Computer engineering, University of Sfax, Tunisian Republic

Gopal Sakarkar, Shri. Ramdeobaba College of Engineering and Management, Republic of India

V.V. Krishna Maddinala, Assistant Professor, GD Rungta College of Engineering and Technology, Republic of India

Anand N. Khobragade, Scientist, Maharashtra Remote Sensing Applications Centre, Republic of India

Abdallah Handoura, Assistant Prof., Computer and Communication Laboratory, Telecom Bretagne, France

Technical Program Committee Members

Ivo Bukovsky
Miroslaw Ochodek
Bronislav Chramcov
Eric Afful Dazie

Michal Bliznak
Donald Davendra
Radim Farana
Zuzana Kominkova Oplatkova
Martin Kotyrba
Erik Kral
David Malanik
Michal Pluhacek
Zdenka Prokopova
Martin Sysel
Roman Senkerik
Petr Silhavy
Radek Silhavy
Jiri Vojtesek
Eva Volna
Janez Brest
Ales Zamuda
Roman Prokop
Boguslaw Cyganek
Krzysztof Okarma
Monika Bakosova
Pavel Vaclavek
Olga Brovkina
Elarbi Badidi

Organizing Committee Chair

Radek Silhavy, Ph.D., Tomas Bata University in Zlín, Faculty of Applied
Informatics, e-mail: rsilhavy@fai.utb.cz

Conference Organizer (Production)

OpenPublish.eu s.r.o.
Web: http://www.openpublish.eu
e-mail: csoc@openpublish.eu

Conference Website, Call for Papers

http://www.openpublish.eu

Contents

Part I
Automation Control Theory Perspectives in Intelligent Systems

A Novel Color Image Encryption Algorithm Using Chaotic Map and Improved RC4

Cong Jin and Zhengwu Tu

Abstract In this paper, color image encryption algorithm based on improved RC4 and chaotic maps is proposed. In proposed algorithm, the classic RC4 algorithm in cryptography is improved, and then applied to proposed encryption algorithm. Firstly, the original image is divided into some sub-blocks. Then, the improved RC4 algorithm is applied to the operation between the adjacent two sub-blocks, thereby changes the value of pixels. Finally, the image is scrambled by logistic map. Experimental results show that the original image has a big change and a flat histogram after encrypted, and that the proposed algorithm has an enough large key space and a very high sensitivity to the key.

Keywords Chaotic maps · Image encryption · RC4 algorithm · Logistic map

1 Introduction

A good image encryption algorithm should have a high security, good encryption effect and spend a little time. Especially in the transmission of network, the encryption speed is an important indicator in the real-time requirements of image encryption. The common problem in traditional image encryption is that the process of image encryption need a long time and has a poor real-time in image transmission. In order to improve the speed of encryption, we will divide the image into many pixel blocks and consider pixel block as the minimum operation unit. Recently, a variety of image encryption based on blocks is proposed [1]. These encryption algorithms have a fast speed of encryption. The traditional encryption algorithm in modern cryptography has a high security for text data. In encryption theory, digital image can be encrypted by encryption algorithm of modern cryptography. But digital image is a kind of special data, which is featured with huge data capacity, two-dimensional data, and high redundancy and so on. Those

C. Jin (✉) · Z. Tu
School of Computer, Central China Normal University, Wuhan 430079, Hubei, China
e-mail: jincong@mail.ccnu.edu.cn

© Springer International Publishing Switzerland 2016
R. Silhavy et al. (eds.), *Automation Control Theory Perspectives in Intelligent Systems*, Advances in Intelligent Systems and Computing 466,
DOI 10.1007/978-3-319-33389-2_1

encryption algorithms of modern cryptography, such as Data Encryption Standard
(DES), International Data Encryption Algorithm (IDEA) and Advanced Encryption
Standard (AES) [2], etc., are designed for text data which is one-dimension, and
doesn't combine with the feature of image data. So it is very difficult to satisfy the
image encryption. But we can modify the traditional cipher algorithms to be suitable for the image encryption. The proposed algorithm improves the classic RC4
algorithm in cryptography to encrypt the digital image.

The chaos system possesses of some characteristics, such as high sensitivity to
initial values and system parameter, the statistical property of white noise and
sequence ergodicity, and the characteristic of diffusion, permutation and randomness which conform to the requirements of cryptography. The chaotic maps can
export the pseudo-random sequences whose structures are very complex and difficult to be analyzed and predicted. Researchers encrypt the digital image by the
chaotic map, which reflects on three aspects: (1). Change the pixel value of original
image by the pseudo-random sequences which are exported by the chaotic maps
[3]; (2). Scramble the pixel position of original image by the chaotic maps [4]; (3).
Combine two methods above. So far, many image encryption based on chaotic
maps have been proposed [5]. Specific to these characteristics of the chaos system,
the proposed algorithm applies the chaotic maps to change the pixel value and
scramble the pixel position. Combined with the above points, color image
encryption algorithm based on modified RC4 and chaotic maps is proposed.
Proposed algorithm modifies the classical RC4 algorithm in cryptography and is
applied to the encryption process of digital image. So this makes that the pixel value
of the image changes largely, and conforms to the security and reliability in the
process of transmission. Firstly, the original image is divided into blocks whose size
is 8 × 8. Otherwise, a set of pseudo-random number are got by the iteration of
logistic map and consist of a pseudo-random block whose size is 8 × 8. Secondly,
the improved RC4 algorithm is operating between the adjacent blocks. Finally, the
image is scrambled by the logistic map to generate the encrypted image.

2 Encryption Algorithm

The encryption algorithm is composed by five parts. These parts are divided into
blocks, generating the pseudo-random block, operation between blocks, modified
RC4 and scrambling the image. The architecture is shown in Fig. 1.

2.1 Divide Image into Several Blocks

As is assumed, the size of original image I is $M \times N$, so the pixel value of original
image I is $I(i, j)$, $i = 0, 1,..., M - 1$; $j = 0, 1,..., N - 1$. The original image I is
divided into blocks, whose size is 8 × 8. So the original image can be divided into

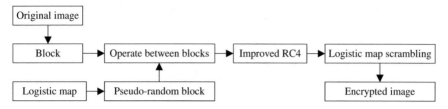

Fig. 1 Architecture of the encryption algorithm

$m \times n$ sub-blocks, which are represented as $I_{s,t}$, $s = 0, 1, \ldots, m - 1$; $t = 0, 1, \ldots,$ $n - 1$ and in which the pixel can be represented as $I_{s,t}(p, q)$, $p = 0, 1, \ldots, 7$; $q = 0,$ $1, \ldots, 7$. $\begin{cases} m = M/8 + \lceil M\%8 \rceil \\ n = N/8 + \lceil N\%8 \rceil \end{cases}$. Where $\lceil \rceil$ is the ceil operation and $\%$ is the mod operation. The formula is the conversion between the original image and sub-block $I_{s,t}(p, q) = I(i, j)$. Where $s = i/8, t = j/8, p = i\%8, q = j\%8$.

2.2 Generate Pseudo-Random Block

This pseudo-random block is composed of a set of pseudo-random numbers which is got by the iteration of the logistic map. The logistic map is described by $x_{n+1} = 4x_n(1 - x_n)$, $x_0 \in (0, 1)$, $x_0 \neq 0.5$.

Step1 The initial value x_0 of the logistic map is got by the following formula $x_0 = key_1/10^8$. Where, the range of the key key_1 is 1-99999999.

Step2 Iterate the logistic map for N_0 times. In the experiment, N_0 is 100.

Step3 We continue to iterate the logistic map 8×8 times, and obtain a pseudo-random sequence $X = \{x_1, x_2, \ldots, x_{64}\}$ from the state value. We can get a new sequence $L = \{l_1, l_2, \ldots, l_{64}\}$ by the formula $l_i = x_i \times 2^{24}$.

Step4 The pseudo-random block J is composed of the sequence $L = \{l_1, l_2, \ldots, l_{64}\}$ by the formula $J(i, j) = l_{j+8 \times i}$.

2.3 Operation Between Blocks

In the encryption algorithm, the operation between blocks uses a kind of improved RC4 algorithm. The operator of this operation is recorded as $\otimes$. In this paper, every sub-block of the image can be represented as $I_{s,t}$, which is same as I_u, $u = 0, 1, \ldots,$ $mn-1$. $I_u = I_{t+8 \times s} = I_{s,t}$ In the operation between blocks, the first sub-block I_0 operates with the pseudo-random block J, and generates a new first block, which is still recorded as I_0. Then I_1 operates with I_0, and generates a new second block, which is still recorded as I_1. It goes on as so until the last sub-block. The calculating formula is as follows

$$\begin{cases} I_u = I_u \otimes J, when\, u = 0 \\ I_u = I_u \otimes I_{u-1}, when\, 0 < u < mn \end{cases}$$

2.4 Improved RC4 Algorithm

The operation of the improved RC4 algorithm which is recorded as $\otimes$ between adjacent sub-blocks $I_u(p, q)$, $I_{u-1}(p, q)$ is described as follows:

Step1 The two-dimensional sub-block $I_u(p, q)$, $I_{u-1}(p, q)$ is converted into one-dimension by $I_u(p,q) = I_u(q + 8p) = I_u(k); I_{u-1}(p,q) = I_{u-1}(q + 8p) = I_{u-1}(k)$.

Step2 We can define an array s[64] as S-box, then initialize S-box by the $s[i] = i, i = 0, 1, \ldots, 63$.

Step3 We randomly exchange two value of the S-box by the string key key_3 to disturb the S-box. The string key key_3 is converted into the byte array k[i], then exchange s[i] and s[j] by $k[i] = key_3[i\%length], j = (j + s[i] + k[i])\%64$. Where $length$ is length of the string key key_3 and i is the index of char in the string key key_3. The range of every char in the string key key_3 is 0-255, and this string key key_3 has at least 8 chars.

Step4 After finishing the exchange of S-box in Step2, we continue to exchange s[i] and s[j] of S-box, and get the index r which is the sum of i and j by the $j = (j + s[i])\%64, r = (s[i] + s[j])\%64$.

Step5 We can get a sequence $R = \{r_1, r_2, \ldots, r_{64}\}$, and XOR the r_ith pixel of the front sub-block with the ith pixel of the behind sub-block by the $I_u(i) = I_u(i) \oplus I_{u-1}(r_i)$.

Finally, we can merge all of the sub-blocks after XOR into the image E_1.

2.5 Scramble Image

We can scramble the pixel position of the image E_1 by the logistic map to get the encrypted image E_2. We can get a pseudo-random sequence by the logistic map, which is described as $x_{n+1} = 4x_n(1 - x_n); x_0 \in (0, 1), x_0 \neq 0.5$. In the scrambling process, the image is scrambled by the pseudo-random sequence, is described as follows:

Step1   We can get the initial value x_0 of the logistic map by $x_0 = key_2/10^8$. Where, the range of the key key_2 is 1-99999999.

Step2   Iterate the logistic map for N_0 times. In our experiments, N_0 is 100.

Step3   Traverse every pixel of the image E_1 in order, and iterate the logistic map for one time when accessing a pixel $E_1(i, j)$. So we can get a random number x_i, and a sequence number m_i by the following formula $m_i = \lfloor x_i \times (M \times N) \rfloor$. Where $M \times N$ is the size of the encrypted image E_2 and $\lfloor \rfloor$ is the floor operation.

Step4   If the pixel $E_2(p, q)$ is occupied, we will continue to carry out step3 until the $E_2(p, q)$ isn't occupied. So the pixel $E_2(p, q)$ is replaced $E_1(i, j)$, which is described as $E_2(p,q) = E_1(i, j)$. Where $p = m_i/M$, $q = m_i\%M$.

Finally, the image E_2 is the encrypted image which we get after the image E_1 is scrambled.

3   Decryption Algorithm

Compared with the encryption, the decryption is the inverse process of the encryption algorithm. The decryption algorithm is different from the encryption algorithm in only operation between blocks and scrambling the image. Firstly, the encrypted image E_2 is scrambled for the image E_1 by the logistic map, and the image E_1 is divided into 8×8 sub-blocks. Secondly, the pixels between adjacent sub-blocks have a XOR operation. Finally, the sub-blocks are merged into the image I.

3.1   Inversing Operation of Scrambling Image

In the process of decryption, the encrypted image E_2 is scrambled for the image E_1. In the process of scrambling the image, step1, step2 and step3 are same as the encryption algorithm, but step4 has a little difference. So we only show the process of step4, which is described as

Step4   If the pixel $E_2(p, q)$ is occupied, we will continue to carry out step3 until the $E_2(p, q)$ isn't occupied. So the pixel $E_1(i, j)$ is replaced $E_2(p, q)$, which is described as $E_1(i,j) = E_2(p,q)$.

3.2   Inversing Operation Between Blocks

As same as the encryption algorithm, the image E_1 is divided into sub-blocks, which is described as $I_u, u = 0, 1,\ldots, mn-1$. In the process of decryption, we begin to operate from the last sub-block I_{mn-1}. The last sub-block I_{mn-1} operates with its front sub-block I_{mn-2} to get the last sub-block, which is still recorded as I_{mn-1}. The sub-block I_u operates with its front sub-block I_{u-1} until the first sub-block I_0 in order. Finally, the first sub-block I_0 operates with the pseudo-random block Logistic which is got by the logistic map. This is the opposite order of encryption. The operation is described as

$$\begin{cases} I_u = I_u \otimes J, when\, u = 0 \\ I_u = I_u \otimes I_{u-1}, when\, 0 < u < mn \end{cases}$$

4 Experiment Results

The experiment of the encryption algorithm is conducted under eclipse using java in a computer with an Intel Core i5 2.5 GHz and 4 GB RAM running Windows 8.0 operating system. In the experiment, we select *Airplane* with size 128 × 128, *Sailboat* with size 256 × 256 and *Pens* with size 512 × 512 as the original image. In encrypting the image we select a set of key with $key_1 = 88888888$, $key_2 = 88888888$, $key_3 = abcdefgh$. After the image is encrypted, we compare the original image with the encrypted image and the recovered image, which is shown as Fig. 2.

Fig. 2 Comparison after encrypted and recovered **a** *Airplane*, **b** encrypted *Airplane*, **c** recovered *Airplane*, **d** *Sailboat*, **e** encrypted *Sailboat*, **f** recovered *Sailboat*, **g** *Pens*, **h** encrypted *Pens*, **i** recovered *Pens*

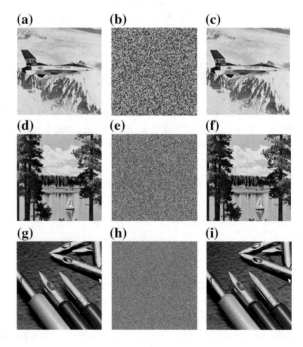

In Fig. 2, the encrypted image is the noise-like image, and it can't be seen any useful information. Besides, the recovered image is same as the original image. So these experiments demonstrate that the proposed algorithm has an ideal performance.

5 Security Analysis

5.1 Key Space Analysis

In experiment, the key of encryption algorithm is composed of the key key_1, key_2 of the logistic map and the string key key_3 of the pixel XOR. The key key_1, key_2 are the decimal integer with 8 bits, and the key key_3 has at least eight chars. So the smallest key space of the proposed algorithm is 10^{32}. This key space is enough large to resist the exhaustive attack.

5.2 Key Sensitive Test

The key sensitive test is shown as Fig. 3. In experiments, we select *Pens* as the original image. This test encrypts the original image (a) to get the encrypted image (b) with the key K_1 (key_1 = 88888888, key_2 = 88888888, key_3 = abcdefgh) and decrypts the encrypted image (b) to get the recovered image (c) with the same key K_1. Then let the key change Key_1 change a little precision to get the key K_2 (key_1 = 88888889, key_2 = 88888888, key_3 = abcdefgh). This test decrypts the encrypted image (b) to get the recovered image (d) which is still a noise-like image with the key K_2.

This sensitive test shows that the original image can be completely recovered only when the correct security key is being utilized and that the decryption result is completely different from the correct recovered image when the key has a little precision change. Therefore, this demonstrates the proposed algorithm is also sensitive to its key changes during the decryption process.

(a)   **(b)**   **(c)**   **(d)**

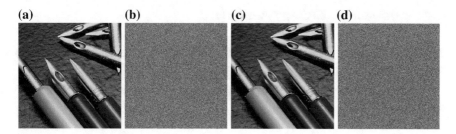

Fig. 3 Key sensitive test

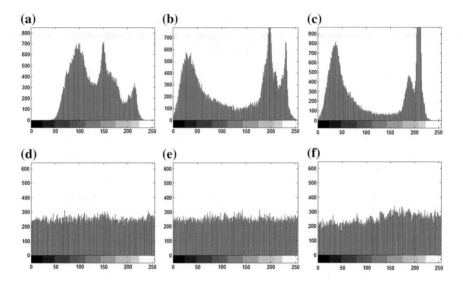

Fig. 4 Comparison of histogram on color components (**a–c**) is the histogram of red, green, blue color component in the original image, respectively; (**d–f**) is the histogram of red, green, blue color component in the encrypted image respectively

5.3 Histogram Analysis

Image histogram represents the intensity distribution of pixels within an image. An encrypted image with a flat histogram is able to resist statistic attacks. In this analysis, we select *Sailboat* image as the original image. Figure 4 compares the encrypted image with the original image on the histogram of red, green, blue color component.

The result of Fig. 4 shows that the histogram of red, green, blue color component in the encrypted image have a big change, which becomes flat and uniform. This change can resist statistic attacks, and the encrypted image is difficult to be decoded from the histogram of the encrypted image.

5.4 Information Entropy Analysis

Information entropy (IFE) is designed to evaluate the uncertainty in a random variable as shown in the equation $H_L = \sum_{l=0}^{F-1} P(L=l) \log_2 \frac{1}{P(L=l)}$.

Where F is the gray level and $P(L=l)$ is the percentage of pixels of which the value is equal to l.

Table 1 Information entropy analysis

File name	Image size	Original image	Encrypted image
Airplane.bmp	128 × 128	6.7754	7.9860
Sailboat.bmp	256 × 256	7.7741	7.9903

The IFE can be used for evaluating the randomness of an image. An IFE score of an image close to the maximum IFE value means the excellent random property. For a grayscale image with a data range of [0, 255], its maximum IFE is 8. For a color image, we can convert the color image into the grayscale image, and calculate the IFE. In the experiment of information entropy analysis, we select *Airplane* and *Sailboat* images as the original images. Table 1 shows the IFE scores of images before and after applying the proposed encryption algorithm.

From these results, the IFE scores of all encrypted images with different sizes are close to 8. The IFE scores of the encrypted image are closer to 8 than that of the original image, which means that the encrypted images after applying the proposed encryption algorithm have better random distributions.

5.5 Correlation Analysis

The original image has high correlations between pixels and their neighboring pixels at horizontal, vertical and diagonal directions. The encryption algorithm aims at breaking these pixel correlations in the original images, and transforming them into noise-like encrypted images with little or no correlations. The correlation values can be calculated by the following equation $C_{xy} = \frac{E[(x-\mu_x)(y-\mu_y)]}{\sigma_x \sigma_y}$. Where μ and σ are the mean value and standard deviation respectively, $E[\cdot]$ is the expectation value.

Hence, a good encrypted image should be unrecognized and has the correlation values close to zero. Correlation analysis here is to test the relationships of adjacent pixels in the original and encrypted images. 3000 pairs of two adjacent pixels are randomly chosen from the original and encrypted images in the horizontal, vertical and diagonal directions to perform this analysis. Table 2 compares correlations of the original image with its encrypted versions generated by different encryption algorithm. We select *Sailboat* image as the original image.

Table 2 Correlation values at the horizontal, vertical and diagonal directions

Image	Encryption algorithm	Horizontal	Vertical	Diagonal
Original image		0.9164	0.9140	0.8773
Encrypted image	Proposed algorithm	7.6182×10^{-4}	8.4534×10^{-4}	0.0081
Encrypted image	[6]'s	0.0141	0.0107	0.0097

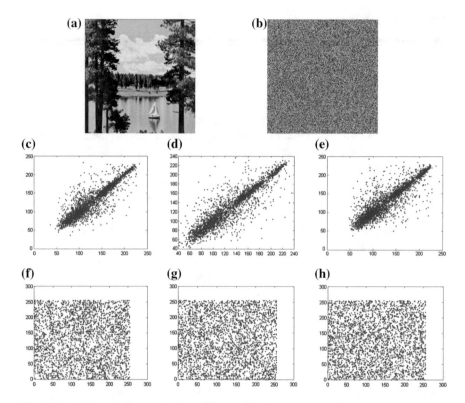

Fig. 5 Correlation of adjacent pixels at different directions

From the results of Table 2, the original image has high correlation values in all directions while all encrypted images have very low correlation values, which shows the excellent performance of image encryption algorithms. Furthermore, compared to the [6]'s algorithm, the proposed algorithm obtains the smaller correlation values in all directions. It out performs [6]'s algorithm with respect to performance of image encryption.

We can set the intensity values of adjacent pixel pairs as the horizontal and vertical axes, respectively. Figure 5 plots their distributions in three directions. In this paper, we select *Sailboat* image as the original image. If two adjacent pixels are equal, their tracks are located in the diagonal line. As can be seen, the distributions of adjacent pixels in the original image are around the diagonal line, which means the original image has high correlation. On the contrary, the distributions of adjacent pixels in the encrypted image disperse in the whole data range of the image, which proves that the encrypted image has extremely low correlation and show high randomness.

Table 3 Comparison of the encryption time of different algorithms

File name	Image size	Proposed algorithm (s)	[7] (s)
Sailboat. bmp	256 × 256	0.273	0.569
Pens.bmp	512 × 512	1.455	2.251

5.6 Speed Analysis

The proposed algorithm is mainly based on modified RC4 and chaotic maps. The basic operation of RC4 algorithm is XOR, whose time-consuming is very little. So the proposed algorithm has a high efficiency on speed. In the speed analysis, we select *Sailboat* and *Pens* images as the original images. Besides, we compare the encryption speed of proposed algorithm with [7]'s. The result is shown as Table 3.

As can be seen from Table 3, the proposed algorithm performs faster than [7]'s, which demonstrates that the proposed algorithm has a fast speed performance and is suitable for real applications.

6 Conclusions

This paper proposes an image encryption algorithm based on modified RC4 and chaotic maps. Through the experiment results and security analysis, the proposed algorithm not only has a high security, but also has a fast speed performance. The main advantages of proposed algorithm are as follows:

(1) The encrypted image is a noise-like image and has a flat and uniform histogram and a good randomness.
(2) The adjacent pixels of the encrypted image have very low correlation.
(3) The proposed algorithm has a large key space and is very sensitive to the key.

Because the main operation of the encrypted algorithm is XOR operation in modified RC4 algorithm and the pixel block is considered as the minimum operation unit, the encryption algorithm has a fast encryption speed.

Acknowledgments This work was supported by the fundamental research funds for the central universities (Grant No.20205001537).

References

1. Liu, H.J., Kadir, A., Niu, Y.J.: Chaos-based color image block encryption scheme using S-box. AEU-Int. J. Electron. Commun. **68**(7), 676–686 (2014)
2. Yin, A.H., Wang, S.K.: A novel encryption scheme based on time stamp in gigabit Ethernet passive optical network using AES-128. Optik **125**(3), 1361–1365 (2014)

3. Kadir, A., Hamdulla, A., Guo, W.Q.: Color image encryption using skew tent map and hyper chaotic system of 6th-order CNN. Optik **125**(5), 1671–1675 (2014)
4. Zhang, Z., Sun, S.L.: Image encryption algorithm based on logistic chaotic system and s-box scrambling. Image Sig. Process. **1**(4), 177–181 (2011)
5. Tong, X.J.: Design of an image encryption scheme based on a multiple chaotic map. Commun. Nonlinear Sci. Numer. Simul. **18**(7), 1725–1733 (2013)
6. Wang, X.Y., Chen, F., Wang, T.: A new compound mode of confusion and diffusion for block encryption of image based on chaos. Commun. Nonlinear Sci. Numer. Simul. **15**(9), 2479–2485 (2010)
7. Liao, X.F., Lai, S.Y., Zhou, Q.: A novel image encryption algorithm based on self-adaptive wave transmission. Sig. Process. **90**(9), 2714–2722 (2010)

Modified Discrete LQ Control Algorithm for Situations with the Scan Period Variance

Jan Cvejn

Abstract Computer-based control systems, especially if they run under general-purpose operating systems, often exhibit variance of the scan period of processing inputs and outputs. Although this fact is usually not taken into account when discrete control algorithms are used, it can cause worse performance of the control loop in comparison to the theoretical case. In this paper we describe a modified discrete LQ control algorithm that takes disturbances of the scan period into account and partially compensates their influence. We also show that such a controller can be implemented even on low-performance hardware platforms, if they are equipped with a sufficient amount of memory.

Keywords Optimal control · LQ controller · Linear systems · Discrete control

1 Introduction

In the control theory analysis and design of control algorithms in the continuous-time domain and in the discrete-time domain are studied separately. Continuous approach is natural for modelling and analysis of real processes and will be always used for designing controllers on the basis of analog components. Discrete approach seems to be natural for technical implementation of control algorithms on microprocessor-based platforms.

Discrete control algorithms rely upon constant period of processing inputs and outputs. However, constant scan period is often not fully guaranteed in real situations. This phenomenon can occur due to handling asynchronous hardware events in computer systems. Although this problem is typical for general-purpose multi-tasking operating systems, even the most robust hardware platforms such as PLCs

J. Cvejn (✉)
Faculty of Electrotechnics and Informatics, University of Pardubice,
Studentská 95, 532 10 Pardubice, Czech Republic
e-mail: jan.cvejn@upce.cz

© Springer International Publishing Switzerland 2016
R. Silhavy et al. (eds.), *Automation Control Theory Perspectives
in Intelligent Systems*, Advances in Intelligent Systems and Computing 466,
DOI 10.1007/978-3-319-33389-2_2

exhibit scan variance. A similar situation could occur at remote control, where the measurements and the control signal are transported over a communication network.

Irregularities of the scan period cause worse performance of the control loop in comparison to the theoretical case. This influence can be neglected if the irregularities occur rarely and the system time constants are large in comparison to the scan period. In the other cases the influence on the closed-loop dynamics can be significant.

In this paper, we describe a modification of the classical linear-quadratic (LQ) discrete control algorithm taking into account irregularities of the scan period. We show that the effect of the scan variance can be partially compensated by mathematical means if a hybrid control law is used, working at discrete steps, but using a continuous-time model for the determination of the control output. In this way the control reliability and performance can be enhanced, especially at time-critical applications.

This problem has been studied already in [1] in a more general form as a stochastic control problem, considering also the optimal estimation. In this paper only the controller part is discussed, but an extended model of the scan period disturbances, which better corresponds to some real situations, is considered. This modification requires a corresponding extension of the control algorithm. The determination of the control action at each step is still not a time-consuming operation, although the enhanced control algorithm needs a table of data stored in the controller memory, which is initialized at the design phase.

2   Motivation

Consider a continuous time-invariant linear system

$$\dot{\mathbf{x}} = \mathbf{A}\mathbf{x}(t) + \mathbf{B}\mathbf{u}(t) \tag{1}$$

where the dimensions of $\mathbf{x}$ and $\mathbf{u}$ are n and m, respectively, $n \geq m$. $\mathbf{A}$ and $\mathbf{B}$ are known matrices of corresponding dimensions. We are looking for a control history such that

$$J = \frac{1}{2} \int_0^\infty \mathbf{x}^T(t)\mathbf{Q}\mathbf{x}(t) + \mathbf{u}^T(t)\mathbf{R}\mathbf{u}(t) \; dt \;\; \rightarrow \;\; \min \tag{2}$$

where $\mathbf{Q}$, $\mathbf{R}$ are given symmetric positive definite matrices. We assume that the current state $\mathbf{x}(t)$, or its estimate, is known.

Although the system nature is continuous, we consider that the measurements and the control actions are taken at discrete-time steps $t_0 < t_1 < \cdots$. Although the scan period T is assumed to be known and constant, due to external factors the actual difference $t_{i+1} - t_i$ can fluctuate. We assume that each scan is provided with its time mark, which can be usually easily technically realized. Then, from the known sequence $t_1, t_2, \ldots, t_k$ of the previous scan instants it is possible to estimate the future sequence $\bar{t}_{k+1}, \bar{t}_{k+2}, \bar{t}_{k+3}, \ldots$, which is considered to be equidistant, i.e.

$$\bar{t}_{k+2} - \bar{t}_{k+1} = \bar{t}_{k+3} - \bar{t}_{k+2} = \cdots = \overline{T}. \tag{3}$$

But note that $\overline{T} = T$ need not hold in general, because the response can be delayed permanently in some time interval. Although the scan interrupts are generated by hardware clock with the period T, which does not depend on the previous scan instants, the delay can be caused by omitting some scan instants, e.g. due to service of hardware events in the operating system (this problem is discussed below in more detail). This delay can be detected from the sequence of the past measurements t_i, $i \leq k$ and this information also can be used to estimate the future sequence (3), i.e. the expected period $\overline{T}$.

There are several possible ways how to estimate the next scan instant $\bar{t}_{k+1}$. In [1] a simplified model was used, which assumed $\overline{T} = T$. If we denote t_c the instant of the next time interrupt closest to t_k, in most cases the following estimate seems to be more adequate:

$$\bar{t}_{k+1} = t_c + \overline{T} - T. \tag{4}$$

But if $t_c - t_k \leq \alpha T$, where $\alpha \in (0, 1)$ is a known parameter, the control algorithm should be designed to omit the next scan, i.e. in this case

$$\bar{t}_{k+1} = \left(t_c + \overline{T} - T\right) + T = t_c + \overline{T}. \tag{5}$$

This behavior indeed depends on implementation of the controller on given platform. For instance, $t_c - t_k \leq 0.25T$ may indicate that the processor looses the ability to process the hardware events at given moment and in such a case the control algorithm should be designed to drop the following scan and wait until the next one to prevent the system from overloading, which would affect the overall functionality. This modification is especially needed in the case of multi-tasking operating systems, where the control algorithm is a high-priority task, and if processing of the measurements and computation of the control action is a time-consuming operation within the interval $[t_k, t_{k+1}]$.

3 Optimal Control Algorithm

To summarize the considerations of the previous section, at given moment t_k we assume that the estimates of the next scan instant $\bar{t}_{k+1} \geq t_c$ and the future scan period $\overline{T} \geq T$ are known, in general different from the next hardware clock instant t_c and the clock period T.

The criterion (2) value from $t = t_k$ can be expressed as

$$J(t_k) = \frac{1}{2} \sum_{i=k}^{N} \int_{t_i}^{t_{i+1}} \mathbf{x}^T(t)\mathbf{Q}\mathbf{x}(t) + \mathbf{u}^T(t)\mathbf{R}\mathbf{u}(t)\, dt \qquad (6)$$

where $N \to \infty$ and $\mathbf{u}(t)$ is constant in each interval $[t_i, t_{i+1})$, $i \geq k$. Note that unlike common practice the criterion includes the information about complete state history in $[0, t_f]$, and not only about the values at the discrete points t_i.

Denote for simplicity $\mathbf{x}_k = \mathbf{x}(t_k)$ and $\mathbf{u}_k = \mathbf{u}(t_k)$. For given $\mathbf{x}_k$ and $t > t_k$

$$\mathbf{x}(t) = \mathbf{\Phi}(t - t_k)\mathbf{x}_k + \mathbf{\Psi}(t - t_k)\mathbf{u}_k \qquad (7)$$

holds, where

$$\Phi(h) = e^{\mathbf{A}h}, \quad \mathbf{\Psi}(h) = \int_0^h \mathbf{\Phi}(h - \tau)\,\mathbf{B}\, d\tau = \int_0^h \mathbf{\Phi}(\tau)\, d\tau\, \mathbf{B}. \qquad (8)$$

The term $\mathbf{x}^T(t)\mathbf{Q}\mathbf{x}(t)$ for given $\mathbf{x}_k$ and $t > t_k$ can be written using (8) as

$$\mathbf{x}^T(t)\mathbf{Q}\mathbf{x}(t) = [\mathbf{x}_k^T, \mathbf{u}_k^T] \begin{bmatrix} \mathbf{\Phi}^T(t - t_k) \\ \mathbf{\Psi}^T(t - t_k) \end{bmatrix} \mathbf{Q}[\mathbf{\Phi}(t - t_k), \mathbf{\Psi}(t - t_k)] \begin{bmatrix} \mathbf{x}_k \\ \mathbf{u}_k \end{bmatrix}. \qquad (9)$$

Let us define

$$\mathbf{U}(h) = \int_0^h \left(\begin{bmatrix} \mathbf{\Phi}^T(\tau) \\ \mathbf{\Psi}^T(\tau) \end{bmatrix} \mathbf{Q}[\mathbf{\Phi}(\tau), \mathbf{\Psi}(\tau)] + \begin{bmatrix} 0 & 0 \\ 0 & \mathbf{R} \end{bmatrix} \right) dh$$

$$= \int_0^h \begin{bmatrix} \mathbf{\Phi}^T(\tau)\mathbf{Q}\mathbf{\Phi}(\tau) & \mathbf{\Phi}^T(\tau)\mathbf{Q}\mathbf{\Psi}(\tau) \\ \mathbf{\Psi}^T(\tau)\mathbf{Q}\mathbf{\Phi}(\tau) & \mathbf{\Psi}^T(\tau)\mathbf{Q}\mathbf{\Psi}(\tau) + \mathbf{R} \end{bmatrix} d\tau = \begin{bmatrix} \mathbf{U}_{11}(h) & \mathbf{U}_{12}(h) \\ \mathbf{U}_{21}(h) & \mathbf{U}_{22}(h) \end{bmatrix} \qquad (10)$$

where $\mathbf{U}_{21}(h) = \mathbf{U}_{12}^T(h)$. Using (10) we can write

$$J(t_k) = \frac{1}{2} [\mathbf{x}_k^T\ \mathbf{u}_k^T] \mathbf{U}(\bar{t}_{k+1} - t_k) \begin{bmatrix} \mathbf{x}_k \\ \mathbf{u}_k \end{bmatrix} + J(\bar{t}_{k+1}) \qquad (11)$$

where

$$J(\bar{t}_{k+1}) = \frac{1}{2} \sum_{i=k+1}^{N} [\mathbf{x}_i^T\ \mathbf{u}_i^T] \mathbf{U}(\bar{T}) \begin{bmatrix} \mathbf{x}_i \\ \mathbf{u}_i \end{bmatrix}. \qquad (12)$$

Let us denote $J^*(t_k)$ the minimal value of $J(t_k)$. By application of Bellman's optimality principle [2–3] we obtain

$$J^*(t_k) = \min_{u_k} \left\{ \frac{1}{2} \begin{bmatrix} \mathbf{x}_k^T & \mathbf{u}_k^T \end{bmatrix} \mathbf{U}(\bar{t}_{k+1} - t_k) \begin{bmatrix} \mathbf{x}_k \\ \mathbf{u}_k \end{bmatrix} + J^*(\bar{t}_{k+1}) \right\} \quad (13)$$

where

$$J^*(\bar{t}_{k+1}) = \min_{\{u_{k+1},...,u_N\}} \left\{ \frac{1}{2} \sum_{i=k+1}^{N} \begin{bmatrix} \mathbf{x}_i^T & \mathbf{u}_i^T \end{bmatrix} \mathbf{U}(\bar{T}) \begin{bmatrix} \mathbf{x}_i \\ \mathbf{u}_i \end{bmatrix} \right\} \quad (14)$$

subject to the dynamic constraints

$$\mathbf{x}_{i+1} = \mathbf{\Phi}(\bar{T})\, \mathbf{x}_i + \mathbf{\Psi}(\bar{T})\mathbf{u}_i, \quad i \geq k+1. \quad (15)$$

Equations (14) and (15) formulate a discrete deterministic linear-quadratic optimal control problem. The minimal cost-function value of this problem for $N \to \infty$ is in the form

$$J^*(\bar{t}_{k+1}) = \frac{1}{2} \mathbf{x}_{k+1}^T \mathbf{S}\, \mathbf{x}_{k+1}$$

$$= \frac{1}{2} \begin{bmatrix} \mathbf{x}_k^T, & \mathbf{u}_k^T \end{bmatrix} \begin{bmatrix} \mathbf{\Phi}^T(\bar{t}_{k+1} - t_k) \\ \mathbf{\Psi}^T(\bar{t}_{k+1} - t_k) \end{bmatrix} \mathbf{S}[\mathbf{\Phi}(\bar{t}_{k+1} - t_k), \ \mathbf{\Psi}(\bar{t}_{k+1} - t_k)] \begin{bmatrix} \mathbf{x}_k \\ \mathbf{u}_k \end{bmatrix} \quad (16)$$

where $\mathbf{S}$ is a positive-definite symmetric matrix. If we define

$$\mathbf{Z}(h) = \begin{bmatrix} \mathbf{Z}_{11}(h) & \mathbf{Z}_{12}(h) \\ \mathbf{Z}_{21}(h) & \mathbf{Z}_{22}(h) \end{bmatrix} = \mathbf{U}(h) + \begin{bmatrix} \mathbf{\Phi}^T(h) \\ \mathbf{\Psi}^T(h) \end{bmatrix} \mathbf{S}[\mathbf{\Phi}(h), \mathbf{\Psi}(h)]. \quad (17)$$

where $\mathbf{Z}_{21}(h) = \mathbf{Z}_{12}^T(h)$, the minimizer of $J(t_k)$ for fixed $\bar{T}$ can be written as

$$\mathbf{u}_k^* = \arg \min_{\mathbf{u}_k} \left\{ \frac{1}{2} \begin{bmatrix} \mathbf{x}_k^T & \mathbf{u}_k^T \end{bmatrix} \mathbf{U}(\bar{t}_{k+1} - t_k) \begin{bmatrix} \mathbf{x}_k \\ \mathbf{u}_k \end{bmatrix} + \frac{1}{2} \mathbf{x}_{k+1}^T \mathbf{S}\, \mathbf{x}_{k+1} \right\}$$

$$= \arg \min_{\mathbf{u}_k} \left\{ \frac{1}{2} \begin{bmatrix} \mathbf{x}_k^T & \mathbf{u}_k^T \end{bmatrix} \mathbf{Z}(\bar{t}_{k+1} - t_k) \begin{bmatrix} \mathbf{x}_k \\ \mathbf{u}_k \end{bmatrix} \right\}. \quad (18)$$

The solution can be easily found by differentiation in the form

$$\mathbf{u}_k^* = -\mathbf{Z}_{22}^{-1}(\bar{t}_{k+1} - t_k)\mathbf{Z}_{21}(\bar{t}_{k+1} - t_k)\mathbf{x}_k = \mathbf{C}(\bar{t}_{k+1} - t_k)\mathbf{x}_k. \quad (19)$$

Note that the matrix inversion in (19) always exists, since $\mathbf{Z}_{22}(h)$ is positive definite.

If we put $\bar{t}_{k+1} - t_k = \overline{T}$ in (13) and (19), we have to obtain for optimal $\mathbf{u}_k$

$$J^*(t_k) = \frac{1}{2}\mathbf{x}_k^T \mathbf{S}\,\mathbf{x}_k. \tag{20}$$

This condition can be used to determine $\mathbf{S}$. By substituting (20) into (13) and (16) we obtain:

$$
\begin{aligned}
J^*(t_k) &= \frac{1}{2}\begin{bmatrix}\mathbf{x}_k^T & \mathbf{u}_k^T\end{bmatrix}\mathbf{Z}(\overline{T})\begin{bmatrix}\mathbf{x}_k \\ \mathbf{u}_k\end{bmatrix} \\
&= \frac{1}{2}\mathbf{x}_k^T\begin{bmatrix}\mathbf{I} & -\mathbf{Z}_{22}^{-1}(\overline{T})^{-1}\mathbf{Z}_{21}(\overline{T})\end{bmatrix}\mathbf{Z}(\overline{T})\begin{bmatrix}\mathbf{I} \\ -\mathbf{Z}_{22}^{-1}(\overline{T})^{-1}\mathbf{Z}_{21}(\overline{T})\end{bmatrix}\mathbf{x}_k \\
&= \frac{1}{2}\mathbf{x}_k^T\begin{bmatrix}\mathbf{I} & -\mathbf{Z}_{22}^{-1}(\overline{T})^{-1}\mathbf{Z}_{21}(\overline{T})\end{bmatrix}\begin{bmatrix}\mathbf{Z}_{11}(\overline{T}) - \mathbf{Z}_{12}(\overline{T})\mathbf{Z}_{22}^{-1}(\overline{T})\mathbf{Z}_{21}(\overline{T}) \\ \mathbf{0}\end{bmatrix}\mathbf{x}_k \\
&= \frac{1}{2}\mathbf{x}_k^T\left(\mathbf{Z}_{11}(\overline{T}) - \mathbf{Z}_{12}(\overline{T})\mathbf{Z}_{22}^{-1}(\overline{T})\mathbf{Z}_{21}(\overline{T})\right)\mathbf{x}_k.
\end{aligned}
\tag{21}
$$

This shows that

$$\mathbf{S} = \mathbf{Z}_{11}(\overline{T}) - \mathbf{Z}_{12}(\overline{T})\mathbf{Z}_{22}^{-1}(\overline{T})\mathbf{Z}_{21}(\overline{T}) \tag{22}$$

must hold. By substituting for $\mathbf{Z}_{ij}(\overline{T})$ from (17) it is easily seen that (22) can be rewritten into an algebraic matrix Riccati equation [4, 3].

4 Implementation of the Controller

Obtained expressions are indeed rather complicated to be computed at each control step. The controller matrix $\mathbf{C}(\bar{t}_{k+1} - t_k)$ in (20) is dependent on the expected distance of the next scan $\bar{t}_{k+1} - t_k$ and on $\mathbf{S}$, while $\mathbf{S}$ depends on $\overline{T}$.

For given $\overline{T}$ the solution to the Riccati Eq. (22) can be obtained off-line as a part of the controller design. Denote $\mathbf{Z}(\overline{T}; \mathbf{S}_i)$ the value of $\mathbf{Z}(\overline{T})$ for $\mathbf{S} = \mathbf{S}_i$. A basic method of obtaining $\mathbf{S}$ consists in solving

$$\mathbf{S}_{i+1} = \mathbf{Z}_{11}(\overline{T}; \mathbf{S}_i) - \mathbf{Z}_{12}(\overline{T}; \mathbf{S}_i)\mathbf{Z}_{22}^{-1}(\overline{T}; \mathbf{S}_i)\mathbf{Z}_{21}(\overline{T}; \mathbf{S}_i) \tag{23}$$

iteratively until $\|\mathbf{S}_{i+1} - \mathbf{S}_i\| < \varepsilon$, where ε is sufficiently small [4]. More sophisticated methods, preferable both from numerical point of view and for improved efficiency, were proposed in [5, 6].

If we assume $\overline{T} \in [T, T_{max}]$, $T_{max} \geq 2T$, the solutions to (23) can be obtained for the values in this interval with a sufficiently small discrete step $\Delta\overline{T}$ at the initialization phase. These solutions can be stored in the controller memory and in the real-time operation the values corresponding to the current estimate of $\overline{T}$ are being picked from this table. Although it may seem that the initialization phase could be computationally very demanding, if the value of $\mathbf{S}$ is known of some $\overline{T}$, it can be used as a very good estimate of $\mathbf{S}$ for the iterative computation based on (23) corresponding to $\overline{T} + \Delta \overline{T}$, because $\mathbf{S}$ depends only moderately on $\overline{T}$. Therefore, the computation of the whole table of the values of $\mathbf{S}$ for $\overline{T} \in [T, T_{max}]$ is not a time-demanding operation.

In the same way, the values of the matrices $\mathbf{\Phi}(h)$, $\mathbf{\Psi}(h)$ and $\mathbf{U}(h)$ have to be computed for $0 < h \leq h_{max}$, where $h_{max} \geq 2T$ is known, with a sufficiently small discrete step of h and stored in the controller memory. Formally written, it is needed to solve the following set of differential equations in the interval $h \in [0, h_{max}]$:

$$\frac{d}{dh} \mathbf{\Phi}(h) = \mathbf{A} \, \mathbf{\Phi}(h) \tag{24}$$

$$\frac{d}{dh} \mathbf{\Psi}(h) = \mathbf{\Phi}(h) \, \mathbf{B} \tag{25}$$

$$\frac{d}{dh} \mathbf{U}(h) = \begin{bmatrix} \mathbf{\Phi}^T(\tau)\mathbf{Q}\mathbf{\Phi}(\tau) & \mathbf{\Phi}^T(\tau)\mathbf{Q}\mathbf{\Psi}(\tau) \\ \mathbf{\Psi}^T(\tau)\mathbf{Q}\mathbf{\Phi}(\tau) & \mathbf{\Psi}^T(\tau)\mathbf{Q}\mathbf{\Psi}(\tau) + \mathbf{R} \end{bmatrix} \tag{26}$$

with the initial conditions

$$\mathbf{\Phi}(0) = \mathbf{I}_n, \quad \mathbf{\Psi}(0) = \mathbf{0}, \quad \mathbf{U}(0) = \mathbf{0}. \tag{27}$$

It is important to mention here that the controller implementation is significantly more efficient in the simplified version where $\overline{T} = T$ is fixed. In this case $\mathbf{S}$ is constant and the controller matrix $\mathbf{C}(h)$ in (19) can be computed in forward and stored, so the matrix inversion in (19) need not be computed in real time.

5 Example

Consider the double-integrator system in the form (1) where

$$\mathbf{A} = \begin{bmatrix} 0 & 0 \\ 2 & 0 \end{bmatrix}, \, \mathbf{B} = \begin{bmatrix} 1 \\ 0 \end{bmatrix} \tag{28}$$

with the initial condition

$$\mathbf{x}(0) = \begin{bmatrix} 1 \\ 1 \end{bmatrix}. \tag{29}$$

The criterion (6) parameters were chosen as

$$\mathbf{Q} = \mathbf{I}_2, \ \mathbf{R} = 1. \tag{30}$$

The scan period is $T = 2\,s$, but each fourth scan is delayed of 50 % and the following scan is omitted. In the initialization phase it was needed to obtain the solution to (24)–(26) for $h \in [0, 2T]$ and to solve (23) for the sequence of values of $\overline{T} \in [T, 2T]$ with the step size of $\Delta \overline{T} = 0.01$. The initialization was not a time-demanding operation.

Figure 1 shows the response of the system when the standard LQ controller is used, i.e. if $\bar{t}_{k+1} = t_k + T$ and $\overline{T} = T$, while Fig. 2 shows the responses if the modified controller (19) is used for $\overline{T} = 5T/4$. This estimate of $\overline{T}$ corresponds to the fact that each fifth scan is omitted.

It can be seen that the control loop behavior was significantly enhanced, although a similar effect indeed could be achieved by decreasing the scan period, if

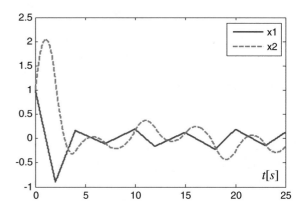

Fig. 1 The history of the state variables—standard LQ controller

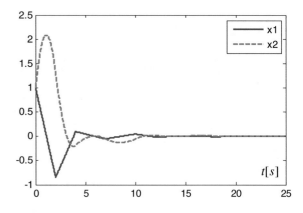

Fig. 2 The history of the state variables—the modified LQ controller, full version

Fig. 3 The history of the state variables—the modified LQ controller, simplified version

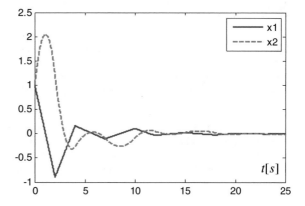

it was technically possible. Even bigger differences can be observed if the controller is equipped with the optimal state estimator, which can be also designed so that it takes the variance of the scan period into account, as described in [1].

Figure 3 shows that the responses obtained for the simplified version where $\overline{T} = T$ are similar to the full version. This indicates that the simplified version, which is preferable from the implementation point of view, would be usually sufficient and recommendable for practical use.

6 Conclusions

The modification of the LQ control algorithm described in this paper tries to reduce the influence of the scan-period variance, which can occur in computer-based control systems, on the closed-loop control performance. Although obtained expressions for the control output may be rather complicated to be computed in real time, if sufficient memory in the control system is available, it is possible to carry out most of these computations in forward and the determination of the control output is not a complicated or time-consuming operation. Consequently, such a control algorithm can be implemented even on low-performance hardware platforms. In the simplified version, which seems to be sufficient for practical purposes, the computation of the control action is as demanding as a matrix-vector product, like in the case of standard LQ control algorithm. However, the controller has to be equipped with sufficient amount of memory to store a table of the control matrices, dependent on a single parameter.

References

1. Cvejn, J.: Compensation of the scan-period irregularities in lqg control systems. Arch. Control Sci. **18**, 3 (2008)
2. Bryson, A.E., Ho, Y.C.: Applied Optimal Control. Hemisphere Corp, New York (1975)
3. Fleming, W.H., Rishel, R.W.: Deterministic and Stochastic Optimal Control. Springer, Berlin (1995)
4. Stengel, R.F.: Stochastic Optimal Control: Theory and Application. Wiley, New York (1986)
5. Pappas, T., Laub, A.J., Sandell Jr, N.R.: On the numerical solution of the discrete-time algebraic riccati equation. IEEE Trans. Autom. Control **25**, 631–641 (1980)
6. Kleinman, D.L.: Stabilizing a discrete, constant linear system with application to iterative methods for solving the riccati equation. IEEE Trans. Autom. Control **19**, 252–254 (1974)

Polynomial Approximation of Quasipolynomials Based on Digital Filter Design Principles

Libor Pekař and Pavel Navrátil

Abstract This contribution is aimed at a possible procedure approximating quasipolynomials by polynomials. Quasipolynomials appear in linear time-delay systems description as a natural consequence of the use of the Laplace transform. Due to their infinite root spectra, control system analysis and synthesis based on such quasipolynomial models are usually mathematically heavy. In the light of this fact, there is a natural research endeavor to design a sufficiently accurate yet simple engineeringly acceptable method that approximates them by polynomials preserving basic spectral information. In this paper, such a procedure is presented based on some ideas of discrete-time (digital) filters designing without excessive math. Namely, the particular quasipolynomial is subjected to iterative discretization by means of the bilinear transformation first; consequently, linear and quadratic interpolations are applied to obtain integer powers of the approximating polynomial. Since dominant roots play a decisive role in the spectrum, interpolations are made in their very neighborhood. A simulation example proofs the algorithm efficiency.

Keywords Approximation · Bilinear transformation · Digital filter · MATLAB · Polynomials · Pre-warping · Quasipolynomials

1 Introduction

Mainly due to that delay appears in many real-word systems, such as economical, biological, networked, mechanical, electrical etc. [1, 2], this phenomenon have been intensively studied during recent decades [3, 4]. The most specific feature of time

L. Pekař (✉) · P. Navrátil
Faculty of Applied Informatics, Tomas Bata University in Zlín,
Zlín, Czech Republic
e-mail: pekar@fai.utb.cz

P. Navrátil
e-mail: pnavratil@fai.utb.cz

© Springer International Publishing Switzerland 2016   25
R. Silhavy et al. (eds.), *Automation Control Theory Perspectives*
in Intelligent Systems, Advances in Intelligent Systems and Computing 466,
DOI 10.1007/978-3-319-33389-2_3

delay systems (TDS) and models can be viewed in the fact that they own infinite spectra; thus, they are included in the family of infinite-dimensional systems. Linear time-invariant TDS, considered in this paper, can primarily be described by ordinary difference-differential (or shifted-argument) equations [5] which can be subjected to the Laplace transform [6]. As a consequence, the corresponding transfer function (matrix) is obtained in which its denominator quasipolynomial (QP) called also as the characteristic QP dominantly decides about dynamical and stability system properties [7]. QP roots (or zeros) agree with system poles, except for some special cases. Hence, the knowledge about the spectrum of QP roots is a crucial matter for TDS analysis.

Various tools and methods have been developed and designed for TDS spectrum computation or estimation, mainly in the state space domain, see e.g. [8]. These results represent a certain kind of TDS discretization as well. In the input-output Laplace space, the Quasi-Polynomial mapping Rootfinder (QPmR) has proved to be a very effective and practically usable when computing QP zeros within the determined region of the complex plane [9–11]. This method omits any QP simplification or approximation and a special software package was developed for its practical usability. TDS model reduction or rationalization represents another way how to cope with the problem [12–14]. However, expect for the well-known Padé approximation of exponential terms, these methods have been designed primarily for approximation of the complete TDS model, not only of the QP itself, and their applicability is mostly worsen due to a high mathematical knowledge level required out of the user. Delta models [15] represent an easy-to-handle rationalization and discretization methodology for practitioners, usable for TDS as well [16], which are closely related to the notion of the bilinear transformation [17] and give a polynomial discrete-time approximation representation of the system model.

The goal of this paper is to design a sufficiently simple, fast and practically usable technique for polynomial approximation of a QP without the necessity of advanced mathematical knowledge or using uncommon software tools. It is based on two main principles adopted from digital filter designing: As first, exponential terms in the QP are subjected to the innate time shifting and consequently to linear or quadratic interpolation such that the eventual shifts are integer multiplies of a basic time period. As second, s-powers representing derivatives are recursively put through the bilinear transformation the efficiency of which is further enhanced by pre-warping [17] that preserves the particular selected frequency under discretization. Since the dominant (i.e. the rightmost) pole (or the pair) has the decisive impact to system dynamics, all the interpolations and extrapolations are performed in the neighborhood of a close dominant QP root estimation.

The rest of the paper is organized as follows: Basic properties of zeros of a retarded QP and herein utilized techniques and tools are provided in the next, preliminary, section. Afterward, the reader is acquainted with the approximating procedure in details. A numerical example is given to illustrate the accuracy and efficiency of the technique; then the paper is concluded.

2 Preliminaries

Prior to the description of the approximation algorithm, retarded quasipolynomials and their spectral features ought to be introduced. This section, moreover, provides the reader with necessary mathematical tool and techniques that are utilized during the approximation procedure.

2.1 Retarded Quasipolynomial and Its Spectrum

A retarded QP can be expressed as

$$X(s, \mathbf{x}, \boldsymbol{\tau}) = s^n + \sum_{i=0}^{n-1} \sum_{j=1}^{h_i} x_{ij} s^i \exp\left(-s \sum_{k=1}^{L} \lambda_{ij,k} \tau_k\right) \tag{1}$$

where $\boldsymbol{\tau} = [\tau_1, .. \tau_L] \in \mathbb{R}_+^L, \tau_i > 0$ represents independent delays, $\lambda_{ij,k} \in \mathbb{N}_0$, $\mathbf{x} = [x_{01}, .. x_{n-1,h_{n-1}}] \in \mathbb{R}^{\sum_{i=0}^{n-1} h_i} \neq 0$.

Let $\Sigma := \{s_i\}$ be the spectrum of roots (zeros) of (1).

Property 1 [2, 5]. *For (1) it holds that*

1. *If there exist nonzero $x_{ij}, \lambda_{ij,k}$ for some positive τ_k and some i, j, k, then the number of QP zeros is infinite.*
2. *For any fixed real $\beta > -\infty$, the number of roots with $\mathrm{Re}\, s_i > \beta$ is finite.*
3. *Isolated roots behave continuously and smoothly with respect to $\boldsymbol{\tau}$ on $\mathbb{C}$.*

Definition 1 The *spectral abscissa*, $\alpha(\boldsymbol{\tau})$, is the function

$$\alpha(\boldsymbol{\tau}) := \boldsymbol{\tau} \mapsto \sup \mathrm{Re}\Sigma \tag{2}$$

Property 2 [18]. *For function $\alpha(\boldsymbol{\tau})$, it holds that:*

1. *It may be nonsmooth, and hence not differentiable, e.g. in points with more than one real root or conjugate pairs with the same maximum real part.*
2. *It is non-Lipschitz, for instance, at points where the maximum real part has multiplicity greater than one.*

To sum up main findings from Properties 1 and 2, although the rightmost part of the spectrum contains isolated roots, the position of which is continuously changed with $\boldsymbol{\tau}$, the abscissa might evince abrupt changes in its value.

Definition 2 The *leading* (dominant) *root*, s_L, or pair $\{s_L, \bar{s}_L\}$ satisfies

$$\alpha(\tau) = \text{Re } s_L = 0 \tag{3}$$

i.e. it represents the rightmost root or the pair from Σ.

2.2 Discretization Techniques and Tools

As mentioned above, the approximation algorithm is designed to be practically implementable by means of the most standard programs without the necessity of the use of special software and the knowledge of advanced math.

Derivatives in (1) are expressed by s-powers. The idea of the derivative approximation is based on the iterative digital-filter-like discretization of the QP depending on the current leading root estimation via the bilinear (or Tustin) transformation

$$s \rightarrow \frac{2}{T} \frac{1-q}{1+q} \tag{4}$$

where q means the shifting operator that agrees with z^{-1} in the z-transform and T is the sampling period.

Let $X(s, \mathbf{x}, \tau)$ be the characteristic quasipolynomial of a system. Since, however, transformation (4) does not preserve frequencies (namely, the system eigenfrequency), it is desirable to find another mapping that keeps "continuous" frequencies (i.e. those in the s-plane) and "discrete" frequencies (i.e. those in the z-plane) identical. It can be derived [17] that this requirement is satisfied if the following modified mapping is used

$$s \rightarrow \frac{\omega}{\tan(\omega T/2)} \frac{1-q}{1+q} \tag{5}$$

where ω stands for the desired frequency. Note that the operation of the frequency preservation is called *pre-warping*. Because of the decisive role of leading roots, we have set $\omega = \text{Im } s_L$.

From the point of view of derivative discretization or delta models, the value of T in (4) or (5) should be sufficiently small; however, the lower T is, the higher resulting approximating polynomial degree is obtained. In the contrary, the z-transform demands significantly lower values; for instance, in [19] the following recommendation for periodic systems is given

$$T = [0.2/\omega_0, 0.5/\omega_0] \tag{6}$$

where ω_0 expresses the frequency of undumped oscillations. Note that $\omega_0 \approx |s_L|$.

Regarding terms expressing delays in (1), i.e. the exponentials, they can be subjected to inherent shifting

$$\exp(-\vartheta s)X(s) \hat{=} x(t - \vartheta) \hat{=} q^{\vartheta/T}x(k) \hat{=} z^{-\vartheta/T}X(z) \tag{7}$$

Nevertheless, in general, delay value ϑ might not be an integer multiple of T; hence, term $z^{-\vartheta/T}$ should be interpolated by a linear combination of integer powers of z. The following lemma gives two possible solutions of this task.

Lemma 1 *Consider a term* $z^{-(\lfloor\vartheta\rfloor+\bar{\vartheta})}, \lfloor\vartheta\rfloor \in \mathbb{N}_0, 0 < \bar{\vartheta} < 1$. *In the vicinity of* $z_0 \in \mathbb{C}$, *the term can be interpolated linearly as (8) or quadratically as (9):*

$$(1 - \lfloor\vartheta\rfloor)z_0^{-\lfloor\vartheta\rfloor}z^{-\bar{\vartheta}} + \lfloor\vartheta\rfloor z_0^{-\lfloor\vartheta\rfloor+1}z^{-(\bar{\vartheta}+1)} \tag{8}$$

$$\begin{aligned} 0.5(2 - \lfloor\vartheta\rfloor)(1 - \lfloor\vartheta\rfloor)z_0^{-\lfloor\vartheta\rfloor}z^{-\bar{\vartheta}} + \lfloor\vartheta\rfloor(2 - \lfloor\vartheta\rfloor)z_0^{-\lfloor\vartheta\rfloor+1}z^{-(\bar{\vartheta}+1)} \\ + 0.5\lfloor\vartheta\rfloor(\lfloor\vartheta\rfloor - 1)z_0^{-\lfloor\vartheta\rfloor+2}z^{-(\bar{\vartheta}+2)} \end{aligned} \tag{9}$$

The proof is omitted due to the limited space. In this study, the value of z_0 is selected as

$$z_0 = z_L = \exp(Ts_L) \tag{10}$$

which agrees with the z-transform and is closely related to mapping (5), and it is consistent with the idea of the leading root importance.

3 Polynomial Approximation Algorithm

Two versions of the algorithm are presented. The former one can be utilized whenever the leading root of a QP sufficiently close to the studied QP. The latter can be used even if no leading root estimation is known and, naturally, it is expected to give less accurate results and is computationally more complex.

Algorithm 1 Input: The QP $X(s, \mathbf{x}, \tau)$ to be approximated.

Step 1: Consider that there exists a QP $X(s, \mathbf{x}_0, \tau_0)$ with $\|X(s, \mathbf{x}, \tau) - X(s, \mathbf{x}_0, \tau_0)\|$ $< \Delta$, for a sufficiently small $\Delta > 0$, the leading root of which, $s_0 = \hat{s}_0$, is known exactly. Set $\varepsilon > 0$.
Step 2: Compute polynomial $P(z^{-1}|\hat{s}_0)$ according to (4)–(9). Define and compute

$$\hat{s}_1 := \{\arg\min|s - \hat{s}_0| : s = T^{-1}\log(z), P(z^{-1}|\hat{s}_0) = 0\} \tag{11}$$

Step 3: While $|\hat{s}_1 - \hat{s}_0| \geq \varepsilon$, set $\hat{s}_0 := \hat{s}_1$ and go to Step 5.

Output: $P(z^{-1})$ and its roots.

Remark 1 The norm in Step 1 of Algorithm 1 can simply be computed as a point norm in s_0, i.e. $\|X(s_0, \mathbf{x}, \tau)\|$. The problem may appear if $|s_1 - s_0| > \delta$ for some $\delta > 0$ and any value of $\|X(s_0, \mathbf{x}, \tau)\|$ due to Property 2.

Algorithm 2 Input: The QP $X(s, \mathbf{x}, \tau)$ to be approximated.

Step 1: Define the mesh grid $\tau_{k,j+1} = \tau_{k,j} + \Delta\tau_{k,j}$, $\tau_{k,0} = 0$, $\tau = \left[\tau_{1,N}, \tau_{2,N}, \ldots, \tau_{L,N}\right]$, $k = 1\ldots L$, $j = 0\ldots N - 1$, and set $\varepsilon > 0$.
Step 2: Compute

$$\hat{s}_{0,\ldots,0} = s_{0,\ldots,0} := \{\arg\max\operatorname{Re} s : X(s, \mathbf{x}, \mathbf{0}) = 0\} \tag{12}$$

exactly.
Step 3: For $(j_1 = 0\ldots N - 1$, for $(j_2 = 0\ldots N - 1,\ldots$ (for $j_L = 0\ldots N - 1$ do: If $\exists j_l \neq 0, l = 1\ldots L$ do Steps 4–6))) (nested loops).
Step 4:   Define $M := \max\{k : j_k \neq 0\}$ and   set $\bar{\tau} = \left[\tau_{1,j_1}, \tau_{2,j_2}\ldots, \tau_{L,j_L}\right]$, $\hat{s}_0 = \hat{s}_{j_1,\ldots,j_{M-1},j_M-1,0\ldots0}$.
Step 5: Compute polynomial $P(z^{-1}|\bar{\tau}, \hat{s}_0)$ according to (4)–(9) and $\hat{s}_1$ by means of (11).
Step 6: While $|\hat{s}_1 - \hat{s}_0| \geq \varepsilon$, set $\hat{s}_0 := \hat{s}_1$ and go to Step 5.

Output: $P(z^{-1})$ and its roots.

4   Numerical Example

Consider a skater on the remotely swaying bow sketched in Fig. 1. The skater controls the power input, $P(t)$, to the servo giving rise to the angle deviation, $u(t)$, from the horizontal position and, consequently, the angle between the skater and the bow symmetry axis, $y(t)$, emerges. If the friction is neglected but the skater's reaction time and servo latency included, the following particular transfer function can be written [20]

Fig. 1 A skater on the remotely controlled swaying bow

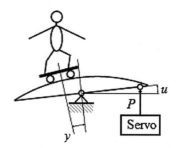

$$G(s) = \frac{0.2\exp(-(\tau_1 + \tau_2)s)}{s^2(s^2 - \exp(-\tau_2 s))} \qquad (13)$$

A particular generalized (third-order) finite-dimensional linear proportional-integrative-derivative controller stabilizes system (13) with $\tau = [\tau_1, \tau_2] = [0.08, 0.08]$, which yields the feedback characteristic quasipolynomial (14), see [21].

$$X(s, \cdot, \tau) = s^2(s^2 - \exp(-0.08s))(s^3 + 469418.6s^2 + 640264.6s + 10560107)$$
$$+ 0.2\exp(-0.16s)(82226506s^3 + 106523134s^2 + 26247749s + 5617613)$$
$$(14)$$

the roots of which decide about control system stability.

Let us test Algorithm 1 for various values of T first. With respect to the upper bound of condition, the sampling period can be written as $T = (k_T|s_L|)^{-1}, k_T \geq 2$, the value of which is updated in every iteration step (see Steps 2 and 3 in Algorithm 1). Assume that the leading pair, $s_0 = -9.9669e - 3 \pm 3.9672704i$, of $X(s, \mathbf{x}_0 = \mathbf{x}, [0.08, 0.07])$ is known exactly and set $\varepsilon = e - 6$. Selected algorithm results are summarized in Table 1. Due to limited space, some observations are further commented rather than being included in the table.

Table 1 Results of Algorithm 1

Method	k_T	Dominant pair of roots		Num. of iter.
(4), (8)	2	−2.11963e–2 + 3.8592406i	−9.1702e–2 – 3.917348i	5
	5	−2.80361e–2 + 3.9172242i	−3.60554e–2 – 3.9223709i	4
	10	−2.90655e–2 + 3.9254363i	−3.30879e–2 – 3.9277296i	3
	20	−2.9325e–2 + 3.927487i	−3.05583e–2 – 3.9281274i	3
	30	−3.8027e–3 + 3.9947106i	−3.504e–3 – 3.9948482i	8
(5), (8)	2	−2.83932e–2 + 3.9282875i	−0.1062955 – 3.9922307i	5
	5	−2.92479e–2 + 3.9281888i	−3.78099e–2 – 3.9336952i	4
	10	−2.93707e–2 + 3.9281745i	−3.34359e–2 – 3.9304912i	3
	20	−2.94013e–2 + 3.92817i	−3.06333e–2 – 3.9288096i	3
	30	−3.77434e–3 + 3.9950239i	−3.47358e–3 – 3.9951627i	14
(4), (9)	2	−2.11963e–2 + 3.8592406i	−6.70344e–2 – 3.8332474i	5
	5	−2.80361e–2 + 3.9172242i	−2.96717e–2 – 3.9154286i	4
	10	−2.90655e–2 + 3.9254364i	−2.93613e–2 – 3.924999i	3
	20	−2.93249e–2 + 3.9274854i	−2.93571e–2 – 3.9274282i	3
	30	−1.29309e–2 + 3.9607762i	−1.29307e–2 – 3.960776i	13
(5), (9)	2	−2.83932e–2 + 3.9282875i	−7.80977e–2 – 3.8996772i	5
	5	−2.92479e–2 + 3.9281888i	−3.09832e–2 – 3.9262835i	4
	10	−2.93707e–2 + 3.9281746i	−2.96686e–2 – 3.9277336i	3
	20	−2.94014e–2 + 3.9281712i	−2.94335e–2 – 3.9281141i	3
	30	−1.29634e–3 + 3.9610884i	−1.29632e–3 – 3.961088i	14

Table 2 Values of dominant pairs for selected values of k_T

Method	k_T	Dominant pair of roots		Num. of iter.
(4), (8)	10	−2.90655e−2 + 3.9254363i	−3.30879e−2 − 3.9277296i	3
	20	−2.93249e−2 + 3.927486i	−3.05583e−2 − 3.9281274i	3.05
(5), (8)	10	−2.93707e−2 + 3.9281745i	−3.34359e−2 − 3.9304912i	3
	20	−2.94013e−2 + 3.9281699i	−3.06333e−2 − 3.9288096i	3.01
(4), (9)	10	−2.90655e−2 + 3.9254363i	−2.93613e−2 − 3.924999i	3
	20	−2.9325e−2 + 3.927487i	−2.93572e−2 − 3.9274298i	3
(5), (9)	10	−2.93707e−2 + 3.9281745i	−2.96686e−2 − 3.9277336i	3
	20	−2.94014e−2 + 3.9281706i	−2.94334e−2 − 3.9281135i	3.05

The exact leading pair of roots of $X(s, \mathbf{x}, \tau)$ found by the QPmR reads $s_L = -0.0294116 \pm 3.9281699i$. However, in fact, leading roots of the approximating polynomial do not constitute a conjugate pair because of its complex coefficients. We have observed from the test that higher values of k_T (limited by an upper bound of approx. $k_T \approx 30$) give less number of iterations within Steps 2 and 3, better approaching of leading roots and higher leading root estimation accuracy. The beneficial impact of pre-warping (5) can be seen in significantly better leading root estimation (mainly in the imaginary part of the root); however, such an improvement is not confirmed in the case of less dominant roots. The advantage of the quadratic interpolation (9) compared to the linear one (8) can also be observed. It distinctively improves less-dominant pairs mutual approaching and has a slight impact to their loci estimation and the approaching of the leading pair. In the contrary, it has no effect on the leading poles estimation.

However, the discretization procedure gives rise to "parasitic" high-frequency roots not included in the original QP. These polynomial roots are located very close to the imaginary axis with a high imaginary part value. Such an observation has been made by Vyhlídal and Zítek [16] as well; they have utilized a delta models in their work.

Algorithm 2 applied to (14) starts with $s_{0,...,0} = 0.1214757 \pm 4.5573833i$, see (12), and let $\Delta \tau_{k,j} = 0.01$. Eventual values of dominant pairs for very selected values of k_T (to be concise) are displayed in Table 2. Apparently, the results are very close to those in Table 1, which implies that the polynomial approximation based on the information about the dominant (leading) root is sufficiently robust with respect to successive procedure of delay values shifting introduced in Algorithm 2.

5   Conclusion

The presented paper has been aimed at the possible polynomial approximation of a quasipolynomial by means of tools and techniques used for digital filters design; namely, the bilinear transformation with/without pre-warping and a specific linear and quadratic interpolation for the acquisition of commensurate delays. This activity is useful mainly for stability and dynamical analysis of time delay systems when the characteristic quasipolynomial, as the transfer function denominator, is analyzed. Since the decisive information is contained in a small number of the rightmost, i.e. leading, quasipolynomial zeros, the approximating polynomial has been found iteratively based on the leading root estimation. A rather tricky step is a proper choice of the discretization step and the sampling period.

The presented simulation example has indicated the beneficial impact of the use of pre-warping and more complex, i.e. quadratic, interpolation to the leading root estimation accuracy and the approaching of both dominant roots in the complex conjugate pair, respectively. Besides the leading pair, a small number of less-dominant pairs must also be observed. This feature might play a leading role in the process of the determination of our future research in this field.

Acknowledgments The work was performed with the financial support by the Ministry of Education, Youth and Sports of the Czech Republic within the National Sustainability Programme project No. LO1303 (MSMT-7778/2014) and also by the European Regional Development Fund under the project CEBIA-Tech No. CZ.1.05/2.1.00/03.0089.

References

1. Chiasson, J., Loiseau, J.J.: Applications of Time Delay Systems. Springer, New York (2007)
2. Sipahi, R., Vyhlídal, T., Niculescu, S.-I., Pepe, P.: Time Delay Systems: Methods, Applications and New Trends. LNCIS, vol. 423. Springer, New York (2012)
3. Richard, J.P.: Time-Delay systems: an overview of some recent advances and open problems. Automatica **39**, 1667–1694 (2003)
4. Loiseau, J.J., Michiels, W., Niculescu, S.-I., Sipahi, R.: Topics in Time Delay Systems: Analysis, Algorithm and Control. LNCIS, vol. 388. Springer, Berlin (2009)
5. Hale, J.K., Verduyn Lunel, S.M.: Introduction to Functional Differential Equations. Applied Mathematical Sciences, vol. 99. Springer, New York (1993)
6. Zítek, P., Víteček, A.: Control Design of Time-Delay and Nonlinear Subsystems. CTU Publishing (1999) (in Czech)
7. Gu, K., Kharitonov, V.L., Chen, J.: Stability of Time-Delay Systems. Birkhäuser, Boston (2003)
8. Breda, D., Maset, S., Vermiglio, R.: Pseudospectral differencing methods for characteristic roots of delay differential equations. SIAM J. Sci. Comput. **27**, 482–495 (2005)
9. Vyhlídal, T., Zítek, P.: Quasipolynomial mapping algorithm rootfinder for analysis of time delay systems. In: Proceedings of the 4th IFAC Workshop on Time-Delay Systems (TDS 2003). Rocquencourt, France (2003)
10. Vyhlídal, T., Zítek, P.: Mapping based algorithm for large-scale computation of quasipolynomial zeros. IEEE Trans. Autom. Control **54**, 171–177 (2009)
11. Vyhlídal, T., Zítek, P.: QPmR—Quasi-Polynomial Root-Finder: Algorithm Update and Examples. In: Vyhlídal, T., Lafay, J.-F., Sipahi, R. (eds.) Delay Systems: From Theory to Numerics and Applications, pp. 299–312. Springer, New York (2014)
12. Partington, J.R.: Some frequency-domain approaches to the model reduction of delay systems. Ann. Rev. Control **28**, 65–73 (2004)
13. Pekař, L.: On a controller parameterization for infinite-dimensional feedback systems based on the desired overshoot. WSEAS Trans. Syst. **12**, 325–335 (2013)

14. Seuret, A., Özbay, H., Bonnet, C., Mounier, H.: Low Complexity Controllers for Time Delay Systems. Advances in Delays and Dynamics, vol. 2. Springer, New York (2014)
15. Middleton, R.H., Goodwin, G.C.: Digital Control and Estimation: A Unified Approach. Prentice Hall, Detroit (1990)
16. Vyhlídal, T., Zítek, P.: Discrete Approximation of a Time Delay System and Delta Model Spectrum. In: Proceedings of the 16th IFAC World Congress, p. 636. IFAC, Prague (2005)
17. Oppenheim, A.: Discrete Time Signal Processing. Pearson Higher Education, Upper Saddle River, NJ (2010)
18. Vanbiervliet, T., Verheyden, K., Michiels, W., Vandewalle, S.: A nonsmooth optimization approach for the stabilization of time-delay systems. ESIAM Control Optim. Ca. **14**, 478–493 (2008)
19. Baláté, J.: Automatic Control. BEN Publishing, Prague (2004). (in Czech)
20. Zítek, P., Kučera, V., Vyhlídal, T.: Meromorphic observer-based pole assignment in time delay systems. Kybernetika **44**, 633–648 (2008)
21. Pekař, L.: A Simple DDS Algorithm for TDS: An Example. In: Proceedings of the 29th European Conference on Modelling and Simulation (ECMS 2015), pp. 246–251. European Council for Modelling and Simulation (ECMS), Varna, Bulgaria (2015)

An Implementation of a Tilt-Compensated eCompass

Martin Sysel

Abstract This paper describes implementation of an electronic compass and calibration method. Firstly describes used hardware and then focus on the method of sensor calibration parameters. The eCompass uses a three-axis accelerometer and three-axis magnetometer. The compass heading is a function of all three accelerometer readings and all three magnetometer readings. The accelerometer measures the components of the earth gravity and provide pitch and roll angle information which is used to correct the magnetometer data. The magnetometer measures the components of earth's magnetic field (called geomagnetic field) to determine the heading angle to the magnetic north.

Keywords eCompass · MEMS · Accelerometer · Magnetometer · Calibration

1 Introduction

The first compass was probably a magnetized stone, that when suspended, would always point the same way. No one knows who first discovered the compass. The Chinese understood its use 3,000 years before Europeans learnt to travel without using the sun or the stars. Marco Polo is reputed to have brought the compass back to Europe on his return from Cathay in 1260. Todays all non-electronics compasses use a magnetized steel needle, supported in the middle. The important points of a compass are North, East, South and West; always read clockwise around the circle. North is found at 0° (which is also 360°), East becomes magnetic azimuth 090 (90°), South becomes magnetic azimuth 180 (180°), West becomes magnetic azimuth 270 (270°). More accurate directions are given by using all the numbers in between these. Always give readings in degrees, it is more accurate [1].

M. Sysel (✉)
Faculty of Applied Informatics, Tomas Bata University in Zlín, Zlín,
Czech Republic
e-mail: Sysel@fai.utb.cz

© Springer International Publishing Switzerland 2016   35
R. Silhavy et al. (eds.), *Automation Control Theory Perspectives*
in Intelligent Systems, Advances in Intelligent Systems and Computing 466,
DOI 10.1007/978-3-319-33389-2_4

Fig. 1 Observed north dip
poles

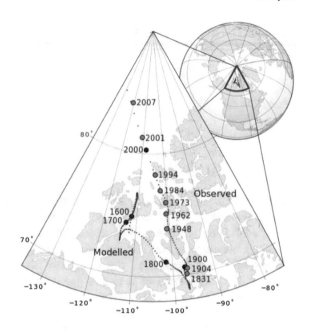

The compass points to the magnetic north, not true geographic North Pole. The magnetic north pole is actually south of the geographic North Pole. The compass therefore points a little to the side of the geographic North Pole, and this varies depending on where you are on Earth, and also with time. In fact, the magnetic pole migrate over time, it moves about 55 km a year, and the movement has speeded up recently [2]. Figure 1 shows observed north dip poles during 1831–2007 which are marked as yellow circles. Modeled pole locations from 1590 to 2020 are blue circles.

The Earth acts like a large spherical magnet: it is surrounded by a magnetic field. That changes with time and location. At any point and time, the Earth's magnetic field is characterized by a direction and intensity which can be measured.

The intensity of the magnetic field is about 0.25–0.65 gauss (25000–65000 nT) and has a component parallel to the earth surface that always points toward the magnetic North Pole. In the northern hemisphere this field point down. At the equator it points horizontally and in the southern hemisphere it points up. This angle between the earth's magnetic field and horizontal plane is defined as an inclination angle. Another angle between the earth's magnetic north and geographic north is defined as a declination angle in the range of ±20° depending on geographic location [3]. A magnetic compass needle tries to align itself with the magnetic field lines. However, near the magnetic poles, the fields are vertically converging. The strength and direction tend to "tilt" the compass needle up or down into the Earth.

Fig. 2 eCompass coordinate system [3]

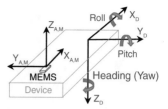

A tilt compensated electronics compass system uses a three-axis accelerometer sensor and three-axis magnetic sensor. The accelerometer is used to measure the tilt angles of pitch and roll for compensation. The magnetic sensor is used to measure the earth's magnetic field. Then it is possible to calculate the compass heading angle (yaw) with respect to the magnetic north. The declination angle at the current geographic location should be used for compensation of the heading to obtain true geographic North Pole. The Fig. 2 shows coordinate system which uses the industry standard "NED" (North, East, Down) to label axes on the device. The X_D axis of the device is the eCompass pointing direction, the Y_D axis points to the right and Z_D axis points downward. Positive yaw angle is defined to be a clockwise rotation about the positive Z_D axis.

Similar, positive pitch and roll angles are defined as clockwise rotations. When MEMS is installed in a device as shown in Fig. 2 then the sign of $Y_{A,M}$ and $Z_{A,M}$ from the sensor measurements needs to be reversed to make the sensing axes the same direction as the device axes [3, 4].

2 Accelerometer

2.1 Physical Principles and Structure

An accelerometer is an electromechanical device that measures acceleration forces. These forces may be static, like the gravity, or they could be dynamic—caused by moving or vibrating. There are many types of accelerometers and there are many different ways to make an accelerometer. Some accelerometers contain microscopic crystal structures that get stressed by accelerative forces use the piezoelectric effect. Another sensing changes in capacitance. Capacitive sensing has excellent sensitivity.

Typical MEMS accelerometer is composed of movable proof mass with plates that is attached through a mechanical suspension system to a reference frame, as shown in Fig. 3. A MEMS accelerometer differs from integrated circuits in that a proof mass is machined into the silicon. Any displacement of the component causes this mass to move slightly according to Newton's second law, and that change is

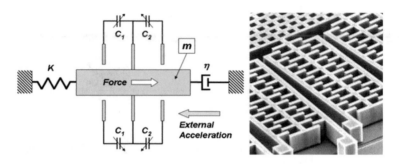

Fig. 3 Schematic and internal structure of a capacitive accelerometer [6]

detected by sensors. Usually the proof mass disturbs the capacitance of a nearby node; that change is measured and filtered. Movable plates and fixed outer plates represent capacitors. The deflection of proof mass is measured using the capacitance difference [5]. The free-space (air) capacitances between the movable plate and two stationary outer plates C_1 and C_2 are functions of the corresponding displacements [6].

The most important specification is the number of axis. The MEMS proof mass can measure only one parameter in each available axis, so a one axis device can sense acceleration in a single direction. Three axis units return sensor information in the X, Y, and Z directions.

Accelerometer sensors measure the difference between any linear acceleration in the accelerometer's reference frame and the earth's gravitational field vector. The earth's gravitational field is defined by a force vector that points directly down towards the earth's core. In the absence of linear acceleration, the accelerometer output is a measurement of the rotated gravitational field vector and can be used to determine the accelerometer pitch and roll orientation angles. The orientation angles are dependent on the order in which the rotations are applied. The most common order is the aerospace sequence of yaw then pitch and finally a roll rotation [7].

Three-axis accelerometers supplied for the consumer market are typically calibrated by the sensor manufacturer using a six-element linear model comprising a gain and offset in each of the three axes. This factory calibration will change slightly as a result of the thermal stresses during soldering of the accelerometer to the circuit board. Rotation of the accelerometer package relative to the circuit board and misalignment of the circuit board to the final product will also add small errors. The original factory accelerometer calibration is adequate for the majority of consumer applications. However, own calibration improve accuracy for high-accuracy applications (tilt applications, accuracies below 2°).

2.2   Accelerometer Calibration

The relation between corrected accelerometer output and accelerometer raw measurements [3, 5] can be expressed as

$$
\begin{bmatrix} A_x \\ A_y \\ A_z \end{bmatrix} = \begin{bmatrix} Am_{xyz} \end{bmatrix}_{3x3} \begin{bmatrix} \frac{1}{S_x} & 0 & 0 \\ 0 & \frac{1}{S_y} & 0 \\ 0 & 0 & \frac{1}{S_z} \end{bmatrix} \cdot \begin{bmatrix} R_x - O_x \\ R_y - O_y \\ R_z - O_z \end{bmatrix} \tag{1}
$$

where

A is the corrected reading in X, Y and Z axes.
Am is the 3×3 misalignment matrix between sensor axes and device axes.
S is the sensitivity of each channel.
R is the raw data from the accelerometer.
O is the accelerometer's zero-g level.

The sensitivities and offsets are constants, so the matrix can be simplified to:

$$
\begin{bmatrix} A_x \\ A_y \\ A_z \end{bmatrix} = \begin{bmatrix} C_{11} & C_{12} & C_{13} \\ C_{21} & C_{22} & C_{23} \\ C_{31} & C_{32} & C_{33} \end{bmatrix} \cdot \begin{bmatrix} R_x \\ R_y \\ R_z \end{bmatrix} + \begin{bmatrix} C_{10} \\ C_{20} \\ C_{30} \end{bmatrix} \tag{2}
$$

The main goal of accelerometer calibration is to determine 12 parameters [8, 9] from C_{10} to C_{33}. This is typically done using the least squares method [3, 4, 7].

The original 6-parameter (gain and offset in each channel) factory calibration can be recomputed to correct for thermal stresses introduced in the soldering process. Used 12 parameter linear calibration model can correct for accelerometer package rotation on the circuit board and for cross-axis interference between the accelerometer's x, y and z channels. Another recalibration, a 15 parameter model includes cubic nonlinearities in the accelerometer response can grow up accuracy. The orientation angles used for the recalibration must be carefully selected to provide the best calibration accuracy from the limited number of measurement orientations available [10].

It is possible rewrite (2) as

$$
[A_x \quad A_y \quad A_z][R_x \quad R_y \quad R_z \quad 1] \cdot \begin{bmatrix} C_{11} & C_{21} & C_{31} \\ C_{12} & C_{22} & C_{32} \\ C_{13} & C_{23} & C_{33} \\ C_{10} & C_{20} & C_{30} \end{bmatrix} \tag{3}
$$

and then

$$Y = w \cdot X \tag{4}$$

where

Matrix X is calibration parameters.
Matrix w is raw data from the accelerometer collected at 6 stationary positions (when device lays on each side of the device and normalized gravity vector is zero in two axis). For better accuracy is possible add more stationary positions.
Matrix Y is the known normalized gravity vector.

Finally, calculation of 12 calibration parameters by least squares method can be written as

$$X = \left[w^T \cdot w\right]^{-1} \cdot w^T \cdot Y \tag{5}$$

3   Magnetometer

A magnetometer is a type of sensor that measures the strength and direction of the local magnetic field. The magnetic field measured is a combination of both the earth's magnetic field and any magnetic field created by nearby objects. The accuracy is highly dependent on the calculation and subtraction in software of stray magnetic fields both within, and in the vicinity of, the magnetometer on the PCB (Printed Circuit Board). By convention, these fields are divided into those that are fixed (termed Hard-Iron effects) and those that influence or distort geomagnetic field (termed Soft-Iron effects), see Fig. 4.

Hard-iron interference magnetic field is normally generated by ferromagnetic materials with permanent magnetic fields. These materials could be permanent

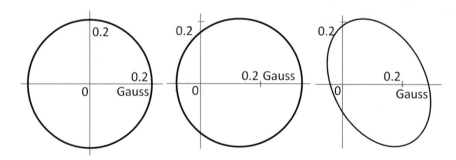

Fig. 4 Magnetic field distortions

magnets or magnetized iron or steel, for example the metal chassis or frame of the device, any actual magnets attached such as speakers etc. This interference pattern is unique to the device and is constant (time invariant). The unwanted magnetic fields are superimposed on the output of the magnetic sensor measurements of the earth's magnetic field. The effect of this superposition is to bias the magnetic sensor outputs. If we consider the magnetic data circle (the magnetic field strength in one horizontal direction), hard iron interference shift the entire circle away from the origin by some amount [11] (see Fig. 4). The amount is dependent on different factors and can greatly exceed the geomagnetic field.

A soft-iron interference magnetic field is generated by the items inside the device. They could be current carrying traces on the PCB or magnetically soft materials. They generate a time varying magnetic field that is superimposed on the magnetic sensor output in response to the earth's magnetic field. The effect of the soft-iron distortion is to make a full round rotation circle become a tilted ellipse. The distortion in this case depends on the direction that the compass is facing. Because of this the distortion cannot be calibrated out with a simple offset. Figure 4 shows the "Undistorted Field" diagram, the hard iron distortion and an additional soft iron distortion results an ellipse. Devices without strong soft-iron interference can simplify calculation, by comprising only the hard-iron offsets and the geomagnetic field strength.

The relationship between the normalized vector and the magnetic sensor raw data can be expressed as [3]:

$$\begin{bmatrix} M_x \\ M_y \\ M_z \end{bmatrix} = \begin{bmatrix} Mm_{xyz} \end{bmatrix}_{3x3} \begin{bmatrix} \frac{1}{S_x} & 0 & 0 \\ 0 & \frac{1}{S_y} & 0 \\ 0 & 0 & \frac{1}{S_z} \end{bmatrix} \cdot \begin{bmatrix} Ms_{xyz} \end{bmatrix}_{3c3} \cdot \begin{bmatrix} R_x - O_x \\ R_y - O_y \\ R_z - O_z \end{bmatrix} \tag{6}$$

where

M is the corrected reading in X, Y and Z axes.
Mm is the 3×3 misalignment matrix between sensor axes and device axes.
S is the sensitivity of each channel.
R is the raw data from the magnetometer.
Ms is the 3×3 matrix caused by soft iron distortion.
O is the offset caused by hard iron distortion.

The main goal of the magnetometer sensor calibration is to determine parameters to the normalized values could be obtained at any measurement position [3, 7, 10].

4 eCompass Heading Calculation

The accelerometer readings provide pitch and roll angle information which is used to correct the magnetometer data. This allow accurate calculation of the compass heading (yaw) when the device is not held flat. The pitch and roll angles are computed on assumption that the accelerometer readings result entirely from the compass orientation in the earth's gravitational filed. Fast acceleration (or low-g conditions) influence pitch and roll calculation, so the accuracy of an electronic compass means static accuracy when the device is without high acceleration. If the PCB remains flat, then the compass heading could be computed from the arctangent of the ratio of the two horizontal magnetic field components. In general, the PCB has an arbitrary orientation, the compass heading is a function of all three accelerometer readings and all three magnetometer readings [4].

Pitch and roll can be calculated by equations [3, 4]:

$$Pitch = \rho = \arcsin(-A_x) \tag{7}$$

$$Roll = \gamma = \arcsin\left(\frac{A_y}{\cos \rho}\right) \tag{8}$$

Pitch is defined as the angle between the X_D axis and the horizontal plane. Roll is defined as the angle between Y_D axis and the horizontal plane. The acrsin function has good linearity between about $-45°$ to $+45°$, so accuracy degrades when pitch or roll angles exceed this range [3]. All accelerometers are completely insensitive to rotations about the gravitational field vector and cannot be used to determine such a rotation. The unknown yaw angle represents the device rotation from magnetic north. Its determination requires a magnetometer sensor.

For he heading calculation, magnetometer measurements need to be normalized by applying calibration parameters and then by tilt compensation. Tilt compensated magnetic sensor measurements can be obtained as [3, 4]:

$$M_{x2} = M_x \cos \rho + M_z \sin \rho \tag{9}$$

$$M_{y2} = M_x \sin \gamma \sin \rho + M_y \cos \gamma - M_z \sin \gamma \cos \rho \tag{10}$$

$$M_{z2} = -M_x \cos \gamma \sin \rho + M_y \sin \gamma + M_z \cos \gamma \cos \rho \tag{11}$$

where M_x, M_y and M_z are normalized magnetic sensor measurements with applied calibration correction (6).

Subsequently, the x and y components can be used for heading angle calculation (yaw angle) which is computed to magnetic north [3, 4]:

$$Heading = \Psi = \arctan\left(\frac{M_{y2}}{M_{x2}}\right) \tag{12}$$

Equation (12) is valid for positive values of M_{x2} and M_{y2}. When M_{x2} is positive and M_{y2} component is negative then a negative result is summed with 360°. For the negative value of M_{x2} need to be add 180° to heading. Zero value of the M_{x2} means that device points to East (90°—M_{y2} is negative value) or West (270°—M_{y2} is positive value). Finally, value should be modified by declination which is valid at the current location.

5 Conclusion

An implementation of a tilt compensated eCompass using accelerometer and magnetometer has been described. The accelerometer sensor output is used by the tilt-compensated eCompass algorithms to compute the roll and pitch angles. These allow tilt correction of the magnetometer data. The magnetometer readings must be corrected for hard-iron and soft-iron effects.

Although, all MEMS are factory calibrated for common using, it is recommended to perform own calibration to reach a better accuracy. It is possible to say that eCompass can be calibrated with no a priory information about the direction of magnetic North Pole. Techniques of calibration can be extended to include temperature dependence by performing the recalibration at several temperatures and interpolating the fitted calibration parameters to the actual temperature [9].

References

1. Scouts New Zealand, Compass Basics, New Zealand (2012)
2. NOAA, Wandering of the Geomagnetic poles, National Oceanic And Atmospheric Administration. http://www.ngdc.noaa.gov/geomag/GeomagneticPoles.shtml
3. STMicroelectronics, AN3192 Application note: using LSM303DLH for a tilt compensated electronic compass, 34p. (2010). https://www.pololu.com/file/0J434/LSM303DLH-compass-app-note.pdf
4. Ozyagcilar, T.: AN4248 Application note: Implementing a Tilt-Compensated eCompass using Accelerometer and Magnetometer Sensors (2013). http://cache.freescale.com/files/sensors/doc/app_note/AN4248.pdf
5. Ganssle, J.: A Designer's Guide to MEMS Sensors (2012). http://www.digikey.com/en/articles/techzone/2012/jul/a-designers-guide-to-mems-sensors
6. Cacchione, F.: Mechanical Characterisation and simulation of fracture processes in polysilicon MEMS, PhD Thesis. Politecnicio di Milano (2007). st.com/web/en/resource/technical/document/white_paper/phd_thesis.pdf
7. Pedley, M.: AN3461 Application note: Tilt Sensing Using a Three-Axis Accelerometer (2013). http://www.freescale.com/files/sensors/doc/app_note/AN3461.pdf
8. Stančin, S., Tomažič, S.: Time-and computation-efficient calibration of MEMS 3D accelerometers and gyroscopes. Sensors **14.8**, 14885–14915 (2014). http://www.mdpi.com/1424-8220/14/8/14885/pdf

9. Fang, B., Chou, W., Ding, L.: An optimal calibration method for a MEMS inertial measurement unit. Int. J. Adv. Robot. Syst. **11.14**, 57516 (2014). http://cdn.intechopen.com/pdfs-wm/46177.pdf
10. Pedley, M.: AN4399 Application note: High Precision Calibration of a Three-Axis Accelerometer (2013). http://cache.freescale.com/files/sensors/doc/app_note/AN4399.pdf
11. Skula, D.: New Method for Magnetometer Offset Compensation, Annals of DAAAM for 2010 and proceedings, Zadar (2010) ISBN 978-3-901509-73-5

Calibration of Triaxial Accelerometer and Triaxial Magnetometer for Tilt Compensated Electronic Compass

Ales Kuncar, Martin Sysel and Tomas Urbanek

Abstract This research paper describes the method for the calibration of accelerometer and magnetometer for tilt compensated electronic compass. The electronic compass is implemented using triaxial MEMS accelerometer and triaxial MEMS magnetometer. The heading of the compass is generally influenced by scale factors, offsets and misalignment errors of these sensors. The proposed calibration method determines twelve calibration parameters in six stationary positions for accelerometer and twelve calibration parameters in 3D rotations for magnetometer.

Keywords Accelerometer · Calibration · Electronic compass · Magnetometer · Micro-electro-mechanical-system

1 Introduction

The electronic compasses are a crucial navigation tool in many areas even in present time of the global positioning system (GPS). They are used as a component for dead reckoning for determining the location. However, the electronic compasses are vulnerable to a variety of external influences such as hard iron and soft iron interferences and accelerations errors which can affect the calculated heading.

The electronic compasses are based on the measurement of the Earth's geomagnetic field. The Earth can be considered as a magnetic dipole with poles near the North Pole and the South Pole. The magnetic vectors points toward the

A. Kuncar (✉) · M. Sysel · T. Urbanek
Faculty of Applied Informatics, Tomas Bata University in Zlin,
Nad Stranemi 4511, Zlín, Czech Republic
e-mail: kuncar@fai.utb.cz

M. Sysel
e-mail: sysel@fai.utb.cz

T. Urbanek
e-mail: turbanek@fai.utb.cz

© Springer International Publishing Switzerland 2016   45
R. Silhavy et al. (eds.), *Automation Control Theory Perspectives in Intelligent Systems*, Advances in Intelligent Systems and Computing 466,
DOI 10.1007/978-3-319-33389-2_5

Fig. 1 Earth's magnetic
field [3]

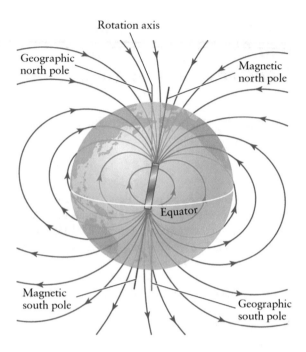

North Pole and near the equator, the magnetic vectors are parallel to the surface
(Fig. 1) [1].

This inclination, angle between the magnetic vector and the horizontal plane, is
highly dependent on the geographical latitude (Fig. 2); same as the intensity. The
intensity of the Earth's magnetic field is approximately between 25 and 65 μT
(0.25 – 0.65 gauss) [2].

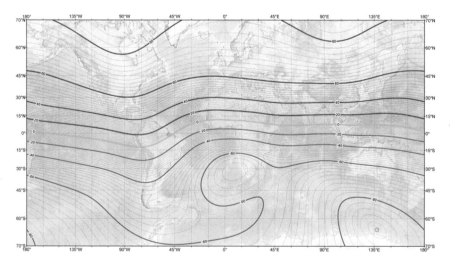

Fig. 2 Magnetic inclination map [4]

The common approach measures two orthogonal components of the magnetic vector from magnetometer to estimate the heading. However, if the sensor is tilted, the measured values of these components change and the calculated heading is not accurate. To avoid this error, a triaxial accelerometer must be coupled to compensate the tilt [1].

In this research paper, we proposed the method for the calibration of accelerometer and magnetometer for the implementation of tilt compensated electronic compass. The electronic compass is implemented using triaxial MEMS accelerometer and triaxial MEMS magnetometer.

The reminder of this paper is organized as follows. The accelerometer calibration and magnetometer calibration is described in Sects. 2 and 3. Section 4 explains the procedure of tilt compensation algorithm.

2 Accelerometer

2.1 Accelerometer Error Model

The accelerometers suffer from a variety of error sources which are slightly different depending on the type of accelerometer.

Ordinarily, the accelerometer measurement (R_X) can be represented in form of the acceleration along its sensitive axis $(\tilde{A}_X)$ and the accelerations along the pendulum and hinge axes, $\tilde{A}_Y$ and $\tilde{A}_Z$ respectively, by equation [5]:

$$R_X = (1 + S_X) \cdot \tilde{A}_X + M_Y \cdot \tilde{A}_Y + M_Z \cdot \tilde{A}_Z + B_f + B_v \cdot \tilde{A}_X \cdot \tilde{A}_Y + n_X \qquad (1)$$

where

S_X is the scale factor,
M_Y, M_Z are the cross-axis coupling factors,
B_f is the measurement bias,
B_v is the vibro-pendulous error coefficient,
n_X is the random noise.

For MEMS accelerometers, it is expected that the cross-axis coupling factors and vibro-pendulous error are negligible [6]. Then Eq. (1) can be simplified to

$$R_X = \tilde{A}_X + S_X \cdot \tilde{A}_X + B_f + n_X \qquad (2)$$

2.2 Accelerometer Calibration

The relation between the raw accelerometer readings and the correct accelerometer outputs is described as [2]:

$$\begin{bmatrix} A_X \\ A_Y \\ A_Z \end{bmatrix} = [Am]_{3x3} \cdot \begin{bmatrix} \frac{1}{S_X} & 0 & 0 \\ 0 & \frac{1}{S_Y} & 0 \\ 0 & 0 & \frac{1}{S_Z} \end{bmatrix} \begin{bmatrix} R_X - O_X \\ R_Y - O_Y \\ R_Z - O_Z \end{bmatrix} \quad (3)$$

where

A_X, A_Y, A_Z are known correct accelerometer outputs,
Am is a 3×3 misalignment matrix between accelerometer axes and device axes,
S_X, S_Y, S_Z are the scale factors,
R_X, R_Y, R_Z are the raw accelerometer readings,
O_X, O_Y, O_Z are the offsets.

Equation (3) can be simplified to

$$\begin{bmatrix} A_X \\ A_Y \\ A_Z \end{bmatrix} = \begin{bmatrix} P_{11} & P_{12} & P_{13} \\ P_{21} & P_{22} & P_{23} \\ P_{31} & P_{32} & P_{33} \end{bmatrix} \cdot \begin{bmatrix} R_X \\ R_Y \\ R_Z \end{bmatrix} + \begin{bmatrix} P_{10} \\ P_{20} \\ P_{30} \end{bmatrix} \quad (4)$$

The goal of this calibration is to determine twelve unknown calibration parameters P_i. These parameters are typically obtained by the least square method [2].

The basic calibration is performed at two stationary positions in each sensitive axis which are showed in Table 1. In order to get more precise output values, we can add more stationary positions.

Equation (4) can be rewritten as

$$[A_X \quad A_Y \quad A_Z] = [R_X \quad R_Y \quad R_Z \quad 1] \cdot \begin{bmatrix} P_{11} & P_{21} & P_{31} \\ P_{12} & P_{22} & P_{32} \\ P_{13} & P_{23} & P_{33} \\ P_{10} & P_{20} & P_{30} \end{bmatrix} \quad (5)$$

or also to

$$Y = w \cdot X \quad (6)$$

where

Y is known correct gravity vector,
w is the raw accelerometer reading at each positon,
X is the matrix containing twelve calibration parameters.

Table 1 Sign definition of raw accelerometer readings	Stationary position	A_X	A_Y	A_Z
	1	0	0	+g
	2	0	0	−g
	3	0	+g	0
	4	0	−g	0
	5	+g	0	0
	6	−g	0	0

3 Magnetometer

3.1 Magnetometer Error Model

The magnetometer data are influenced by wide band measurement noise, stochastic biases, installation errors and magnetic interferences in the vicinity of the sensors.

The magnetic interference can be divided up into two groups. The first group, known as hard iron interference, consists of fixed or slightly time-varying field generated by ferromagnetic materials [7].

In the second group, soft iron interference, the magnetic field is generated within the device itself.

$$R = C \cdot [M_M \cdot S \cdot SI \cdot (M + B + n)] \tag{7}$$

where

C is the matrix of misalignment between magnetometer axes and device axes,
M_M is the matrix of misalignment errors,
S is the matrix of scale factors,
SI is the matrix of soft iron biases
M is vector of magnetic field along sensitive axis,
B is vector of hard iron biases,
n is the wideband noise.

3.2 Magnetometer Calibration

The relation between the correct magnetometer data and the raw measurements of magnetometer is expressed as [1, 2]

$$
\begin{bmatrix} M_X \\ M_Y \\ M_Z \end{bmatrix} = [Mm]_{3x3} \cdot \begin{bmatrix} \frac{1}{S_X} & 0 & 0 \\ 0 & \frac{1}{S_Y} & 0 \\ 0 & 0 & \frac{1}{S_Z} \end{bmatrix} \cdot [SI]_{3x3} \cdot \begin{bmatrix} R_X - O_X \\ R_Y - O_Y \\ R_Z - O_Z \end{bmatrix} \tag{8}
$$

where

M_X, M_Y, M_Z are the correct magnetometer outputs,
Mm is a 3×3 misalignment matrix between the magnetometer axes and the device axes,
S_X, S_Y, S_Z are the scale factors,
R_X, R_Y, R_Z are the raw magnetometer readings,
SI is a 3×3 matrix of offsets caused by soft-iron interference,
O_X, O_Y, O_Z are the offsets caused by hard-iron interference.

Equation (8) can be simplified to [2]

$$\begin{bmatrix} M_X \\ M_Y \\ M_Z \end{bmatrix} = \begin{bmatrix} P_{11} & P_{12} & P_{13} \\ P_{21} & P_{22} & P_{23} \\ P_{31} & P_{32} & P_{33} \end{bmatrix} \cdot \begin{bmatrix} R_X - P_{10} \\ R_Y - P_{20} \\ R_Z - P_{30} \end{bmatrix} \qquad (9)$$

The goal of magnetometer calibration is to determine twelve unknown calibration parameters P_i to known correct magnetometer outputs which can be obtained at random positions [2].

4   Electronic Compass

The tilt compensated electronic compass uses triaxial accelerometer and triaxial magnetometer. The accelerometer measures components of gravity and magnetometer measures parts of geomagnetic field. The accelerometer readings provide roll and pitch angles which are used for magnetometer correction. That allows accurate calculation of compass heading. The tilt compensated electronic compass is not operating under freefall, low-g and high-g accelerations [2, 8].

4.1   Tilt Compensation Algorithm

The electronic compass coordinate system is represented on Fig. 3, where X_B, Y_B, Z_B are the device body axes and X, Y, Z are the accelerometer and magnetometer sensing axes [2, 9].

Rotations can be expressed as quaternions, Euler angles, roll-pitch-heading and rotations matrix. We used the rotations matrix. This method defines the roll γ as the angle between the Y_B axis and the horizontal plane. The pitch ρ is defined as the

Fig. 3 Electronic compass coordinate system

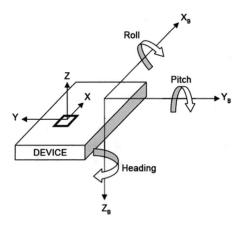

angle between X_B axis and horizontal plane and the heading ψ is the angle with respect to the magnetic north pole [9].

To obtain the rotation matrixes, we need to measure the rotations around each sensitive axis. These rotations can be described by the roll matrix R_γ (Eq. 10), the pitch matrix R_ρ (Eq. 11) and the heading matrix R_ψ (Eq. 12) [2, 7, 8].

$$R_\gamma = \begin{bmatrix} 1 & 0 & 0 \\ 0 & \cos\gamma & \sin\gamma \\ 0 & -\sin\gamma & \cos\gamma \end{bmatrix} \tag{10}$$

$$R_\rho = \begin{bmatrix} \cos\rho & 0 & -\sin\rho \\ 0 & 1 & 0 \\ \sin\rho & 0 & \cos\rho \end{bmatrix} \tag{11}$$

$$R_\psi = \begin{bmatrix} \cos\psi & \sin\psi & 0 \\ -\sin\psi & \cos\psi & 0 \\ 0 & 0 & 1 \end{bmatrix} \tag{12}$$

The relationship between the arbitrary position X_B', Y_B', Z_B' and the device body axes is [2, 7, 8]

$$\begin{bmatrix} X_B' \\ Y_B' \\ Z_B' \end{bmatrix} = R_\gamma \cdot R_\rho \cdot R_\psi \cdot \begin{bmatrix} X_B \\ Y_B \\ Z_B \end{bmatrix} \tag{13}$$

In the horizontal plane $X_B = Y_B = 0, Z_B = +1g$ and X_B', Y_B', Z_B' are raw accelerometer readings. Then we can rewrite Eq. (13) to [2, 7, 8]

$$\begin{bmatrix} A_X \\ A_Y \\ A_Z \end{bmatrix} = R_\gamma \cdot R_\rho \cdot R_\psi \cdot \begin{bmatrix} 0 \\ 0 \\ 1 \end{bmatrix} \tag{14}$$

Therefore, the roll and the pitch angle can be computed as

$$Pitch = \rho = \arcsin(-A_X) \tag{15}$$

$$Roll = \gamma = \arcsin\left(\frac{A_Y}{\cos\rho}\right) \tag{16}$$

The equations for tilt compensated x and y axis are expressed on Eq. (17), where M_X, M_Y, M_Z are the components of the magnetometer measurements [2, 7, 8].

$$\begin{aligned} X_H &= M_X \cdot \cos\rho + M_Z \sin\rho \\ Y_H &= M_X \cdot \sin\gamma \cdot \cos\rho + M_Y \cdot \cos\gamma - M_Z \sin\gamma \cdot \cos\rho \end{aligned} \tag{17}$$

The heading angle is calculated as [1, 2]

$$Heading = \psi = \arctan\frac{Y_H}{X_H} \qquad (18)$$

5 Conclusion

This research paper has presented calibration methods for accelerometer and magnetometer. These sensors are implemented in tilt compensated electronic compass. The roll and pitch angles are calculated from accelerometer readings that provides tilt compensation for the compass and magnetometer measures the components of the geomagnetic field.

The proposed calibration methods for accelerometer and magnetometer follow simple steps. After the calibration, the tilt compensated electronic compass is capable of providing precise accuracy which is below 2°.

Acknowledgment This work was supported by Internal Grant Agency of Tomas Bata University in Zlin under the project No. IGA/FAI/2016/035.

References

1. Grygorenko, V.: Application note: Sensing—Magnetic Compass with Tilt Compensation
2. STMicroelectronics: Application note: Using LSM303DLH for a tilt compensated electronic compass (2010)
3. Topic 6: Fields and Forces (2007). https://www.patana.ac.th/secondary/science/anrophysics/ntopic6/images/magnetic_field_earth.jpg
4. World Magnetic Model (2010). http://ngdc.noaa.gov/geomag/WMM/
5. Titterton, D., Weston, J.: Strapdown Inertial Navigation Technology. Institution of Electrical Engineers, Stevenage (2004)
6. Allen, J.J., Kinney, R.D., Sarsfield, J., Daily M.R., Ellis, J.R., Smith, J.H., Montague, S., Howe, R.T., Boser, B.E., Horowitz, R., Pisano, A.P., Lemkin, M.A., Clark, W.A., Juneau, C.T.: integrated micro-electro mechanical sensor development for inertial application. IEEE AES Syst. Mag. 36–40 (1998)
7. Ozyagcilar, T.: Application note: Calibrating an eCompass in the Presence of Hard and Soft-Iron Interference (2013)
8. Ozyagcilar T.: Application note: Implementing a Tilt-Compensated eCompass using Accelerometer and Magnetometer Sensors (2012)
9. Pedley, M.: Application note: Tilt Sensing Using a Three-Axis Accelerometer (2013)

Multivariable Gain Scheduled Control of Four Tanks System: Single Scheduling Variable Approach

Adam Krhovják, Stanislav Talaš and Lukáš Rušar

Abstract Motivated by the special class of nonlinear systems, we introduce a scheduling technique that aims at extending the region of validity of linearization by designing an extra scheduling mechanism. Specifically, we show, how a simplification of a control problem may results in a considerably difficult scheduling procedure. In particular, the scheduling problem in the context of a nonlinear model of four tanks is addressed. The main innovation consists in the use of auxiliary scheduling variables dealing with the problem of limited number of output variables. This allows to construct a linear feedback controller at each operating point. Additionally, an integral control which ensures desired stability and performance requirements is presented. The resulting method has been integrated into the gain scheduled control design of the four tanks system and has shown a great performance through the operating range.

Keywords Gain scheduled design · Integral control · Four tanks system

1 Introduction

Today's world is full of highly complex systems that exhibit nonlinear behavior. Engineers facing control of these systems are required to design such mechanisms that would satisfy desired characteristics through the operating range. In many situations, a tracking problem may involve multiple variables interacting with each other. Dealing with multivariable nonlinear systems, superposition principle known from linear systems does not hold any longer and we are faced difficult situations.

However, since linear equations are so much easier to solve than nonlinear ones, much research across a range of disciplines is devoted to finding linear approximations of nonlinear phenomena. Because of the powerful tools which are known

A. Krhovják (✉) · S. Talaš · L. Rušar
Faculty of Applied Informatics, Department of Process Control, Tomas Bata University in Zlin, Zlin, Czech Republic
e-mail: krhovjak@fai.utb.cz

53

from linear systems, there is no question that whenever it is possible, we should take an advantage of linearization. However, one must bear in mind the basic limitation associated with this approach. The fact that linearization represents an approximation in the neighborhood of an operating point. In other words, linearization cannot be viewed globally since it can only predict the local behavior of a nonlinear system. Consequently, classical controllers with fixed parameters can only guarantee performance in the vicinity of that point. Interestingly enough, in many situations, it is possible to capture how the dynamic of a system change in its equilibrium points by introducing a family of linear models. Moreover, it may be even possible to find one or more variables that parameterizes equilibrium points so that each member of a family is prescribed by them. In such cases, engineers does not have to absorb a wide range of nonlinear analysis tools. It is intuitively reasonable to linearize a nonlinear model about selected operating points, capturing key states of a system, design a linear controller at each point, and interpolate the resulting family of linear controllers by monitoring scheduling variables. There has been a significant research in gain scheduling (GS). We refer the interest reader to [5, 7] for deeper and more insightful understanding of the gain scheduling procedure. Gain scheduling principles has been successfully implemented for sophisticated control design of many technological systems [1–3]. As one can expect the most challenging point of the design is to find appropriate scheduling variables. Several references, including [8–10] discuss this crucial point. Despite these intensive efforts, we are still far from finding a general answer to the question of scheduling variable. In order to address those needs we have stressed to illustrate a single scheduling scenario for multivariable nonlinear system of four tanks. As we go over the model of the multivariable systems, it will determined that simplification of the control problem by minimizing output variables does not reduce the number of scheduling variables. Thus, an additional scheduling mechanism has to be implemented. Throughout the paper we gradually reveal the analytical solution of the scheduling procedure satisfying a tracking problem for a multivariable nonlinear system of four tanks (FT).

2   Four Tanks System

A simplified model of the FT system taken from [6] is schematically shown in Fig. 1. Here h_i represents the liquid level of the corresponding tank and q_a, q_b are the manipulated inflows. Finally, γ_1 and γ_2 are constants which are responsible for the inflow separation to the lower and upper level tank. There are no reactants or reaction kinetics and stoichiometry to consider. The model also includes hydraulic relationship for the tank outlet streams. Both parameters of the tanks and initial liquid levels are captured in Table 1.

In this case study we have used $\gamma_1 = \gamma_2 = 0.4$, $q_a = 43.4\,\text{ml/s}$, $q_b = 35.4\,\text{ml/s}$ and g as the gravitational acceleration $981\ \text{cm}^{-2}$.

Fig. 1 Four tanks system

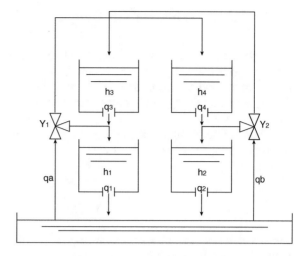

Table 1 Model parameters

Tank	F_i (cm^2)	k_i (cm^2)	h_i (cm)
1	50.27	0.233	14
2	50.27	0.242	14
3	28.27	0.127	14.25
4	28.27	0.127	21.43

The only step needed to develop the model of FT is to write conservation equations, representing material balance for a single material. Recalling the general form of a mass balance is given, it is easy to see the simplified system of FT is modeled by

$$q_a\gamma_1 = F_1\frac{dh_1}{dt} + q_1 - q_3$$
$$q_b\gamma_2 = F_2\frac{dh_2}{dt} + q_2 - q_4$$
$$q_b(1 - \gamma_2) = F_3\frac{dh_3}{dt} + q_3 \qquad (1)$$
$$q_a(1 - \gamma_1) = F_4\frac{dh_4}{dt} + q_4$$

where

$$q_i = k_i\sqrt{2gh_i} \quad \text{for} \quad i = 1, \dots, 4 \qquad (2)$$

and k_i is positive constant

3   Model Structure for GS Design

In the process of designing and implementing a gain scheduled controller for a nonlinear system, we have to find its approximations about the family of operating (equilibrium) points. Thus, the coupled nonlinear first-order ordinary differential Eq. (1) capturing the dynamics of the four tanks system have to be transformed into its linearized form.

In view of our example, we shall deal with multi-input multi-output linearizable nonlinear [3] system represented by

$$\dot{x} = f(x, u) \tag{3}$$

$$y = g(x) \tag{4}$$

where $\dot{x}$ denotes derivative of x with respect to time variable and u are specified input variables. We call the variable x the state variable and y the output variable. We shall refer to (3) and (4) together as the state-space model.

To obtain the state-space model of the FT system, let us take us take $x = h$ as state variables and $u_1 = q_a$, $u_2 = q_b$ as control inputs. Then the state equations take form

$$
\begin{aligned}
\frac{dx_1}{dt} &= \frac{1}{F_1}\left[-k_1\sqrt{2gx_1} + u_1\gamma_1 + k_3\sqrt{2gx_3}\right] \\
\frac{dx_2}{dt} &= \frac{1}{F_2}\left[-k_2\sqrt{2gx_2} + u_2\gamma_2 + k_4\sqrt{2gx_4}\right] \\
\frac{dx_3}{dt} &= \frac{1}{F_3}\left[-k_3\sqrt{2gx_3} + u_2(1-\gamma_2)\right] \\
\frac{dx_4}{dt} &= \frac{1}{F_4}\left[-k_4\sqrt{2gx_4} + u_1(1-\gamma_1)\right]
\end{aligned}
\tag{5}
$$

For clarity of scheduling design, and our convenience, we restrict our attention to the case of the single output variable h_1. Then the output equations is given by

$$y = x_1 \tag{6}$$

One can easily sketch the trajectory of steady-state characteristic by setting $\dot{x} = 0$ and solving for unknown vector x.

Expanding the right hand side of (3) about point $(\bar{x}, \bar{u})$, we obtain

$$f(x, u) \approx f(\bar{x}, \bar{u}) + \frac{\partial f(\bar{x}, \bar{u})}{\partial x} + \frac{\partial f(\bar{x}, \bar{u})}{\partial u} + \text{H.O.T.} \tag{7}$$

If we restrict our attention to a sufficiently small neighborhood of the equilibrium point such that the higher-order terms are negligible, then we may drop these terms and approximate the nonlinear state equation by the linear state equation

$$
\begin{bmatrix} x_1 \\ x_2 \\ x_3 \\ x_4 \end{bmatrix} = \underbrace{\begin{bmatrix} -\frac{1}{T_1} & 0 & \frac{F_3}{F_1 T_3} & 0 \\ 0 & -\frac{1}{T_2} & 0 & \frac{F_4}{F_2 T_2} \\ 0 & 0 & -\frac{1}{T_3} & 0 \\ 0 & 0 & 0 & -\frac{1}{T_4} \end{bmatrix}}_{A} \begin{bmatrix} x_1 \\ x_2 \\ x_3 \\ x_4 \end{bmatrix} + \underbrace{\begin{bmatrix} \frac{\gamma_1}{F_1} & 0 \\ 0 & \frac{\gamma_2}{F_2} \\ 0 & \frac{1-\gamma_2}{F_3} \\ \frac{(1-\gamma_1)}{F_4} & 0 \end{bmatrix}}_{B} \begin{bmatrix} u_1 \\ u_2 \end{bmatrix} \qquad (8)
$$

where

$$
T_i = \frac{F_i}{k_i} \sqrt{\frac{2x_i}{g}} \qquad (9)
$$

4 Parametrization via Scheduling Variable

Before we present a parametrization via scheduling variable, let us first examine configuration of the gain scheduled control system captured in Fig. 2. From the figure, it can be easily seen that controller parameters are automatically changed in open loop fashion by monitoring operating conditions. From this point of view, presented gain scheduled control system can be understand as a feedback control system in which the feedback gains are adjusted using feedforward gain scheduler.

Then it comes as no surprise that first and the most important step in designing a controller is to find an appropriate scheduling strategy. Once the strategy is found, it can be directly embedded into the controller design.

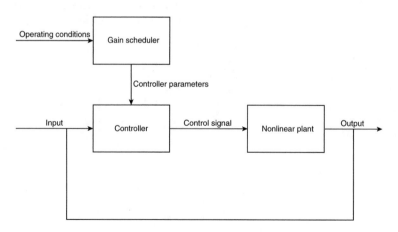

Fig. 2 Gain scheduled control scheme

Toward the goal, let us rewrite the nonlinear system (3)–(4) as

$$\dot{x} = f(x, u, \alpha) \tag{10}$$

$$y = g(x, \alpha) \tag{11}$$

We can see that the nonlinear system (10)–(11) is basically same as the system that we have introduced by Eqs. (3)–(4). The only difference is that both state and output equation are now parametrized by a new variable α, which represents operating conditions. Interested reader will find a comprehensive treatment in [7–10].

Even if we have succeeded in identifying the scheduling variable, we would face the second non-trivial problem. For this control to make any sense, we must have α of the same dimension as output vector. Since our control objective is only to control level h_1 we need to have the same number reference variables. The interesting way how to deal with this complication is to carry out an addition scheduling.

Assume that it is possible find a scheduling variable satisfying the condition

$$\frac{d\alpha}{dt} = 0 \tag{12}$$

over the operating range.

Since one of the variables is known and derivative of the second one is supposed to be zero it remains to display an additional scheduling for unknown variables.

To illustrate this motivating discussion let us consider this crucial point in the context of our example.

Suppose the system is operating at steady state and we want to design controller such that x tracks a reference signal y_r. In order to maintain the output of the plant at the value $\bar{x}_1$, we have to generate the corresponding input signal to the system at $\bar{u}_1 = k_1\sqrt{2g\bar{x}_1}$. Thus it means, that we can directly schedule on a reference trajectory.

Having identified a scheduling variable, the common scheduling scenario would usually takes this form

$$
\begin{bmatrix} x_1 \\ x_2 \\ x_3 \\ x_4 \end{bmatrix} =
\begin{bmatrix}
-\frac{1}{T_1(\alpha_1)} & 0 & \frac{F_3}{F_1 T_3(\alpha_3)} & 0 \\
0 & -\frac{1}{T_2(\alpha_2)} & 0 & \frac{F_4}{F_2 T_2(\alpha_2)} \\
0 & 0 & -\frac{1}{T_3(\alpha_3)} & 0 \\
0 & 0 & 0 & -\frac{1}{T_4(\alpha_4)}
\end{bmatrix}
\begin{bmatrix} x_1 \\ x_2 \\ x_3 \\ x_4 \end{bmatrix} +
\begin{bmatrix}
\frac{\gamma_1}{F_1} & 0 \\
0 & \frac{\gamma_2}{F_2} \\
0 & \frac{1-\gamma_2}{F_3} \\
\frac{(1-\gamma_1)}{F_4} & 0
\end{bmatrix}
\begin{bmatrix} u_1 \\ u_2 \end{bmatrix}
\tag{13}
$$

where now

$$T_i = \frac{F_i}{k_i}\sqrt{\frac{2\alpha_i}{g}} \tag{14}$$

However, since the system is viewed as single output system, it is impossible to directly schedule on α_2, α_3 α_4 and we have to introduce auxiliary functions, reflecting the absence of corresponding reference trajectories in the form

$$\dot{\alpha}_4 = 0 \tag{15}$$

$$\alpha_3 = \frac{\left(k_1\sqrt{2}\sqrt{g\alpha_1} - k_4\sqrt{2}\gamma_1\sqrt{g\alpha_4}\right)^2}{2gk_3^2} \tag{16}$$

$$\alpha_3 = \frac{\left(k_4\sqrt{2}\sqrt{g\alpha_4} + \frac{\gamma_2 k_3\sqrt{2}\sqrt{g\alpha_3}}{1-\gamma_2}\right)^2}{2gk_2^2} \tag{17}$$

Notice that newly introduced expressions play and essential role in (13) because they stand for missing reference variables.

An important feature of our analysis is that even if α_1 represents reference variable the Eqs. (13)–(17) still capture the behavior of the system around equilibria.

5 Regulation via Integral Control

Since the previous sections resulted in a family of parametrized linear models we want to design state feedback control such that

$$y \to y_r \text{ as } t \to \infty \tag{18}$$

Further, we assume that we can physically measure the controlled output y. In order to ensure zero steady-state tracking error in the presence of uncertainties, we want to use integral control. The regulation task will be achieved by stabilizing system at an equilibrium point where $y = y_r$.

To maintain the system at that point it must be true, that there exists a pair of $(\bar{x}, \bar{u})$ such that

$$0 = f(\bar{x}, \bar{u}) \tag{19}$$

$$0 = g(\bar{x}) - y_r \tag{20}$$

Note, that for Eqs. (19)–(20) we assume a unique solution $(\bar{x}, \bar{u})$.

Toward the goal, we have integrate the tracking error $e = y - y_r$

$$\dot{\sigma} = e \tag{21}$$

Having defined the integrator of the tracking error let now augment the system (3) to obtain

$$\dot{x} = f(x, u) \tag{22}$$

$$\dot{\sigma} = g(x) - y_r \tag{23}$$

It follows the control u will be designed as a feedback function of (x, σ). For such control the new system has an equilibrium point $(\bar{x}, \bar{\sigma})$.

To proceed with the design of the controller, we now linearize (22)–(23) about $(\bar{x}, \bar{\sigma})$ to obtain augmented state space model as

$$\dot{\xi} = \begin{bmatrix} A & 0 \\ C & 0 \end{bmatrix} \xi + \begin{bmatrix} B \\ 0 \end{bmatrix} \upsilon \overset{\text{def}}{=} A\xi + B\upsilon \tag{24}$$

where

$$\xi = \begin{bmatrix} x - \bar{x} \\ \sigma - \bar{\sigma} \end{bmatrix}, \upsilon = u - \bar{u}$$

Now we have to design a matrix K such that $A + BK$ is Hurwitz.

Partition K as $K = \begin{bmatrix} K_1 & K_2 \end{bmatrix}$ implies that the state feedback control should be taken as

$$u = K_1(x - x) + K_2(\sigma - \sigma) + \bar{u} \tag{25}$$

And by applying the control (25), we obtain the closed-loop system

$$\dot{x} = f(x, K_1(x - x) + K_2(\sigma - \sigma) + \bar{u}) \tag{26}$$

$$\dot{\sigma} = g(x) - y_r \tag{27}$$

So far, the basic idea of the control problem has been formed. For a complete coverage of the integral control, readers may consult [4]. All that remains now is to simulate the performance of the gain scheduling procedure with the help of the integral control.

The simulation results of tracking control are presented in Figs. 3 and 4. From a pole placement viewpoint, the matrix K is designed to assign the eigenvalues of $A + BK$ at $-0.0222, -0.0215, -0.0261, -0.0353 + 0.0360i, -03530 - 0.0360i$

Figure 3 shows the responses of the control system to sequences of step changes in reference signals. However, to appreciate what we gain by gain scheduling, Fig. 4 illustrates responses of the closed-loop system to the same sequence of

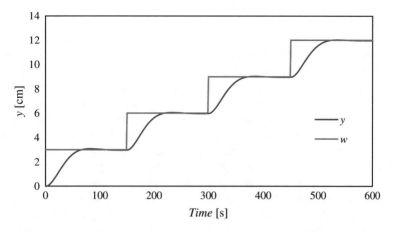

Fig. 3 The responses of the closed-loop system to a sequence of step changes

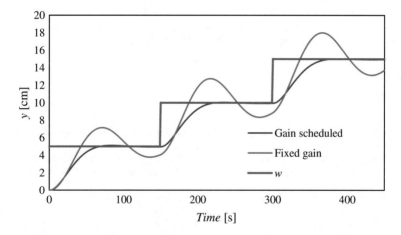

Fig. 4 Gain scheduled versus fix gain controller

changes. In the first case, a gain scheduled controller is applied, while in the second case a fixed-gain controller evaluated at $a_1 = 1$ cm is used. As the reference signal is far from the operating point, the performance deteriorates and system may go unstable. In some situations, it may be possible to overcome this difficulty by sequence of small step changes, allowing enough time for the system to settle down as is shown in Fig. 3. In other words, it is necessary to move slowly from one point to another. These observations are consistent with a common gain scheduling rule-of-thumb about the behavior of gain scheduled controller under slowly varying scheduling variable.

6 Conclusion

This paper addressed the problem of scheduling variables for a nonlinear multi-variable four tanks system. First, the simplified model of four tanks system has been detailed. Based on this a control strategy utilizing a single variable to capture the dynamics of the system through the range of operating conditions has been presented. It has been shown that the model reduction complicates the main scheduling task. Consequently, we initiated the promising solution that aims at extending the validity of scheduling by introducing an auxiliary scheduling variables.

An important feature of this approach is that linear design methods can be applied to the linearized system at each operating point. Thanks to this feature, the presented procedure leaves room for many other linear control methods. It has been demonstrated that a proposed approach integrated to the gain schedule control scheme has the potential to respond rapidly changing operating conditions.

This work is intended to be foundation for applying a unified scheduling approach to reduced models of multivariable nonlinear systems.

Acknowledgments This article was created with support of the Ministry of Education of the Czech Republic under grant IGA reg. n. IGA/FAI/2016/006.

References

1. Jiang, J.: Optimal gain scheduling controllers for a diesel engine. IEEE Control Syst. Mag. **14** (4), 42–48 (1994)
2. Krhovják, A., Dostál, P., Talaš, S. Rušar, L.: Nonlinear gain scheduled controller for a sphere liquid tank. s.l. In: European Council for Modelling and Simulation (ECMS), pp. 327–332 (2015)
3. Krhovják, A., Dostál, P., Talaš, S., Rušar, L.: Multivariable gain scheduled control of two funnel liquid tanks in series. In: Strbske Pleso, s.n., pp. 60–65 (2015)
4. Khalil, H.K.: Nonlinear Systems, 3rd edn. Prentice Hall, Upper Saddle River (2002)
5. Lawrence, D.A., Rugh, W.J.: Gain scheduling dynamic linear controllers for a nonlinear plant. Automatica **31**, 381–390 (1995)
6. Raff, T., Hubber, S., Nagy, Z.K., Allglöwer, F.: Nonlinear Model Predictive Control of a Four Tank System: An Experimental Stability Study, pp. 237–242. IEEE, Munich (2006)
7. Rugh, W.J.: Analytical framework for gain scheduling. IEEE Control Syst. Mag. **11**(1), 79–84 (1991)
8. Shamma, J.S., Athans, M.: Analysis of gain scheduled control for nonlinear plants. IEEE Trans. Autom. Control **35**(8), 898–907 (1990)
9. Shamma, J.S., Athans, M.: Gain scheduling: potential hazards and possible remedies. IEEE Control Syst. Mag., 101–107 (1991)
10. Shamma, J.S., Athans, M.: Guaranteed properties of gain scheduled control of linear parameter-varying plants. Automatica **27**(4), 559–564 (1991)

Inverted Pendulum Optimal Control Based on First Principle Model

František Dušek, Daniel Honc, K. Rahul Sharma and Libor Havlíček

Abstract This paper describes the design procedure of nonlinear dynamical model of a real system—inverted pendulum—cart with pendulum. The aim of the paper is to create a mathematical model based on known constructional, mechanical and electrical characteristics of the physical system. Such a model is linearized into standard linear time-invariant state-space model where the input is motor power voltage and the outputs are cart position and pendulum angle. A linear model is used for discrete-time LQ controller design—state variables are estimated and the cart position is controlled with pendulum in upright metastable position.

Keywords Inverted pendulum · First principle model · Optimal control · LQ controller

1 Introduction

Mathematical model of inverted pendulum can be found in the literature quite often, especially in connection with the control of two-wheeled inverted pendulum self-balancing mobile robot. Review of modelling and control of two-wheeled robots is in [1]. Three possible approaches for creating nonlinear dynamical model are by applying: Newton's laws, Euler-Lagrange equation or Kane's method. The most commonly used methods are Euler-Lagrange equations. On the other hand, motion equations concept is closest to its physical meaning—the variables have natural physical interpretation. Unfortunately, even models based on Newton's equations [2, 3] do not start with initial form like balances of forces and moments but frequently as resulting equations of motion. Another common feature of such models is that the properties of the motors are neglected. The torque acting directly on the wheels of the robot are inputs of the model.

F. Dušek · D. Honc (✉) · K. Rahul Sharma · L. Havlíček
Faculty of Electrical Engineering and Informatics, Department of Process Control, University of Pardubice, nám. Čs. legií 565, 532 10 Pardubice, Czech Republic
e-mail: Frantisek.dusek@upce.cz

© Springer International Publishing Switzerland 2016
R. Silhavy et al. (eds.), *Automation Control Theory Perspectives in Intelligent Systems*, Advances in Intelligent Systems and Computing 466,
DOI 10.1007/978-3-319-33389-2_7

The article deals with a simple device that allows only longitudinal movement of the cart and pendulum. Similar problem is solved in [4, 5] where the model is based on Newton's equations but does not include the description of the motor. Mathematical model for the needs of the control design must include description of all parts influencing the behaviour of the controlled system between the control action (input) and the controlled variable (output). Model in this article uses Newton's equations—it is strictly based on the basic balance of forces and moments and includes the dynamic behaviour of DC motor with permanent magnets.

The laboratory system is designed and realized at Department of process control, Faculty of Electrical Engineering and Informatics, University of Pardubice for teaching and research purpose in the areas of modelling, identification, simulation and control. The mathematical model describes behaviour of three elementary parts of the system (pendulum, cart and motor). Analytical nonlinear model is the starting point for experimental identification where corrections of theoretical parameters will be estimated so that the dynamic behaviour of the model corresponds to measurements on real system. Analytically linearized model can be used for large number of modern control methods—in this paper LQ optimal controller is designed and applied.

2 Laboratory System

The entire device has been mechanically and electrically designed for laboratory use from the beginning. The stand is made from aluminium sections with two rods leading the cart with pendulum rod (see Fig. 1). The pendulum rotates freely and its position (angle) is sensed by the rotary 10 bits (1024 pulses per revolution) magnetic encoder AS5040 AMS AG [6] (see Fig. 1). Cart movement is controlled via a toothed belt, mounted directly on the pulley of a DC brush motor with permanent magnets Mabuchi C2162. Speed of the motor is measured by the second AS5040 encoder mounted on the motor axis. The motor is powered by 18 V power supply and its speed is controlled by a control power unit with the control signal between 0

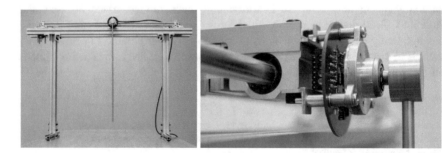

Fig. 1 Inverted pendulum stand and cart with encoder

and 5 V, where zero speed corresponds to 2.5 V, zero voltage causes the maximum speed in one direction and 5 V maximum speed in the opposite direction. Extreme cart positions are equipped with limit switches.

3 Mathematical Model

The mathematical description consists of three separate parts which influence each other mutually. The model describes the ideal behaviour of individual components based on the balance of forces and moments. Real behaviour is achieved by including mechanical resistances with simple approximation—resistance force is directly proportional to the linear or angular velocity. The correction factors will be determined experimentally on the basis of measurements carried out on the device.

Balance is based on a fundamental scheme as in Fig. 2. It shows equivalent circuit of motor with winding resistance R and inductance L including source voltage with internal resistance R_z. The motor is connected through transmission P with the cart of weight m_v. For the simplicity it is assumed that this connection realized by pulley of radius r and is ideally rigid (elasticity is neglected). We also assume that the pendulum is a rod with constant cross area, weight m_d and length d with weight mass m_k at the end.

3.1 Balanced Forces—Cart

Only forces acting in axis of movement (x axis) affect the movement of the cart. Position of the cart is labelled as x. The centre of gravity of the rod (or mass) of pendulum is moving relatively in direction of the x axis to the position of the cart

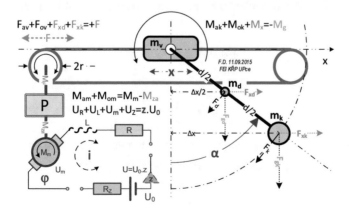

Fig. 2 Fundamental scheme

with distance $\Delta x/2$ (or Δx). Forces acting to the cart are: acceleration force F_{av}, the resistive force F_{ov} and forces induced by the motion of the pendulum F_{xd} (rod) and F_{xk} (mass). All these forces must be in balance with the force F—belt acting on the cart. Dependence of cart position x on pendulum angle α and the applied force F can be described by the equation

$$F_{av} + F_{ov} + F_{xd} + F_{xk} = F \tag{1}$$

where

F	external force (belt tension),
$F_{av} = m_v \frac{d^2 x}{dt^2}$	cart inertial force,
$F_{ov} = k_v \frac{dx}{dt}$	cart resistive force,
$F_{xd} = m_d \frac{d^2}{dt^2} \left(x + \frac{1}{2}\Delta x \right)$	pendulum rod inertial force,
$F_{xk} = m_k \frac{d^2}{dt^2} \left(x + \Delta x \right)$	pendulum mass inertial force and
$\Delta x = d.\sin(\alpha)$	pendulum centre of gravity relative change.

3.2 Balanced Torques—Pendulum

Torques must be considered by describing the rotational movement of the pendulum. The angle of rotation of the pendulum is marked as α and its positive value indicates turning counter clockwise. Torques M_{ak} (due to the acceleration of the cart), M_{ok} (induced by resistance) and M_x (induced by forces F_{xd} and F_{xk} acting due to the change in position of the pendulum to the cart along the x axis) must be balanced with the effects of torque M_g coming from the force of gravity. The gravitational force is oriented downwards and therefore negative sign in the formulation (2) must be considered. The balance of torques (i.e. the dependence of rotation angle of the pendulum α on position of the cart x) can be described by the equation

$$M_{ak} + M_{ok} + M_x = -M_g \tag{2}$$

where

$M_{ak} = J_k \frac{d^2 \alpha}{dt^2}$	pendulum inertial torque, $J_k = \frac{1}{3} m_d d^2 + m_k d^2$,
$M_{ok} = k_k d \frac{d\alpha}{dt}$	pendulum resistive torque,
$M_g = \frac{d}{2} m_d g . \sin(\alpha) + d.m_k g.\sin(\alpha)$	pendulum torque (gravitation) and
$M_x = \frac{d}{2} F_{xd} \cos(\alpha) + d.F_{xk} \cos(\alpha)$	pendulum torque (cart).

3.3 Balanced Voltages and Torques—Motor

Description of the dynamics of a DC brush motor with permanent magnets is based on the scheme as in Fig. 2 and consists of an electric part—balance of voltages i.e. by Kirchhoff's laws (3) and the mechanical part—balance of torques (4). Both parts are interconnected by the magnetic field Φ which is generally a function of the flowing current i. For simplification, it is assumed that the magnetic field is constant for the current i in the operating area. Specific values are included in the rate constant k_u and torque constant k_m which can determined or calculated from the information about the motor supplied by the manufacturer. The balance of voltages and moments which is the dependence of the motor current i and angle of the rotor φ on the motor supply voltage $z \cdot U_0$ and motor load M_{Za} can be expressed as

$$U_R + U_L + U_m + U_z = z \cdot U_0 \tag{3}$$

where

$U_R = R \cdot i$ voltage across the winding resistance,
$U_L = L \frac{di}{dt}$ voltage across the winding inductance,
$U_m = k_u \Phi \frac{d\varphi}{dt}$ back EMF,
$U_{Rz} = R_z i$ voltage across the internal source resistance and
Z motor control signal ($-1 \leq z \leq 1$).

$$M_{am} + M_{om} + M_{za} = M_m \tag{4}$$

where

$M_{am} = J_m \frac{d^2\varphi}{dt^2}$ inertial torque of all the rotating parts (motor, belt),
$M_{om} = k_o \frac{d\varphi}{dt}$ resistive torque of all the rotating parts,
M_{za} load torque and
$M_m = k_m \Phi \cdot i$ motor torque.

3.4 Linkage Between Motor and the Cart Movement

Transfer of the motor rotation to the movement of the cart is realized by belt and pulley of radius r located on the motor axis. Ideally rigid connection is assumed (the flexibility of the belt is not considered). Connection between angle of the motor rotation φ and cart position x is described by (5). Force F acting on the cart depends on the load torque M_{Za} according to (6). The transmission gear ratio P is considered as well. In our case the transmission gear ratio, $P = 1$.

$$\varphi = \frac{P}{r}x \tag{5}$$

$$F = \frac{P}{r}M_{za} \tag{6}$$

3.5 Nonlinear Dynamical Model

After identifying individual forces, torques and voltages in equations from (1) to (6), neglecting the coupling forces between all the parts, evaluating highest derivatives and introducing substitutions for the constant parameters, we obtain a system of three differential equations in a form suitable for simulation. The resulting model is a 5th order system of following nonlinear differential equations

$$\frac{di}{dt} = -c_3 i - c_4 \frac{dx}{dt} + c_1 z$$

$$\frac{d^2 x}{dt^2} = a_3 \frac{b_2 + \cos^2(\alpha)}{den} i - a_4 \frac{b_2 + \cos^2(\alpha)}{den} \frac{dx}{dt} + b_4 \frac{\cos(\alpha)}{den} \frac{d\alpha}{dt} + b_2 d \frac{\sin(\alpha)}{den} \left(\frac{d\alpha}{dt}\right)^2$$

$$+ b_1 \frac{\sin(\alpha)\cos(\alpha)}{den} g$$

$$\frac{d^2 \alpha}{dt^2} = -a_3 b_1 \frac{\cos(\alpha)}{d.den} i + a_4 b_1 \frac{\cos(\alpha)}{d.den} \frac{dx}{dt} - a_1 b_4 \frac{1}{d.den} \frac{d\alpha}{dt} + (a_1 - b_1) \frac{\cos(\alpha)\sin(\alpha)}{den} \left(\frac{d\alpha}{dt}\right)^2$$

$$- a_1 b_1 \frac{\sin(\alpha)}{d.den} g \tag{7}$$

where, $den = a_1 b_2 + (a_1 - b_1)\cos^2(\alpha)$

$$a_1 = \frac{m_v + m_d + m_k + \frac{P^2}{r^2}\left(J_m + J_p\right)}{\frac{1}{2}m_d + m_k}, \quad a_3 = \frac{\frac{P}{r}k_m \Phi}{\frac{1}{2}m_d + m_k}, \quad a_4 = \frac{k_o \frac{P^2}{r^2} + k_v}{\frac{1}{2}m_d + m_k}$$

$$b_1 = \frac{\frac{1}{2}m_d + m_k}{\frac{1}{4}m_d + m_k}, \quad b_2 = \frac{\frac{1}{3}m_d + m_k}{\frac{1}{4}m_d + m_k}, \quad b_4 = \frac{k_k}{\frac{1}{4}m_d + m_k}$$

$$c_1 = \frac{U_0}{L}, \quad c_3 = \frac{R + R_z}{L}, \quad c_4 = \frac{k_u}{L}\Phi\frac{P}{r}$$

Differential equations (7) describe the time evolution of five state variables (x, $v = x'$, α, $\omega = \alpha'$ and i) according to the initial conditions and the motor power expressed by the control variable z. All the variables have physical meaning and can be directly compared with measured data.

4 Controller Design

An optimal LQ controller based on state-space model is applied. Not all components of the state vector are measurable, so estimator must be a part of the controller design. Two variables are measured—the position of the carriage (distance from the centre position x) and the angle of the pendulum α.

4.1 Linear Approximation of Nonlinear Model

The nonlinear model is approximated with linear form around steady state in the upright metastable position $\alpha_0 = \pi$. This corresponds to the control operating point—position of the cart will be controlled to desired position and the pendulum will be stabilized in the upright position. The nonlinearity of the model (7) is given by trigonometric functions of pendulum angle and squared angular velocity of the pendulum. We replace trigonometric functions with their first derivatives in the selected operating point. New state variable β as a deviation of pendulum angle α from the operating point is introduced. The corresponding relationships are given as,

$$
\begin{aligned}
&\alpha - \alpha_0 = \beta \\
&\cos(\alpha) = \cos(\pi + \beta) \approx -1 \quad \sin(\alpha) = \sin(\pi + \beta) \approx -\beta \\
&\left(\frac{d\alpha}{dt}\right)^2 = \left(\frac{d\beta}{dt}\right)^2 \approx 0
\end{aligned}
\tag{8}
$$

After substituting (8) into (7) and introducing new state variables—cart speed v and speed of pendulum rotation ω, we can express the standard state-space model in matrix form as

$$
\begin{bmatrix} \frac{di}{dt} \\ \frac{dx}{dt} \\ \frac{dv}{dt} \\ \frac{d\beta}{dt} \\ \frac{d\omega}{dt} \end{bmatrix} = \begin{bmatrix} -c_3 & 0 & -c_4 & 0 & 0 \\ 0 & 0 & 1 & 0 & 0 \\ \frac{(b_2+1)a_3}{a_1 b_2 + a_1 - b_1} & 0 & -\frac{(b_2+1)a_4}{a_1 b_2 + a_1 - b_1} & \frac{b_1 g}{a_1 b_2 + a_1 - b_1} & -\frac{b_4}{a_1 b_2 + a_1 - b_1} \\ 0 & 0 & 0 & 0 & 1 \\ \frac{a_3 b_1}{(a_1 b_2 + a_1 - b_1)d} & 0 & -\frac{a_4 b_1}{(a_1 b_2 + a_1 - b_1)d} & \frac{a_1 b_1 g}{(a_1 b_2 + a_1 - b_1)d} & -\frac{a_1 b_4}{(a_1 b_2 + a_1 - b_1)d} \end{bmatrix} \begin{bmatrix} i \\ x \\ v \\ \beta \\ \omega \end{bmatrix} + \begin{bmatrix} i \\ x \\ v \\ \beta \\ \omega \end{bmatrix} z
\tag{9}
$$

5 State Estimator and Controller

The existence of a linear state-space model allows simple LQ controller design and state estimator as well. For practical realization, the controller (11) and estimator (13) is designed in discrete-time domain. The position of the cart x and angle of the pendulum α are measured and the motor voltage is controlled with control variable z on real system. The discrete-time state-space model (10) equivalent to continuous-time model (9) with sampling period T can be expressed as

$$\mathbf{x}(k+1) = \mathbf{A} \cdot \mathbf{x}(k) + \mathbf{B} \cdot \mathbf{u}(k)$$
$$\mathbf{y}(k) = \mathbf{C} \cdot \mathbf{x}(k) \tag{10}$$

where $\mathbf{x}^T = [\,i \quad x \quad v \quad \beta \quad \omega\,]$, $\mathbf{y}^T = [\,x \quad \beta\,]$ and $\mathbf{u} = z$.

State-space LQ controller is used in the form (11) with the reference (desired) state. This formulation together with the physical meaning of state variables allows inclusion of different control requirements in a simple and natural way. In addition to the basic requirements—tracking the desired position of the cart $x_w(k)$ and zero positional deviation of the pendulum angle $\beta(k) = 0$, it is possible to include a request to cart speed $v_w(k)$. Zero values in the reference state means requirement for minimum value of the appropriate state variables (absolute value). Non-zero values indicate a requirement for minimum deviation of the relevant variable from the set point—the position of the cart $x_w(k)$, the angle of the pendulum $\beta(k)$, the speed of the cart v_w (k) etc.

$$\mathbf{u}(k) = -\mathbf{L} \cdot [\mathbf{x}(k) - \mathbf{x}_w(k)] \tag{11}$$

where $\mathbf{x}_w^T(k) = [\,0 \quad x_w(k) \quad v_w(k) \quad 0 \quad 0\,]$.

Controller gain vector $\mathbf{L}$ (MATLAB function $\mathbf{L}$ = lqr(A, B, Q, R)) minimizes the cost function (11) subject to the restrictions (10)

$$J(\mathbf{L}, \mathbf{A}, \mathbf{B}) = \sum_{i=0}^{\infty} \left[\mathbf{x}^T(i)\mathbf{Q} \cdot \mathbf{x}(i) + \mathbf{u}^T(i)\mathbf{R} \cdot \mathbf{u}(i) \right]$$
$$\mathbf{L} = \arg\min_{\mathbf{L}} J(\mathbf{L}, \mathbf{A}, \mathbf{B}) \tag{12}$$

where $\mathbf{Q}$ and $\mathbf{R}$ are symmetric positive definite weighing matrices. Values on the diagonal elements represent "contribution" of corresponding state variable to the total value of the cost function and it is possible with this choice to set preferences to the minimization of deviation of that variable from the reference value.

Discrete state-space estimator is in the form

$$\mathbf{x}_e(k+1) = \mathbf{A} \cdot \mathbf{x}_e(k) + \mathbf{B} \cdot \mathbf{u}(k) + \mathbf{K}[\mathbf{y}(k) - \mathbf{C} \cdot \mathbf{x}_e(k)] \tag{13}$$

where the observer gain vector $\mathbf{K}$ is determined as a solution of the dual problem to the controller design problem. The vector $\mathbf{K}$ (MATLAB functions $\mathbf{K}^T$ = dlqr($\mathbf{A}^T$, $\mathbf{C}^T$, $\mathbf{Q}_e$, $\mathbf{R}_e$)) minimizes the cost function (14) subject to the restrictions (10)

$$J(\mathbf{K}, \mathbf{A}, \mathbf{C}) = \sum_{i=0}^{\infty} \left[\Delta \mathbf{x}^T(i) \mathbf{Q}_e \cdot \Delta \mathbf{x}(i) + \mathbf{y}^T(i) \mathbf{R}_e \cdot \mathbf{y}(i) \right]$$

$$\mathbf{K} = \arg \min_{\mathbf{K}} J(\mathbf{K}, \mathbf{A}, \mathbf{C})$$

(14)

where $\mathbf{Q}_e$ and $\mathbf{R}_e$ are symmetric positive definite weighing matrices and $\Delta \mathbf{x}(k) = \mathbf{x}_e(k) - \mathbf{x}(k)$ is a vector of deviations between the state vector $\mathbf{x}$ and its estimation $\mathbf{x}_e$.

6 Inverted Pendulum Control Simulation

The model parameters used in the simulation experiments are summarized in Table 1. The coefficient of resistive force against cart movement k_v and against the rotation of the pendulum k_k are estimated, so that the settling time approximately reflects the behaviour of the real device. The value of inertia of rotating parts of the

Table 1 Simulation experiments—parameters

Symbol	Units	Value	Meaning
g	m s^{-2}	9.81	Constant of gravity
m_v	kg	0.396	Cart weight
k_v	kg m s^{-1}	0.6	Cart movement resistance coefficient
m_d	kg	0.04	Pendulum rod weight
m_k	kg	0.1	Pendulum mass weight
d	m	0.525	Pendulum rod length
k_k	kg m s^{-1} rad^{-1}	0.01	Pendulum rotation resistance coefficient
J_k	kg m^2	0.003675	Pendulum moment of inertia
U_0	V	18	Supply voltage
R_z	Ω	0.05	Voltage source internal resistance
L	H	0.1×10^{-3}	Motor windings inductance
ω_0	rad s^{-1}	476.47	No load motor speed ($M_{za} = 0$)
i_0	A	**0.15**	No load motor current ($M_{za} = 0$)
M_s	N m	0.287	Stall torque ($\omega = 0$)
i_s	A	**2.5**	Stall motor current ($\omega = 0$)
R	Ω	7.2	Motor windings resistance, $R = U_0/i_s$
k_u	V s rad^{-1}	0.0355	Back EMF constant, $k_u\emptyset = \frac{U_0 - R.i_0}{\omega_0}$
k_m	N m A^{-1}	0.1148	Motor torque constant, $k_m\emptyset = \frac{M_s}{i_s}$
k_o	N m s^{-1} rad^{-1}	$3.614 \cdot 10^{-5}$	Motor rotation resistance coefficient, $k_o = \frac{M_s}{i_s} \frac{i_0}{\omega_0}$
J_m	kg m^2	45.0×10^{-7}	Motor moment of inertia
J_p	kg m^2	0.00015	Other drive components moment of inertia
r	m	7×10^{-3}	Pulley radius

drive except motor Jp is estimated similarly. These values are in Table 1 with red colour and will be determined by the experimental identification.

The motor parameters (green values) like winding resistance R, back EMF constant k_u, torque constant k_m and rotational resistance constant k_o are calculated from the Eqs. (3) and (4) so they match with the engine characteristics taken from the documentation of the motor (current i_0 and speed ω_0 at no load, current i_s and torque M_s at zero speed—bold values). The value of motor winding inductance L was measured.

The estimator (weighing matrices $\mathbf{Q_e}$, $\mathbf{R_e}$) and controller parameters (weighing matrices $\mathbf{Q}$, $\mathbf{R}$) are chosen so that the tracking error of the cart and the pendulum angle from the upper metastable position is minimized—design is based on model (9)

$$\mathbf{Q} = \mathbf{Q}_e = \begin{bmatrix} 1 & 0 & 0 & 0 & 0 \\ 0 & 100 & 0 & 0 & 0 \\ 0 & 0 & 1 & 0 & 0 \\ 0 & 0 & 0 & 100 & 0 \\ 0 & 0 & 0 & 0 & 1 \end{bmatrix} \mathbf{R} = 0.01 \quad \mathbf{R}_e = \begin{bmatrix} 1 & 0 \\ 0 & 1 \end{bmatrix} \quad (15)$$

Following simulation experiments show the course of selected variables on a request to move the cart from a central position to the rightmost position then to the left and back to the starting position with discrete LQ controller with a sampling period $T = 0.25$ s. Minimal cart tracking error and the minimal angle deviation from upright position of the pendulum is preferred—see matrix $\mathbf{Q}$ (15). Nevertheless there is an offset in the tracking error of the cart as seen in Fig. 3. The improvement can be achieved by including a requirement for the cart velocity $x_w(k)$ to desired state vector (11)—see Fig. 4.

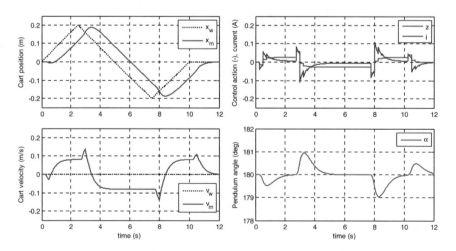

Fig. 3 Control experiment—without cart speed requirement

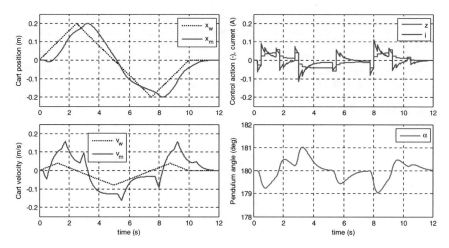

Fig. 4 Control experiment—with cart speed requirement

7 Conclusion

All variables in the presented mathematical model of the inverted pendulum have physical meaning and the values of 15 model parameters from 18 can be directly determined based on the known specifications. This simplifies experimental identification because we need to specify only three unknown parameters (in addition to the initial values—which are able to estimate). Using a state-space model with variables that have physical meaning allows inclusion of additional control requirements. Simulation with continuous nonlinear model and discrete linear controller with the state estimation allows to get the idea about the controller behaviour under ideal conditions and to estimate the default values of some other parameters (e.g. the sample time) which will be used in the control experiments on the real device.

Acknowledgments This research was supported by Institutional support of The Ministry of Education, Youth and Sports of the Czech Republic and SGS project at FEI.

References

1. Chan, R.P.M., Stol, K.A., Halkyard, C.R.: Review of modelling and control of two-wheeled robots. Ann. Rev. Control **37**(1), 89–103 (2013)
2. Gao, L.J.X., Huang, Q., Du, Q., Duan, X.: Mechanical design and dynamic modeling of a two-wheeled inverted pendulum mobile robot. In: 2007 IEEE International Conference on Automation and Logistics, pp. 1614–1619. IEEE (2007)

3. Ruan, X., Cai, J.: Fuzzy Backstepping controllers for two-wheeled self-balancing robot. In: 2009 International Asia Conference on Informatics in Control, Automation and Robotics, pp. 166–169. IEEE (2009)
4. Haugen, F.: Mathematical Model of a Pendulum on a Cart. TechTeach, Skien (2011)
5. Patenaude, J.: Simple Inverted Pendulum Cart Dynamics Classical Development. Kansas State University, Mechanical and Nuclear Engineering, Manhattan (2007)
6. Technical Guide: Mabuchi Motor. Mabuchi Motor Co., Ltd http://www.mabuchi-motor.co.jp/en_US/technic/index.html

A Cross-Layer Routing Metric for Multihop Wireless Networks

I.O. Datyev, M.G. Shishaev and V.A. Putilov

Abstract Multihop wireless networks are the promising direction of communication networks. The paper presents an attempt of different authors past experience generalization of routing metrics development. These mertrics based on multihop wireless networks data transmission features. In particular, characteristics of wireless transmission that are recommended to be considered during the design of routing metrics are reviewed. The cross-layer routing metric for multihop wireless networks is proposed. In addition, discuss the advantages and disadvantages of the well-known routing metrics created for multihop wireless networks.

Keywords Multihop wireless networks · Routing metric · Load balancing · Interference

1 Introduction

Identifying the best data packet transmission path is one of the main problems of wireless multihop networks. The routing metric is intended to solve this problem by routing path estimation. Various characteristics can be taken into account. The most

The work was supported by RFBR grant № 15-29-06973—Development of methodology, model tool kit and information technologies for system risk assessment of new exploration of the Arctic region.

I.O. Datyev (✉) · M.G. Shishaev · V.A. Putilov
Institute for Informatics and Mathematical Modelling of Technological Processes
of the Kola Science Center RAS, 184209 Apatity, Russia
e-mail: datyev@iimm.ru

M.G. Shishaev
e-mail: shishaev@iimm.ru

V.A. Putilov
e-mail: putilov@iimm.ru

common are: hops count, bandwidth of the channel, cost of data transmission over a channel, reliability, delay and etc.

Thus, the metric is used by routing algorithms to determine the best path to the data receiver. To simplify searching for a path, routing algorithms create and regularly update the routing tables, which contain route information. The routing information is changed in accordance with the rules laid down in the used routing algorithm.

In the case of static routing, the metric value typically does not change within a session. If the network is a set of moving nodes, a mobile ad hoc network (MANET) [1], for example, along with frequently changing routes, changing its metric values.

The routing metric should be easily calculated to reduce the service load on the network. On the other hand, the metric should be a measure of the most important characteristics of the path. It is obvious that the set of such features and estimating their significance should be generated based on the communication network features. For example, capacity reduction caused by interference is a major problem faced by WMNs [2]. Therefore, our work focuses on the routing metrics for wireless multihop networks.

This paper is organized as follows. In Sect. 2, we review the previous works on routing metrics for WMNs and discuss their advantages and drawbacks. Section 3 provides our IALB routing metric. Our conclusions and future work are presented in Sect. 4.

2   Related Works

In this section, we overview some well-known routing metrics proposed for WMNs and we also provide a detailed analysis of their advantages and drawbacks.

2.1   Expected Transmission Count (ETX)

The ETX metric [3] was presented to address the problems faced by Hop-Count metric. ETX, proposed by de Couto et al. [4], is defined as the expected number of MAC layer transmissions that is needed for successfully delivering a packet through a wireless link. The weight of a path is defined as the summation of the ETX's of all links along the path. In mathematical terms, it can be written:

$$P = 1 - (1 - P_f) \times (1 - P_r) \tag{1}$$

where,

P probability of unsuccessful transmission of packet in a link from node a to node b.
P_f probability of path loss in forward direction.
P_r probability of path loss in reverse direction.

The expected number of transmissions to successfully deliver a packet in one hop can be represented by

$$ETX = \sum_{k=1}^{\infty} kp^k(1-p)^{k-1} = \frac{1}{1-p} \tag{2}$$

The ETX metric for a single link is measured in terms of forward and reverse delivery ratio

$$ETX = \frac{1}{M_f \times M_r} \tag{3}$$

where,

M_f forward delivery ratio is $(1 - P_f)$.
M_r reverse delivery ratio is $(1 - P_r)$.

ETX measures the packet loss rate. Every one second, probe packets are sent to all neighboring nodes. On getting the probe packet, the neighboring node sums the number of packets received. Based on this information, every ten seconds packet loss rate is computed. It finds paths, which have a high throughput with the minimum hops because a long path will have less throughput because of intra-flow interference. It indirectly handles inter-flow interference.

In addition, ETX is also an isotonic routing metric, which guarantees easy calculation of minimum weight paths and loop-free routing under all routing protocols. However, the drawbacks of ETX is that it does not consider interference or the fact that different links may have different transmission rates.

2.2 Expected Transmission Time (ETT) Metric

ETT metric [5] was an enhancement over ETX as it took into account the bandwidth of different links. ETT is the time taken to communicate a packet successfully to the MAC layer.

$$ETT_l = ETX_l \times \frac{S_l}{B_l} \tag{4}$$

where S_l is the average data packet size, B_l is bandwidth of link l.

ETT may choose a path that only uses one channel, even though a patch with more diversified channels has less intraflow interference and hence higher throughput. Similarly to ETX, the chosen route is the one with the lowest sum of ETT values.

Similar to ETX, ETT is also isotonic. However, the remaining drawback of ETT is that it still does not fully capture the intra-flow and inter-flow interference in the network.

2.3 Weighted Cumulative Expected Transmission Time (WCETT) Metric

WCETT [5] was presented to enhance the ETT metric in the multi radio mesh networks by taking into account the diversity of the channels. Given the presence of multiple channels and intra-flow interference, WCETT is defined as:

$$WCETT = (1 - \beta) + \sum_{i=1}^{n} ETT_l + \beta \times \max_{1 \leq j \leq k} X_j \qquad (5)$$

WCETT is the sum of ETTs of all the links in the path p operating on Xj channel j, in a system with total k orthogonal channels. β is a tunable parameter subject to $0 <= \beta <= 1$.

WCETT consists of two components: the first component finds the path with the least sum of ETTs; the second accounts for the bottleneck channel dominating the throughput of the total path.

Over performing ETT, it explicitly accounts for the intra-flow interference, providing support for multi-radio or multi-channel wireless networks. Its two weighted components of it substitute the simple summation of ETT and attempt to strike a balance between throughput and delay. It does not capture inter-flow interference compared with Interference Aware Routing Metric (iAWARE). WCETT is an end-to-end metric because it must consider all channels used along the route to avoid intra-flow interference. It is not isotonic as it can be seen by the presence of the second term. As it does not have isotonicity, it is not easy to use for link state routing protocols. This metric has the same limitations as ETT/ETX metric as it does not estimate actual link presence. Also it does not take into account inter-flow interference effect. As a result, routes with high interference may be established.

2.4   Metric of Interference and Channel Switching (MIC)

MIC metric [6] is an isotonic metric and designed to consider inter and intra-flow interference effects besides providing load balancing. Each node estimates the inter-flow interference, by counting the number of interfering nodes in neighborhood. MIC virtual nodes guarantee minimum-cost routes computation. MIC calculates itself by ETT metric. MIC for a path p is defined as follows:

$$MIC(p) = \frac{1}{N \times \min(ETT)} \times \sum_{i \in p} IRU_i + \sum_{node\ i \in p} CSC_i \qquad (6)$$

where N is the total number of nodes in the network and min(ETT) is the smallest ETT in the network. The two components of MIC, IRU (Interference-aware Resource Usage) is

$$IRU_l = ETT_l \times N_l \qquad (7)$$

and CSC (Channel Switching Cost) is defined as:

$$CSC_i = w1 \quad if \quad CH(prev(i)) \neq CH(i) \qquad (8)$$

$$CSC_i = w2 \quad if \quad CH(prev(i)) = CH(i), 0 \leq w1 \leq w2 \qquad (9)$$

where $0 \leq w1 \leq w2$ and N is the set of neighbors that interfere with the transmissions on link i. CH(i) represents the channel assigned for node i's transmission and prev(i) represents the previous hop of node i along the path p. MIC takes the inter-flow interference only in two consecutive links.

MIC considers interference of a link caused by each interfering node in the neighborhood, counts the amount of interferers on a link only by the position of the interfering nodes no matter whether they are involved in any transmission simultaneously with that link. MIC, therefore, utilizes the measurement of signal power to capture inter-flow and intraflow interference. MIC is not an isotonic but breaking into imaginary nodes through a least weight algorithm like Dijkstra's algorithm [7], it can be made isotonic.

2.5   Load Aware Expected Transmission Time (LAETT Metric)

There are two main aims of LAETT [8]. The aim is to create a path for fulfilling the flow bandwidth demand and to keep a space for the future needs. It is a combination of load estimation and features of wireless access. It comprises of an implementation of ETT metric. LAETTjk is defined by:

$$LAETT_{ij} = ETX_{ij} \times \frac{S}{\left(\frac{RC_i + RC_j}{2\gamma_{ij}}\right)} \tag{10}$$

where ETX_{ij}—expected transmission count on a link (i, j); S—data packet size.

To consider load balancing, remaining capacity (RCi) on each node is introduced and it is given by

$$RC_i = B_i - \sum_{k=1}^{N} f_{ik}\gamma_{ik} \tag{11}$$

where f_{ik} is the rate of transmission of current flow N_i that travels across node i. The flow cost on the leftover bandwidth is weighted by factor γ_{ik}, which is equal to 1 for a very good link. LAETT advantage is that it is isotonic and load aware. It utilizes shortest weighted routing path for balancing the network load. It also captures traffic load and quality of links. The weakness of this metric is that it does not consider intra-flow interference and is not considered as it does not consider inter-flow interference.

2.6 Interference Load Aware (ILA)

ILA metric [9] consists of two components: Channel Switching Cost (CSC) and Metric of channel interference (MTI). CSC component is the same like in MIC. The ILA metric of the route p is defined as:

$$ILA(p) = \alpha \times \sum_{link\ i \in p} MTI_i + \sum_{node\ i \in p} CSC_i \tag{12}$$

where α is scaling factor.

ILA takes into account disadvantages of prevailing metrics like WCETT, ETX, ETT and Hop-Count. It successfully discovers lower congested and a low level interference path which has high throughput and a low packet loss ratio. Inter-flow interference is calculated by ILA. The drawback of this metric is that in two successive links only, the second component CSC can capture intra-flow interference. ILA is not an isotonic. Unlike the ILA metric, Mod-ILA does not have an individual summation component that captures intra-flow interference and it is isotonic but this metric has another drawbacks [10].

2.7 Interference Aware (IAWARE)

iAWARE considers intra-flow and inter-flow interference, medium instability, and data transmission time. This metric uses Signal to Noise Ratio (SNR) and Signal to Interference and Noise Ratio (SINR) to continuously reproduce neighboring interference variations onto routing metrics. In this model [11], a communication between nodes u and v on the link(u, v) is successful, if the SINR at the receiver v is above a certain threshold. Let Pu(v) denotes the signal strength of a packet from node u to node v. iAWARE's first component, finds paths with least path cost and other finds paths with least intra-flow interference (exploiting channel diversity). Moreover, the introduction of SINR is a great breakthrough for inter-flow interference aware routing compared with other ETX-based metric like MIC.

Definition of the link metric iAWARE of a link j as follows:

$$iAWARE_j = \frac{ETT_j}{IR_j} \qquad (13)$$

When IRj for the link j is 1 (no interference), iAWAREj is simply ETTj which captures the link loss ratio and packet transmission rate of the link j. ETTj is weighted with IRj to capture the interference experienced by the link from its neighbors. A link with low ETT and high IRj will have a low iAWARE value. Lower the iAWARE of a link better is the link. Interference ration IRj (u) for node u in a link i = (u, v), where IRj(u) (0 < IRj(u) <=1) can be defined as:

$$IR_i(u) = \frac{SINR_i(u)}{SNR_i(u)} \qquad (14)$$

$$SNR_i(u) = \frac{P_u(v)}{N} \qquad (15)$$

$$SINR_i(u) = \frac{P_u(v)}{N + \sum_{w \in \eta(u)} \tau(w)P_u(w)} \qquad (16)$$

Here η(u) denotes the set of nodes from which node u can hear (or sense) a packet and τ(w) is the normalized rate at which node w generates traffic averaged over a period time. τ(w) is 1 when node w sends out packets at the full data rate supported. τ(w) is used to weight the signal strength from an interfering node w as τ(w) gives the fraction of time node w occupies the channel.

SINR implementation is a big step forward for minimizing inter-flow interference routing in comparison with metrics ETX and WCETT. The weakness of iAware is that it does not possess isotonicity. If a link has higher IRj in comparison to ETTj, iAWARE value is lower as a result a lower ETT path but having greater interference is chosen. The biggest weakness of iAWARE is that it allocates additional weightage to ETT in comparison to interfering links. Unlike the iAWARE metric, Adv-iAWARE does not have an individual summation component that captures intra-flow interference and it is isotonic. But this metric has many another drawbacks [10].

2.8 Interference and Bandwidth Aware ETX (IBETX)

IBETX [12] is designed as threefold metric. Firstly, it directly calculates the Expected Link Delivery (ELD), dexp; that avoids the computational burden, as generated by ETX and bypasses the congested regions in the network like ETX.

Secondly, it provides the nodes with the information of nominal bit rates and makes them able to compute Expected Link Bandwidth (ELB), bexp, of all the wireless links in the same contention domain. Thirdly, long-path penalization by ETX is encountered by calculating the interference, Iexp, named as Expected Link Interference (ELI) also by cross-layered approach. IBETX is defined as follows

$$IBETX = \frac{d_{\exp}}{b_{\exp}} I_{\exp} \tag{17}$$

The 802.11's basic Medium Access Control (MAC) is DCF that besides enabling the nodes to sense the link before sending data, also avoids collisions by employing the virtual carrier sensing. DCF achieves this using Request To Send (RTS) and Clear To Send (CTS) control packets that consequently set the Network Allocation Vector (NAV), i.e., $NAV = \tau_{RTS} + \tau_{CTS}$. The NAV is a counter kept that is and maintained by all nodes in the domain with an amount of time that must elapse until the wireless medium becomes idle. Any node cannot transmit until NAV becomes zero. It stores the channel reservation information to avoid the hidden terminal problem. Using the cross-layer approach, DCF periodically probes the MAC to find the time period for which the link is busy; τ_{busy}. The interference, a node m has to suffer, is expressed as

$$i_m = \frac{\tau_{busy}}{\tau_t} \tag{18}$$

where τ_{busy} is the is the duration for which the medium remains busy; in the case of receiving packets it is Rx state (or communication is going-on with other nodes) and the NAV pending. In the interference expression for node m, τt is the total window time (10 s). If a node n is at the transmitting end, its τ_{busy} is given as: $\tau Rx + \tau Tx + \tau RTS + \tau CTS$. Thus the interferences for sending node n and receiving node m are given as

$$i_m = \frac{\tau_{Rx} + \tau_{RTS} + \tau_{CTS}}{\tau_t} \tag{19}$$

$$i_n = \frac{\tau_{Rx} + \tau_{Tx} + \tau_{RTS} + \tau_{CTS}}{\tau_t} \qquad (20)$$

$$i_{mn} = \max(i_m, i_n) \qquad (21)$$

The link mn formed by nodes m and n are suffering from an interference, i_{mn}, that is the maximum of the interferences calculated in Eqs. (19) and (20), is calculated by Eq. (21). The receiving node m saves the information of interference computed by Eq. (19) and sending node n by Eq. (20). Then the expected interference of the link mn calculated as

$$I_{exp}(mn) = \frac{i_{mn}}{1 + i_{mn}} \qquad (22)$$

Being shared in nature, wireless medium has a problem of interference due to contention. This causes packet loss due to collisions that consequently reduces the bandwidth of links. Therefore, Iexp factor is added, that handles the inter-flow interference among the contending nodes. The longer paths with higher throughputs are ignored by ETX and ETX-based metrics, ELI would not let any path (independent of number of hop-counts) to be ignored while selecting high throughput paths. IBETX value for the end-to-end path P is calculated by

$$IBETX(P) = \sum_{mn=1}^{n} IBETX_{mn} \qquad (23)$$

where mn's are the links on P.

Hence, on directly calculating the loss probability, expected bandwidth and expected interference based on the degree of contention present on the links, IBETX successfully finds the quality links. We consider channel switching delay not to be an effective routing metric, which is non-isotonic.

3 The Proposed Metric

Our metric is consisting of two components. First component Iexp is intended to take into account the interference of the links. Interference helps to consider the longer paths ignored by ETX and all those ETX-based metrics that do not calculate directly the interference among the neighbor links. To exactly measure the congestion in the medium and collisions due to hidden nodes, interference also finds the optimal paths in the wireless network. This component is the same as in IBETX. In our opinion this is the best way to take into account the interference based on features of the MAC level, i.e., using cross-level approach.

Second component is for load-balancing. Load Balancing is defined as the capability of a routing metric to balance the traffic load so that overloading of gateways in wireless mesh networks can be avoided and the network resources are used fairly [13]. To reduce the unbalanced load we use the LBC (Load balancing component) in our routing metric. LBC is based on the mathematical equations like in LAETT metric. Our metric for link mn can be defined as:

$$IALB(mn) = I_{\exp}(mn) \times LBC(mn) \tag{24}$$

$$LBC(mn) = \frac{S}{\left(\frac{RC_m + RC_n}{2\gamma_{mn}}\right)} \tag{25}$$

where $I_{\exp}(mn)$—interference on link (m, n) is the same as in IBETX.

$LBC(mn)$—load balancing component on link (m, n) is the same as the second part in LAETT but it is calculated for all pair of nodes along the path p.

Our metric for path p can be defined as:

$$IALB(p) = \sum_{mn=1}^{MN} (I_{\exp}(mn) \times LBC(mn)) \tag{26}$$

where MN is a quantity of links of path p.

This metric does not consider the cost of channel switching delay because we consider it not to be an effective routing metric, which is non-isotonic. Advantages:

1. IALB is isotonic and load aware.
2. IALB directly calculating the expected interference based on the degree of contention present on the links.
3. Relative simple to calculating.
4. Can be easily modified to account significant characteristics for concrete network (e.g. battery capacity).

4   Conclusions and Future Work

In this survey, we presented a comprehensive analysis of the various routing metrics that have been proposed for routing protocols in Wireless Multi-hop Networks. After minimum hop count which usually selects lossy links, ETX is the most widely used routing link metric (in the presence of least mobility of nodes and availability of links). Along with development of routing metrics, it is needed to pay attention to the problem of measurement and calculation of characteristics used in the metrics. It is impossible to develop routing metric, excluding the effect of MAC and PHY layers, since it is necessary to consider features of a wireless transmission. Using cross-layer approach, IALB metric has provided with the MAC layer information.

Future work goals are to simulate the proposed metric with the most widely used protocols, as AODV, OLSR, etc. and to analyze their performance over recently proposed routing metrics, and design an enhance IALB which take into account the features of wireless multihop networks based on non-stationary nodes with low power.

References

1. Khan, S., Loo, J.: Cross layer secure and resource-aware ondemand routing protocol for hybridwireless mesh networks. Wireless Pers. Commun. **62**(1), 201–214 (2012)
2. Ding, Y., Pongaliur, K., Xiao, L.: Channel allocation and routing in hybrid multichannel multiradio wireless mesh networks. IEEE Trans. Mob. Comput. **12**(2), 206–218 (2013)
3. Draves, R., Padhye, J., Zill, B.: Routing in multi radio, multi hop wireless mesh networks. In: Proceedings of ACM MOBICOM, pp. 114–128 (2004)
4. de Couto, D.S.J.: High-throughput routing for multi-hop wireless networks. Ph.D. dissertation. MIT (2004)
5. Yang, Y., Wang, J., Kravets, R.: Designing routing metrics for mesh networks. In: Proceedings of the WiMesh (2005)
6. Yang, Y., Wang, J., Kravets, R.: Interference-aware loop-free routing for mesh networks (2006)
7. Dijkstra, E.W.: A note on two problems in connection with graphs. Numer. Math., 269–271 (1959)
8. Aiache, H., Conan, V., Lebrun, L., Rousseau, S.: A load dependent metric for balancing internet traffic in wireless mesh networks. In: Mobile Ad Hoc and Sensor Systems, pp. 629–634 (2008)
9. Li, J., Blake, C., Couto, D.S.J., Lee, H.I., Morris, R.: Capacity of ad hoc wireless networks. In: Proceedings of 7th ACM International Conference on Mobile Computing and Networking (2001)
10. Venkat Mohan, S. Dr., Kasiviswanath, N.: ETX based routing metrics. (IJCSIT) Int. J. Comput. Sci. Inf. Technol. **2**(4), 1537–1548 (2011)
11. Subramanian, A.P., Buddhikot, M.M., Miller, S.: Interference aware routing in multi-radio wireless mesh networks. Technical Report, Computer Science Department, Stony Brook University (2007)
12. Javaid, N., Bibi, A., Djouani, K.: Interference and bandwidth adjusted ETX for wireless multi-hop networks (2010)
13. Siraj, Mohammad: A survey on routing algorithms and routing metrics for wireless mesh networks. World Appl. Sci. J. **30**(7), 870–886 (2014)

Mean Square Stability and Dissipativity of Split-Step Theta Method for Stochastic Delay Differential Equations with Poisson White Noise Excitations

Haiyan Yuan, Jihong Shen and Cheng Song

Abstract In this paper, a split-step theta (SST) method is introduced and analyzed for nonlinear neutral stochastic differential delay equations (NSDDEs). The asymptotic mean square stability of the split-step theta (SST) method is considered for nonlinear neutral stochastic differential equations. It is proved that, under the one-sided Lipschitz condition and the linear growth condition, for all positive stepsizes, the split-step theta method with $\theta \in (1/2, 1]$ is asymptotically mean square stable. The stability for the method with $\theta \in [0, 1/2]$ is also obtained under a stronger assumption. It further studies the mean square dissipativity of the split-step theta method with $\theta \in (1/2, 1]$ and proves that the method possesses a bounded absorbing set in mean square independent of initial data.

Keywords Split-step theta method · Nonlinear neutral stochastic differential delay equations · Mean square stability · Dissipativity

1 Instruction

Stochastic functional differential equations (SFDEs) play an important role in science and engineering applications, especially for systems whose evolution in time is influenced by random forces as well as its history information. When the time delay in SFDEs is a constant, it turns into stochastic delay differential equations (SDDEs). Both the theory and numerical methods for SDDEs have been well developed in the

H. Yuan (✉) · J. Shen
College of Automation, Harbin Engineering University,
Harbin 150001, China
e-mail: yhy82_47@163.com

H. Yuan
Department of Mathematics, Heilongjiang Institute of Technology,
Harbin 150050, China

C. Song
School of Management, Harbin Institute of Technology, Harbin 150001, China

© Springer International Publishing Switzerland 2016
R. Silhavy et al. (eds.), *Automation Control Theory Perspectives
in Intelligent Systems*, Advances in Intelligent Systems and Computing 466,
DOI 10.1007/978-3-319-33389-2_9

recent decades, see [1–6]. Recently, many dynamical systems not only depend on present and past states but also involve derivatives with delays. The neutral functional differential equations are often used to describe such systems. Taking the environmental disturbances into account, Kolmanovskii and Myshkis [7] introduced neutral stochastic functional differential equations and gave their applications in chemical engineering and aeroelasticity. Subsequently, many authors have studied the numerical methods for neutral stochastic equations, see [8]. Luo [9] studied the exponential stability for stochastic neutral partial functional differential equations by employing the fixed point principle. Li and Cao [10] considered mean-square stability of two-step Maruyama methods for nonlinear NSDDEs and Wu et al. [11] and Huang [12] had studied the exponential mean square stability of the theta approximations for NSDDEs.

To best of our knowledge, no results on mean-square stability and dissipativity of numerical methods for nonlinear NSDDEs have been presented in the literatures.

The aim of our paper is to further study the mean square stability and dissipativity of the split-step theta method under some conditions and the step constrained.

The paper is organized as follows: in Sect. 2, the exponential stability of analytic solution is introduced for NSDDEs. Some necessary notations and preliminaries are also presented in this section. In Sect. 3, the split-step theta method is used to solve the NSDDEs, and the asymptotic stability results are derived.

Section 4 is to illustrate the mean-square dissipativity of the split-step theta method for nonlinear NSDDEs. Finally, in Sect. 5, we give some numerical experiments to illustrate the mean-square stability and dissipativity of the split-step theta method for a given nonlinear NSDDEs.

2 Exponential Mean-Square Stability of Analytic Solution

Let $\{\Omega, F, \{F_t\}_{t \geq 0}, P\}$ be a complete probability space with a filtration $\{F_t\}_{t \geq 0}$ satisfying the usual condition (i.e., it is increasing and right continuous, and F_0 contains all P-null sets). Let $w(t) = (w_1(t), w_2(t), \ldots, w_l(t))^{\mathrm{T}}$ be standard l-dimensional Brownian motion defined on the probability space.

Let $N : R^d \mapsto R^d$, $f : R_+ \times R^d \times R^d \mapsto R^d$ and $g : R_+ \times R^d \times R^d \mapsto R^{d \times l}$ be Borel measurable functions, where $R_+ = [0, +\infty)$. Consider d-dimensional NSDDE of the form

$$\begin{cases} dy(t) = f(t, y(t), y(t - \tau))dt + g(t, y(t), y(t - \tau))dw(t) \\ \qquad\quad + h(t, y(t), y(t - \tau))dN(t), t \geq 0, \\ y(t) = \varphi(t), t \in [-\tau, 0]. \end{cases} \qquad (1)$$

where the delay τ is a positive constants, and $\varphi(t)$ is an F_0-measurable, $C([-\tau, 0]; R^d)$-valued random variable satisfying

$$\sup_{-\tau \leq t \leq 0} E[\varphi^{\mathrm{T}}(t)\varphi(t)] < +\infty \tag{2}$$

with the notation E denoting the mathematical expectation with respect to P. The following conditions (a1) and (a2) are standard for the existence and uniqueness of the solution for Eq. (1).

(a1) (The local Lipschitz condition). There exists constants $K_L > 0$ and $L > 0$ such that

$$|f(t,x_1,y_1) - f(t,x_2,y_2)|^2 \vee |g(t,x_1,y_1) - g(t,x_2,y_2)|^2 \vee |h(t,x_1,y_1) - h(t,x_2,y_2)|^2$$
$$\leq K_L(|x_1 - x_2|^2 + |y_1 - y_2|^2), \tag{3}$$

for all $|x_1| \vee |x_2| \vee |y_1| \vee |y_2| \leq L$ and $t \in R_+$.

(a2) (The linear growth condition)

$$|f(t,x,y)|^2 \vee |g(t,x,y)|^2 \vee |h(t,x,y)|^2 \leq K_G(1 + |x|^2 + |y|^2), \tag{4}$$

for all $(t,x,y) \in R_+ \times R \times R$, where $K_G > 0$ is a constant.

Definition 1 ([6]) The trivial solution of Eq. (1) is said to be exponentially mean-square stable, if there exists a pair of constants $r > 0$ and $C > 0$, such that, whenever $\sup_{-\tau \leq t \leq 0} E[\varphi^{\mathrm{T}}(t)\varphi(t)] < +\infty$,

$$E[y^{\mathrm{T}}(t)y(t)] \leq C \sup_{-\tau \leq t \leq 0} E[\varphi^{\mathrm{T}}(t)\varphi(t)]e^{-rt}, t \geq 0. \tag{5}$$

Lemma 1 *Assume that there exist a symmetric, positive definite $d \times d$ matrix Q and positive constants μ_1, μ_2 such that for all $(t,x,y) \in R_+ \times R^d \times R^d$ and*

$$x^{\mathrm{T}}Qf(t,x,y) + \frac{1}{2}\mathrm{trace}[g^{\mathrm{T}}(t,x,y)Qg(t,x,y)] \leq -\mu_1 x^{\mathrm{T}}Qx + \mu_2 y^{\mathrm{T}}Qy, \tag{6}$$

with $-\mu_1 + \mu_2 < 0$ hold, then the trivial solution of Eq. (1) is exponentially mean square stable.

3   Mean-Square Stability of Split-Step Theta Method

An adaptation of the split-step theta (SST) method in problem (1) leads to

$$Y_n - NY_{n-m} = y_n - Ny_{n-m} + \theta\Delta t f(t_n + \theta\Delta t, Y_n, \bar{Y}_n), \tag{7}$$

$$\bar{Y}_n = Y_{n-m}, \tag{8}$$

$$y_{n+1} - Ny_{n+1-m} = y_n - Ny_{n-m} + \Delta t f(t_n + \theta\Delta t, Y_n, \bar{Y}_n) + g(t_n + \theta\Delta t, Y_n, \bar{Y}_n)\Delta w_n \tag{9}$$

where stepsize $\Delta t = \frac{\tau}{m}$ for a integer m, y_i is an approximation to $y(t_i)$ where $t_i = i\Delta t$, $i = 1, 2, \ldots$, and $y_k = Y_k = \varphi(k\Delta t)$ for $k = -m, -m+1, \ldots, 0$. $\theta \in [0, 1]$ is a fixed parameter, and $\Delta w_k := w((k+1)\Delta t) - w(k\Delta t)$ is the Brownian increment.

In the special case of $\theta = 1$, this method is equivalent to the split-step backward Euler method. The reason why we consider scheme (7)–(9) is that we can establish some provable stability results for it. In particular, it possesses a better exponential mean square stability property than the classic SLT method.

Now we recall some stability concepts for numerical methods.

Definition 2 For a given stepsize Δt, a numerical method is said to be exponentially mean square stable if there is a pair of positive constants γ and C such that for any initial data $\varphi(t)$ the numerical solution y_n produced by the method satisfies $E[y_n^T y_n] \leq Ce^{-\gamma t_n} \sup_{-\tau \leq t \leq 0} E[\varphi^T(t)\varphi(t)], \forall n \geq 0$.

Definition 3 For a given stepsize Δt, a numerical method is said to be asymptotically mean square stable if for any initial data $\varphi(t)$ the numerical solution y_n produced by the method satisfies $\lim_{n\to\infty} E[y_n^T y_n] = 0$.

Theorem 1 *Assume that system (1) satisfies (6) with $-\mu_1 + \mu_2 < 0$, then the SST method (7)–(9) with $\theta \in (1/2, 1]$ is asymptotically mean square stable for all $\Delta t > 0$. If we further assume that there exist constants K_1 and K_2 such that*

$$f^T(t, x, y)Qf(t, x, y) \leq K_1 x^T Qx + K_2 y^T Qy, (t, x, y) \in R_+ \times R^d \times R^d \tag{10}$$

Then for any $\theta \in [0, 1/2)$, there exist a constant Δt_0 depending on θ such that the method is asymptotically mean square stable for $\Delta t \in (0, \Delta t_0)$.

Proof From (9) it follows that

$$
\begin{aligned}
(y_{n+1} - Ny_{n+1-m})^{\mathrm{T}}Q(y_{n+1} - Ny_{n+1-m}) &= (y_n - Ny_{n-m})^{\mathrm{T}}Q(y_n - Ny_{n-m}) \\
&+ \Delta t^2 f^{\mathrm{T}}(t_n + \theta\Delta t, Y_n, \bar{Y}_n)Qf(t_n + \theta\Delta t, Y_n, \bar{Y}_n) \\
&+ \Delta w_n^{\mathrm{T}} g^{\mathrm{T}}(t_n + \theta\Delta t, Y_n, \bar{Y}_n)Qg(t_n + \theta\Delta t, Y_n, \bar{Y}_n)\Delta w_n \\
&+ 2(y_n - Ny_{n-m})^{\mathrm{T}}\Delta t Qf(t_n + \theta\Delta t, Y_n, \bar{Y}_n) \\
&+ 2(y_n - Ny_{n-m})^{\mathrm{T}}Qgt_n + \theta\Delta t, Y_n, \bar{Y}_n)\Delta w_n \\
&+ 2\Delta t f^{\mathrm{T}}(t_n + \theta\Delta t, Y_n, \bar{Y}_n)Qg(t_n + \theta\Delta t, Y_n, \bar{Y}_n)\Delta w_n.
\end{aligned}
\tag{11}
$$

since $w(t) = (w_1(t), w_2(t), \ldots, w_l(t))^{\mathrm{T}}$ is a standard l-dimensional Brownian motion we have $E(\Delta w_i) = 0$, $E[(\Delta w_i)^2] = \Delta t$, and

$$
\begin{aligned}
&E[\Delta w_n^{\mathrm{T}} g^{\mathrm{T}}(t_n + \theta\Delta t, Y_n, \bar{Y}_n)Qg(t_n + \theta\Delta t, Y_n, \bar{Y}_n)\Delta w_n] \\
&= \Delta t E[\operatorname{trace} g^{\mathrm{T}}(t_n + \theta\Delta t, Y_n, \bar{Y}_n)Qg(t_n + \theta\Delta t, Y_n, \bar{Y}_n)]
\end{aligned}
$$

Let $x_n = y_n - Ny_{n-m}$, $X_n = Y_n - NY_{n-m}$, $n = 0, 1, \ldots$. Solving (7) for x_n, substituting it into the above equality and then taking expectation on both sides, one receives

$$
\begin{aligned}
E[x_{n+1}^{\mathrm{T}}Qx_{n+1}] \leq E[x_n^{\mathrm{T}}Qx_n] &+ (1 - 2\theta)\Delta t^2 f^{\mathrm{T}}(t_n + \theta\Delta t, Y_n, \bar{Y}_n)Qf(t_n + \theta\Delta t, Y_n, \bar{Y}_n) \\
&+ 2\Delta t E(Y_n - NY_{n-m})^{\mathrm{T}}\Delta t Qf(t_n + \theta\Delta t, Y_n, \bar{Y}_n) \\
&+ \Delta t E[\operatorname{trace} g^{\mathrm{T}}(t_n + \theta\Delta t, Y_n, \bar{Y}_n)Qg(t_n + \theta\Delta t, Y_n, \bar{Y}_n)]
\end{aligned}
\tag{12}
$$

which combined with (6), gives

$$
\begin{aligned}
E[x_{n+1}^{\mathrm{T}}Qx_{n+1}] \leq E[x_n^{\mathrm{T}}Qx_n] &+ 2\Delta t E(-\mu_1 Y_n^{\mathrm{T}}QY_n + \mu_2 \bar{Y}_n^{\mathrm{T}}Q\bar{Y}_n) \\
&+ (1 - 2\theta)\Delta t^2 f^{\mathrm{T}}(t_n + \theta\Delta t, Y_n, \bar{Y}_n)Qf(t_n + \theta\Delta t, Y_n, \bar{Y}_n)
\end{aligned}
\tag{13}
$$

In the case of that $\theta > \frac{1}{2}$, using $\Delta t f(t_n + \theta\Delta t, Y_n, \bar{Y}_n) = \frac{1}{\theta}(X_n - x_n)$, and $2X_n^{\mathrm{T}}Qx_n \leq \frac{2\theta - 1 - (-\mu_1 + \mu_2)\Delta t\theta^2}{2\theta - 1}X_n^{\mathrm{T}}QX_n + \frac{2\theta - 1}{2\theta - 1 - (-\mu_1 + \mu_2)\Delta t\theta^2}x_n^{\mathrm{T}}Qx_n$, then we have

$$
\begin{aligned}
E[x_{n+1}^{\mathrm{T}}Qx_{n+1}] \leq \left(1 + \frac{(-\mu_1 + \mu_2)\Delta t(2\theta - 1)}{2\theta - 1 - (-\mu_1 + \mu_2)\Delta t\theta^2}\right)&E[x_n^{\mathrm{T}}Qx_n] - 2\Delta t\mu_2 E[Y_n^{\mathrm{T}}QY_n] \\
&+ 2\Delta t((1 - \lambda^2)\mu_2 + \lambda^2\mu_1)E[\bar{Y}_n^{\mathrm{T}}Q\bar{Y}_n].
\end{aligned}
\tag{14}
$$

Let

$$k = \max\left\{1 + \frac{(-\mu_1 + \mu_2)\Delta t(2\theta - 1)}{2\theta - 1 - (-\mu_1 + \mu_2)\Delta t\theta^2}, \left(\frac{\mu_2}{((1 - \lambda^2)\mu_2 + \lambda^2\mu_1)}\right)^{\frac{1}{m}}\right\}, \quad (15)$$

then $0 < k < 1$. By induction, we obtain from (14) that

$$\mathrm{E}[x_{n+1}^{\mathrm{T}}Qx_{n+1}] \leq k^{n+1}\mathrm{E}[x_0^{\mathrm{T}}Qx_0] - 2\Delta t\mu_2 \sum_{j=0}^{n} k^{n-j}\mathrm{E}[Y_j^{\mathrm{T}}QY_j]$$

$$+ 2\Delta t((1 - \lambda^2)\mu_2 + \lambda^2\mu_1) \sum_{j=0}^{n} k^{n-j}\mathrm{E}[\bar{Y}_j^{\mathrm{T}}Q\bar{Y}_j] \quad (16)$$

Follow (8) it gives

$$\sum_{j=0}^{n} k^{n-j}\mathrm{E}[\bar{Y}_j^{\mathrm{T}}Q\bar{Y}_j] \leq mk^{n-m+1} \max_{-m \leq j \leq -1} \mathrm{E}[Y_j^{\mathrm{T}}QY_j] + k^{-m} \sum_{j=0}^{n-m+1} k^{n-j}\mathrm{E}[Y_j^{\mathrm{T}}QY_j] \quad (17)$$

Therefore,

$$\mathrm{E}[x_{n+1}^{\mathrm{T}}Qx_{n+1}] \leq k^{n+1}\left(\mathrm{E}[x_0^{\mathrm{T}}Qx_0] + 2\tau((1 - \lambda^2)\mu_2 + \lambda^2\mu_1)k^{-m} \max_{-m \leq j \leq -1} \mathrm{E}[Y_j^{\mathrm{T}}QY_j]\right)$$

$$- 2\Delta t(\mu_2 - ((1 - \lambda^2)\mu_2 + \lambda^2\mu_1)k^{-m}) \sum_{j=0}^{n-m+1} k^{n-j}\mathrm{E}[Y_j^{\mathrm{T}}QY_j], \quad (18)$$

Consider $\mu_1, \mu_2 > 0$, (17) implies $-(\mu_2 - ((1 - \lambda^2)\mu_2 + \lambda^2\mu_1)k^{-m}) \leq 0$, we have

$$\mathrm{E}[x_{n+1}^{\mathrm{T}}Qx_{n+1}] \leq k^{n+1}\left(\mathrm{E}[x_0^{\mathrm{T}}Qx_0] + 2\tau((1 - \lambda^2)\mu_2 + \lambda^2\mu_1)k^{-m} \max_{-m \leq j \leq -1} \mathrm{E}[Y_j^{\mathrm{T}}QY_j]\right)$$

On the other hand, we have

$$\|y_{n+1}\| = \|y_{n+1} - Ny_{n+1-m} + Ny_{n+1-m}\| \leq \|x_{n+1}\| + \|Ny_{n+1-m}\|,$$

then we get

$$E[y_{n+1}^T Q y_{n+1}] \leq 2E[x_{n+1}^T Q x_{n+1}] + 2\lambda^2 E[y_{n+1-m}^T Q y_{n+1-m}] \quad (19)$$

Define $\varepsilon_0 = k^{n+1}\left(E[x_0^T Q x_0] + 2\tau((1-\lambda^2)\mu_2 + \lambda^2\mu_1)k^{-m} \max_{-m \leq j \leq -1} E[Y_j^T Q Y_j]\right)$,
then (19) follows $E[y_{n+1}^T Q y_{n+1}] \leq \frac{2}{1-2\lambda^2}\varepsilon_0 + (2\lambda^2)^{\lfloor\frac{n}{m}\rfloor+1} \max_{-m \leq j \leq -1} E[y_j^T Q y_j]$
i.e., the method is asymptotically mean square stable.

In the case of that $\theta \in [0, \frac{1}{2}]$, using condition (10), we obtain from (13) that

$$E[x_{n+1}^T Q x_{n+1}] \leq E[x_n^T Q x_n] + \Delta t((1-2\theta)\Delta t K_1 - 2\mu_1)E[Y_n^T Q Y_n]$$
$$+ \Delta t((1-2\theta)\Delta t K_2 + 2\mu_2)E[\bar{Y}_n^T Q \bar{Y}]$$

A combination of (7) and (10) gives

$$x_n^T Q x_n \leq L_1 Y_n^T Q Y_n + L_2 \bar{Y}_n^T Q \bar{Y}_n$$

where $L_1 = (1+\theta\Delta t)(2+\theta\Delta t K_1)$, $L_2 = (1+\theta\Delta t)(2\lambda^2+\theta\Delta t K_2)$.

Let

$$\Delta t_0 = \begin{cases} +\infty, & \theta = \frac{1}{2} \\ \frac{-2(-\mu_1+\mu_2)}{(1-2\theta)(K_1+K_2)}, & \theta \in [0, \frac{1}{2}) \end{cases}$$

Then, for any fixed $\Delta t \in (0, \Delta t_0)$, $2(-\mu_1+\mu_2) + \Delta t(1-2\theta)(K_1+K_2) < 0$, and there exists a small positive number ε such that

$$2(-\mu_1+\mu_2) + \Delta t(1-2\theta)(K_1+K_2) + \frac{L_1+L_2}{\Delta t}\varepsilon < 0.$$

Therefore,

$$E[x_{n+1}^T Q x_{n+1}] \leq (1-\varepsilon)E[x_n^T Q x_n] + \Delta t((1-2\theta)\Delta t K_1 - 2\mu_1 + \frac{L_1}{\Delta t}\varepsilon)E[Y_n^T Q Y_n]$$
$$+ \Delta t((1-2\theta)\Delta t K_2 + 2\mu_2 + \frac{L_2}{\Delta t}\varepsilon)E[\bar{Y}_n^T Q \bar{Y}_n] \quad (20)$$

Let

$$\tilde{k} = \max\left\{1-\varepsilon, \left(\frac{(1-2\theta)\Delta t K_2 + 2\mu_2 + \frac{L_2}{\Delta t}\varepsilon}{-((1-2\theta)\Delta t K_1 - 2\mu_1 + \frac{L_1}{\Delta t}\varepsilon)}\right)^{\frac{1}{m}}\right\},$$

then $0 < \tilde{k} < 1$. Similarly to the derivation of the first part, we can prove from (20) that $\mathrm{E}[x_{n+1}^T Q x_{n+1}] \le k^{n+1} \left(\mathrm{E}[x_0^T Q x_0] + \tilde{L} \tilde{k}^{-m} \max_{-m \le j \le -1} \mathrm{E}[Y_j^T Q Y_j] \right)$.

Where $\tilde{L} = \tau((1 - 2\theta)\Delta t K_2 + 2\mu_2 + \frac{L_2}{\Delta t}\varepsilon)$, similar to (19), we can easily prove that the method is asymptotically mean square stable for $\Delta t \in (0, \Delta t_0)$. This completes the proof of the theorem.

4 Mean Square Dissipativity

In this section, we further study long time behavior of numerical solutions under a more general construction assumption. That is, we assume that there exist a symmetric, positive definite $d \times d$ matrix Q and positive constants μ_1, μ_2 and γ such that for all $(t, x, y) \in R_+ \times R^d \times R^d$

$$[x - N(y)]^T Q f(t, x, y) + \frac{1}{2} \mathrm{trace}[g^T(t, x, y) Q g(t, x, y)] \le \gamma - \mu_1 x^T Q x + \mu_2 y^T Q y, \tag{21}$$

instead of (7). Now we state and prove some conclusions.

Definition 4 Assume that system (1) satisfies (21). The numerical method is said to be dissipative if, when the method is applied to problem (1) with constraint $\tau = mh$ there exists a constant C such that, for any initial values, there exists an n_0, dependent only on initial values $\varphi(t)$, such that

$$\mathrm{E}[y_n^T Q y_n] \le C, n \ge n_0 \tag{22}$$

Theorem 2 Assume that system (1) satisfies (21), there exists a constant C suchthat, for any initial values, there exists an n_0, dependent only on initial values $\varphi(t)$, when $n \ge n_0$, the numerical solution y_n generated by SST method (7)–(9) with $\theta \in (1, 2, 1]$ satisfies (22).

Proof A combination of (12) and (21) leads to

$$\begin{aligned}
\mathrm{E}[x_{n+1}^T Q x_{n+1}] &\le \mathrm{E}[x_n^T Q x_n] + 2\Delta t\gamma + 2\Delta t \mathrm{E}(-\mu_1 Y_n^T Q Y_n + \mu_2 \bar{Y}_n^T Q \bar{Y}_n) \\
&\quad + (1 - 2\theta)\Delta t^2 f^T(t_n + \theta\Delta t, Y_n, \bar{Y}_n) Q f(t_n + \theta\Delta t, Y_n, \bar{Y}_n)
\end{aligned} \tag{23}$$

Similarly to the derivation of (18), we can obtain

$$
E[x_{n+1}^T Q x_{n+1}] \leq k^{n+1} \left(E[x_0^T Q x_0] + 2\tau((1 - \lambda^2)\mu_2 + \lambda^2 \mu_1)k^{-m} \max_{-m \leq j \leq -1} E[Y_j^T Q Y_j] \right)
$$
$$
- 2\Delta t(\mu_2 - ((1 - \lambda^2)\mu_2 + \lambda^2 \mu_1)k^{-m}) \sum_{j=0}^{n-m+1} k^{n-j} E[Y_j^T Q Y_j] + 2\Delta t\gamma \sum_{j=0}^{n} k^j
$$

$$(24)$$

where $0 < k < 1$ is the same as defined in (15).

Because $-(\mu_2 - ((1 - \lambda^2)\mu_2 + \lambda^2 \mu_1)k^{-m}) \leq 0$, we have

$$
E[x_{n+1}^T Q x_{n+1}] \leq k^{n+1} \left(E[x_0^T Q x_0] + 2\tau((1 - \lambda^2)\mu_2 + \lambda^2 \mu_1)k^{-m} \max_{-m \leq j \leq -1} E[Y_j^T Q Y_j] \right) + \frac{2\gamma\Delta t}{1 - k}
$$

Define $\varepsilon_1 = k^{n+1} \left(E[x_0^T Q x_0] + 2\tau((1 - \lambda^2)\mu_2 + \lambda^2 \mu_1)k^{-m} \max_{-m \leq j \leq -1} E[Y_j^T Q Y_j] \right) + \frac{2\gamma\Delta t}{1-k}$, considering (19), then we have

$$
E[y_{n+1}^T Q y_{n+1}] \leq 2E[x_{n+1}^T Q x_{n+1}] + 2\lambda^2 E[y_{n+1-m}^T Q y_{n+1-m}]
$$
$$
\leq 2\varepsilon_1 + 2\lambda^2 E[y_{n+1-m}^T Q y_{n+1-m}] \leq C
$$

where $C = \frac{2\varepsilon_1}{1 - 2\lambda^2} + \varepsilon$.

This proves the theorem.

5   The Numerical Experiment

The purpose of this section is to illustrate our theoretical results of the stability and dissipativity obtained in Sects. 3 and 4.

Let us consider the following nonlinear NSDDEs,

$$
\begin{pmatrix} d(y_1(t) - 0.25\cos(y_1(t - \tau))) \\ d(y_2(t) - 0.125 y_2(t)) \end{pmatrix}
$$
$$
= \left[A \begin{pmatrix} y_1(t) \\ y_2(t) \end{pmatrix} + B \begin{pmatrix} \sin(y_1(t - \tau)) \\ \cos(y_2(t - \tau)) \end{pmatrix} \right] dt + C \begin{pmatrix} y_1(t)dW_1(t) \\ y_2(t - \tau)dW_2(t) \end{pmatrix}
$$

$$(25)$$

Fig. 1 Mean square stability of SST method with $\theta = 0.1$

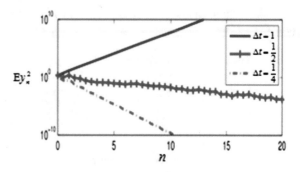

Fig. 2 Mean square stability of SST method with $\theta = 0.6$

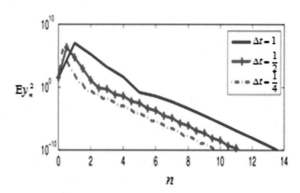

where

$$A = \begin{pmatrix} -28 & 0 \\ 0 & -30 \end{pmatrix}, B = \begin{pmatrix} 2 & -1/2 \\ 1/4 & 1 \end{pmatrix}. C = \begin{pmatrix} 1 & 3/2 \\ 5/2 & -1/2 \end{pmatrix}.$$

The initial condition is given by $\begin{pmatrix} y_1(t) = t + \tau \\ y_2(t) = e^t + 2 \end{pmatrix}$, $t \in [-\tau, 0]$. We take $\tau = 1$.

Consider the SST method (7) for the two-dimensional nonlinear neutral stochastic differential Eq. (25). We take $\Delta t = 1$, $\Delta t = \frac{1}{2}$, $\Delta t = \frac{1}{4}$ and generate 10^3 numerical sample paths using the SST methods. The mean square of the numerical solution is plotted in Figs. 1 and 2. We can see that when $\theta = 0.1$, the SST method is mean square stable for only the step size $\Delta t = \frac{1}{2}$, $\Delta t = \frac{1}{4}$, it is not mean square stable for the step size $\Delta t = 1$, but when $\theta = 0.6$ the SST method is asymptotically mean square stable for all the step size. This is in agreement with our theoretical results.

Acknowledgements This work was supported by the Natural Science Foundation of Heilongjiang Province (A201418) and the Creative Talent Project Foundation of Heilongjiang Province Education Department (UNPYSCT-2015102).

Declare. The authors declare that there is no conflict of interests regarding the publication of this article.

References

1. Baker, C.T.H., Buckwar, E.: Exponential stability in p-th mean of solutions, and of convergent euler-type solutions, of stochastic delay differential equations. J. Comput. Appl. Math. **184**, 404–427 (2005)
2. Buckwar, E.: The θ-maruyama scheme for stochastic functional differential equation with distributed memory term. Monte Carlo Method Appl. **10**, 235–244 (2004)
3. Buckwar, E.: One-step approximations for stochastic functional differential equations. J. Appl. Numer. Math. **56**, 667–681 (2006)
4. Hu, Y., Mohammed, S.A., Yan, F.: Discrete-time approximations of stochastic delay equations: The milstein scheme. Ann. Probab. **32**, 265–314 (2004)
5. Mingzhu, L., Wanrong, C., Zhe, F.: Convergence and stability of the semi-implicit euler method for a linear stochastic differential delay equation. J. Comput. Appl. Math. **170**, 255–268 (2004)
6. Xuerong, M.: Exponential stability of equidistant euler-maruyama approximations of stochastic differential delay equations. J. Comput. Appl. Math. **200**, 297–316 (2007)
7. Kolmanovskii V. B., Myshkis A.: Applied Theory of Functional Differential Equations Kluwer Academic, Norwell, MA(1992)
8. Randjelovic, J., Jankovic, S.: On the pth moment exponential stability criteria of neutral stochastic functional differential equations. J. Math. Anal. Appl. **326**, 266 (2007)
9. Jing. L.: Exponential stability for stochastic neutral partial functional differential equations. J. Math. Anal. Appl. **355**, 414 (2009)
10. Li, X., Wanrong, C.: On mean-square stability of two-step Maruyama methods for nonlinear neutral stochastic delay differential equations. J. Appl. Math. Comput. **261**, 373–381 (2015)
11. Fuke, W., Chengming, H.: Exponential mean square stability of the theta approximations for neutral stochastic differential delay equations. J. Comput. Appl. Math. **286**,172–185 (2015)
12. Chengming, H.: Exponential mean square stability of numerical methods for systems of stochastic differential equations. J. Comput. Appl. Math. **236**, 4016–4026 (2012)

Nonlinearity and Time-Delay Compensations in State-Space Model Based Predictive Control

Stanislav Talaš, Vladimír Bobál, Adam Krhovják and Lukáš Rušar

Abstract In this paper a promising series of modifications of predictive control has been combined in order to extend the functionality of principles of predictive control via linearization. Based on this approach a linear model predictive controller is designed at each point to achieve desired local stability and performance requirements leading to guaranteed functionality through the whole operating range of a nonlinear system. In addition, a compensating technique has been applied in order to deal with the system dynamic burdened with a delayed control input. The improved predictive controller has been implemented and applied on illustrative examples of tank system.

Keywords Model predictive control · Time-delay · Nonlinear system

1 Introduction

A significant number of processes performed in modern industrial sectors tend to be heavily burdened with the time-delay effect. Such cases involve spacious pipe systems, chemical reactors or systems with time-demanding computations. The presence of delayed response in controlled dynamic system results in negative effects such as lowered precision and stability, as well as an increased need of computing power connected with greater hardware requirements.

S. Talaš (✉) · V. Bobál · A. Krhovják · L. Rušar
Faculty of Applied Informatics, Department of Process Control,
Tomas Bata University in Zlin, Zlin, Czech Republic
e-mail: talas@fai.utb.cz

V. Bobál
e-mail: bobal@fai.utb.cz

A. Krhovják
e-mail: krhovjak@fai.utb.cz

L. Rušar
e-mail: rusar@fai.utb.cz

© Springer International Publishing Switzerland 2016
R. Silhavy et al. (eds.), *Automation Control Theory Perspectives in Intelligent Systems*, Advances in Intelligent Systems and Computing 466,
DOI 10.1007/978-3-319-33389-2_10

Considering the large amount of practical areas where time-delay occurs it is not surprising that many efforts were devoted to develop ways to compensate its consequences. A number of techniques that minimize the impact of desynchronized input and output signals are applied in practice. Amongst the most effective is the approach utilizing an internal model of the controlled system to estimate future outcomes as a feedback, therefore it at least partially negates the effect of time-delay in the control structure. As a representative of this strategy the model based predictive control was selected. There has been a significant research in model predictive control; most efforts have been focused on conventional scenarios with linear models where the principle of superposition is applicable [1–3].

From a practical point of view, the basic form of the predictive control cannot be effectively applied for such systems. In other words, the control principle has to be extended by advanced mechanisms. In this paper we have successfully implemented a series of modifications, allowing us to face the problem of both nonlinearities and time-delay. The presented approach will be demonstrated on a system of interconnected liquid tanks simulated in the Matlab environment.

2 Four Tanks System

For testing purposes a simplified model of the four tanks system from [4] was used. The system dynamic behaviour is described by a set of conservation equations, which represents inflow-outflow relations of the liquid inside.

$$q_a\gamma_1 = F_1\frac{dh_1}{dt} + q_1 - q_3$$
$$q_b\gamma_2 = F_2\frac{dh_2}{dt} + q_2 - q_4$$
$$q_b(1 - \gamma_2) = F_3\frac{dh_3}{dt} + q_3$$
$$q_a(1 - \gamma_1) = F_4\frac{dh_4}{dt} + q_4$$
(1)

where

$$q_i = k_i\sqrt{2gh_i} \text{ for } i = 1,\ldots,4$$
(2)

k_i is a positive constant, h_i represents the level of liquid for the corresponding tank, q_a, q_b are manipulated inflows and γ_1, γ_2 are constants representing the inflow separation between tanks. Parameters of tanks and starting liquid levels are noted in Table 1.

For this case study we have considered $\gamma_1 = \gamma_2 = 0.4$, $q_a = 43.4\,\text{ml/s}$, $q_b = 35.4\,\text{ml/s}$ and g as the gravitational acceleration $981\,\text{cm s}^{-2}$. Furthermore, with respect to the speed of valves enabling the control inputs q_a and q_b, all system

Table 1 Model parameters and starting liquid levels

Tank	F_i (cm^2)	k_i (cm^2)	h_i (cm)
1	50.27	0.233	14
2	50.27	0.242	14
3	28.27	0.127	14.25
4	28.27	0.127	21.43

inputs are considered as time-delayed. The length of the delay was estimated to 5 s, representing 5 steps in the sampling period.

Because of the nonlinear behaviour expressed in Eqs. (1), (2), the system was transformed into a linearized incremental form. In order to compensate the inaccuracy for the case of a greater distance from a steady state the strategy of scheduled gain was applied.

The key feature of this approach lies in changing the controller parameters according to the current value of the output variable. With the aim to both simplify the algorithm and satisfy the precision requirements, our assumption is that for the purpose of controller adaptation it applies that

$$y(k) = w(k). \tag{3}$$

Since our goal is to control only output h_1, we need the same number of reference variables [5, 6].

With the scheduling variable identified, the common internal model would take the following form

$$\begin{bmatrix} x_1 \\ x_2 \\ x_3 \\ x_4 \end{bmatrix} = \begin{bmatrix} -\frac{1}{T_1\alpha_1} & 0 & \frac{F_3}{F_1 T_3 \alpha_3} & 0 \\ 0 & -\frac{1}{T_2\alpha_2} & 0 & \frac{F_4}{F_2 T_2 \alpha_2} \\ 0 & 0 & -\frac{1}{T_3\alpha_3} & 0 \\ 0 & 0 & 0 & -\frac{1}{T_4\alpha_4} \end{bmatrix} \begin{bmatrix} x_1 \\ x_2 \\ x_3 \\ x_4 \end{bmatrix} + \begin{bmatrix} \frac{\gamma_1}{F_1} & 0 \\ 0 & \frac{\gamma_2}{F_2} \\ 0 & \frac{1-\gamma_2}{F_3} \\ \frac{1-\gamma_1}{F_4} & 0 \end{bmatrix} \begin{bmatrix} u_1 \\ u_2 \end{bmatrix} \tag{4}$$

where

$$T_i = \frac{F_i}{k_i} \sqrt{\frac{2\alpha_i}{g}}. \tag{5}$$

The system is considered to have a single output and we cannot directly schedule on values of α_2, α_3, α_4, therefore auxiliary functions need to be involved to fill the absence of corresponding reference trajectories as

$$\dot{\alpha}_4 = 0 \tag{6}$$

$$\alpha_3 = \frac{\left(k_1\sqrt{2}\sqrt{g\alpha_1} - k_4\sqrt{2}\gamma_1\sqrt{g\alpha_4}\right)^2}{2gk_3^2} \tag{7}$$

$$\alpha_3 = \frac{\left(k_4\sqrt{2}\sqrt{g\alpha_4} + \frac{\gamma_2 k_3\sqrt{2}\sqrt{g\alpha_3}}{1-\gamma_2}\right)^2}{2gk_2^2} \tag{8}$$

An important feature of the analysis is that even if α_1 represents the reference variable, the Eqs. (4)–(8) still capture the system behaviour around equilibrium.

3   Incremental State-Space Model Based Predictive Control

Use of the state-space description for the internal model makes the control algorithm easier to apply to multi-dimensional systems due to its matrix representation. A slight disadvantage is caused by the need to involve a state observer, which increases computational demands.

We start the predictive system creation with a general form of discrete state-space

$$\begin{aligned} \mathbf{x}(k+1) &= \mathbf{A}\mathbf{x}(k) + \mathbf{B}\mathbf{u}(k+i) \\ \mathbf{y}(k) &= \mathbf{C}\mathbf{x}(k) \end{aligned} \tag{9}$$

With the principle of the predictive control minimizing the difference of control input in mind, Eq. (9) is transformed into model where $\Delta \mathbf{u}$ occurs. The control input is presented as a difference by the definition of the signal increment

$$\Delta u = u(k) - u(k-1) \tag{10}$$

and therefore the entire mathematical model has the following form

$$\begin{bmatrix} \mathbf{x}(k+1) \\ \mathbf{u}(k) \end{bmatrix} = \begin{bmatrix} \mathbf{A} & \mathbf{B} \\ 0 & \mathbf{I} \end{bmatrix} \begin{bmatrix} \mathbf{x}(k) \\ \mathbf{u}(k-1) \end{bmatrix} + \begin{bmatrix} \mathbf{B} \\ \mathbf{I} \end{bmatrix} \Delta \mathbf{u}(k)$$
$$\mathbf{y}(k) = [\mathbf{C} \quad 0] \begin{bmatrix} \mathbf{x}(k) \\ \mathbf{u}(k-1) \end{bmatrix} \tag{11}$$

The new model now contains new matrices $\tilde{\mathbf{A}}$, $\tilde{\mathbf{B}}$, $\tilde{\mathbf{C}}$ and state vector $\tilde{\mathbf{x}}$.

Considering that the available system description is due to presence of nonlinearities expressed in an incremental form, the content needs to be modified into a suitable shape. Equations for state and output predictions are received by transforming (11) by application of a discrete derivative expression $(1 - z^{-1})$ to system state dynamic $\tilde{\mathbf{A}}$ here presented in matrix form [7].

$$
\tilde{\mathbf{x}}(k+i+1) = (\mathbf{I}+\tilde{\mathbf{A}})\tilde{\mathbf{x}}(k+i) - \tilde{\mathbf{A}}\tilde{\mathbf{x}}(k+i-1) + \tilde{\mathbf{B}}\Delta\mathbf{u}(k+i)
$$
$$
\mathbf{y}(k+i) = \tilde{\mathbf{C}}\tilde{\mathbf{x}}(k+i) \tag{12}
$$

for $i = 1, 2, \ldots N$, where N is a size of the prediction horizon determining the extent of predictions performed during the process control.

The objective function of model predictive control is provided as

$$
J = (\mathbf{w} - \mathbf{y})\mathbf{Q}(\mathbf{w} - \mathbf{y}) + \Delta\mathbf{u}\mathbf{R}\Delta\mathbf{u}, \tag{13}
$$

with matrices $\mathbf{Q}$ and $\mathbf{R}$ as weighting parameters dividing the optimization focus between the control divergence and the significance of change in the control input. Length of the vectors $\mathbf{w}$, $\mathbf{y}$ and $\Delta\mathbf{u}$ is determined by the size of the horizon N.

The control law is based on the general prediction of future outputs

$$
\hat{\mathbf{y}} = \mathbf{G}\Delta\mathbf{u} + \mathbf{f}, \tag{14}
$$

where $\mathbf{f}$ is a vector of free response representing the output of the system in case of unchanging control signal.

By combining Eqs. (13) and (14) we receive a way to compute the value of the objective function based on a series of estimated input signals, which tends to be performed by an optimizing algorithm. In case of absence of physical limitations this problem may be simplified by differentiating the objective function with respect to the vector of control inputs $\Delta\mathbf{u}$ and making it equal to zero as an extreme of the function.

$$
\mathbf{u} = -(\mathbf{G}^T\mathbf{Q}\mathbf{G} + \mathbf{R})^{-1}\mathbf{G}^T\mathbf{Q}(\mathbf{f} - \mathbf{w}) \tag{15}
$$

The resulting form of the control law is identical to the declarative model based predictive control. Nonetheless, involved matrices need to be transformed to fit the incremental nature of the controlled system.

$$
\mathbf{G} = \begin{bmatrix} \tilde{\mathbf{C}}\tilde{\mathbf{B}} & 0 & \cdots & 0 \\ \tilde{\mathbf{C}}\sum_{i=0}^{1}\tilde{\mathbf{A}}^i\tilde{\mathbf{B}} & \tilde{\mathbf{C}}\tilde{\mathbf{B}} & & \vdots \\ \vdots & \vdots & & 0 \\ \tilde{\mathbf{C}}\sum_{i=0}^{N}\tilde{\mathbf{A}}^i\tilde{\mathbf{B}} & \tilde{\mathbf{C}}\sum_{i=0}^{N-1}\tilde{\mathbf{A}}^i\tilde{\mathbf{B}} & \cdots & \tilde{\mathbf{C}}\tilde{\mathbf{B}} \end{bmatrix} \quad \mathbf{F} = \begin{bmatrix} \tilde{\mathbf{C}}\sum_{i=1}^{1}\tilde{\mathbf{A}}^i \\ \tilde{\mathbf{C}}\sum_{i=1}^{2}\tilde{\mathbf{A}}^i \\ \vdots \\ \tilde{\mathbf{C}}\sum_{i=1}^{N}\tilde{\mathbf{A}}^i \end{bmatrix} \tag{16}
$$

The free response is consequently calculated from the following expression

$$\mathbf{f} = \mathbf{F}\tilde{\mathbf{x}}. \tag{17}$$

where for the case of a time-delayed system the form of the matrix $\mathbf{F}$ has to be shifted by the corresponding number of steps to provide an estimated output after the interval of the time-delay effect as

$$\mathbf{F}_d = \begin{bmatrix} \tilde{\mathbf{C}} \sum_{i=1}^{1+d} \tilde{\mathbf{A}}^i \\ \tilde{\mathbf{C}} \sum_{i=1}^{2+d} \tilde{\mathbf{A}}^i \\ \vdots \\ \tilde{\mathbf{C}} \sum_{i=1}^{N+d} \tilde{\mathbf{A}}^i \end{bmatrix}. \tag{18}$$

The resulting matrix form suggests an increased necessity of data manipulation, as well as computations connected with the time-delay compensation [8].

4   Regulation Results

The control algorithm containing the time-delay compensating technique, as well as the gain scheduled linearization was tested in the simulation environment Matlab. For the purpose of highlighting main aspects of control dynamics, the reference trajectory was shaped to form a series of sudden step changes.

Figure 1 shows responses of the system controlled with optimization power evenly distributed between precision and input signal change. Simulated results illustrate a fast system response and minimal overdraft in relation to the size of the step in the reference trajectory. Also, with the application of gain scheduled adaptation of system parameters the control precision is same for any system state, despite the nonlinear nature of the system.

To highlight the influence of the time-delay compensation Fig. 2 displays the same procedure without the active suppression of negative effects of the time-delay. As a result, the system output exhibits a significant oscillatory behaviour vastly increasing time required for the controlled variable to settle.

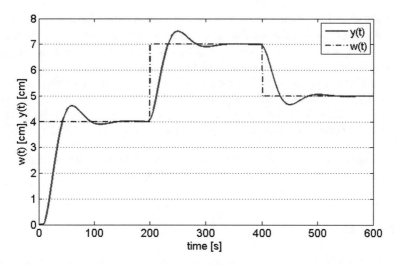

Fig. 1 Control process of the nonlinear time-delayed system of four tanks

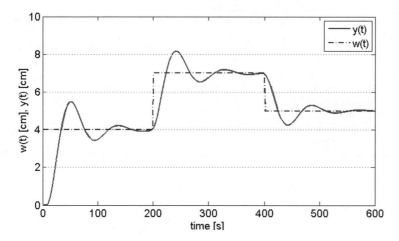

Fig. 2 Regulation procedure without time-delay compensation

5   Conclusion

Impressed by the challenging amount of complications connected with nonlinearity and time-delay compensations we have opened a question of combining available methods dealing with various types of nonlinearities. As an example, a predictive control technique has been designed to control a nonlinear system with time-delay. The presented results have verified our approach combining a gain scheduling control of a nonlinear system with a time-delay compensating procedure.

As the main feat of the control structure we consider the adaptability of the new dynamics in the moment when the present reference value is changed. Time-delay compensating calculations are then performed with greater precision and suppress the instability caused by desynchronized input-output connections.

Further research offers possibilities in the area of a deeper connection of the time-delay compensation and the gain scheduling for the case of known future reference trajectory, as the option of more precise predictions would grant an improved performance, especially for step changes in highly nonlinear system.

Acknowledgments This article was created with support of the Ministry of Education of the Czech Republic under grant IGA reg. n. IGA/FAI/2016/006.

References

1. Qin, S.J., Badgwell, T.A.: A survey of industrial model predictive control technology. Control Eng. Pract. **11**, 733–764 (2003)
2. Haber, R., Bars, R., Schmitz, U.: Predictive Control in Process Engineering: From Basics to the Applications. Willey-VCH Verlag, Weinham (2011)
3. Morari, M., Lee, J.H.: Model predictive control: past, present and future. Comput. Chem. Eng. **23**, 667–682 (1999)
4. Raff, T., Hubber, S., Nagy, Z.K., Allglöwer, F.: Nonlinear Model Predictive Control of a Four Tank System: An Experimental Stability Study, pp. 237–242. IEEE, Munich (2006)
5. Krhovják, A., Dostál, P., Talaš, S., Rušar, L.: Multivariable gain scheduled control of two funnel liquid tanks in series. In: Strbske Pleso, s.n., pp. 60–65 (2015)
6. Shamma, J.S., Athans, M.: Analysis of gain scheduled control for nonlinear plants. IEEE Trans. Autom. Control **35**(8), 898–907 (1990)
7. Lee, J.H., Garcia, C.E.: State-space interpretation of model predictive control. Automatica **30** (4), 707–717 (1994)
8. Normey-Rico, J.E., Camacho, E.F.: Control of Dead-Time Processes. Springer, London (2007)

Plant-Wide Control of a Reactive Distillation Column on Biodiesel Production

Alejandro Regalado-Méndez, Rubí Romero, Reyna Natividad and Sigurd Skogestad

Abstract An economical plant-wide control has been designed and implemented for biodiesel production by reactive distillation. Two available degrees of freedom have been assigned, to control the active constraints. The main conclusions were: (1) the optimal operation with lower value of the cost function (−946.72 USD/min) was established with liquid reflux, vapor stream, and feed flow molar ratio Methanol/Triglycerides of 8.43 kmol/min, 4.52 kmol/min, and 333:25, respectively. Additionally, the Biodiesel European Quality Standards were largely satisfied (Total Glycerol 0.25 wt%), even it exists an over purification, (2) the plant-wide control designed presented an adequate performance, maintaining stable and robust to disturbances and set-point changes, as it always reaches requested reference. Finally, in this study, the Tyreus-Luyben tuning method was the best tuning method, as it takes the least of time to reach the steady state.

Keywords Biodiesel · Daes · Reactive distillation · Plant-wide control

A. Regalado-Méndez (✉)
Universidad Del Mar, Ciudad Universitaria S/N, Oaxaca, Mexico
e-mail: alejandro.regalado33@gmail.com

A. Regalado-Méndez · R. Romero · R. Natividad
Centro Conjunto de Investigación en Química Sustentable UAEMex-UNAM, Unidad San Calletano Km 14.5 Carretera Toluca-Atlacomulco, Toluca, Mexico
e-mail: rromeror@uaemex.mx

R. Natividad
e-mail: rnatividadr@uaemex.mx

S. Skogestad
Norwegian University of Science and Technology, Trondheim, Norway
e-mail: skoge@nt.ntnu.no

© Springer International Publishing Switzerland 2016
R. Silhavy et al. (eds.), *Automation Control Theory Perspectives in Intelligent Systems*, Advances in Intelligent Systems and Computing 466,
DOI 10.1007/978-3-319-33389-2_11

1 Introduction

Ones of the world preoccupations remains in the reduction of petroleum reserves due to the increased use of fuels, and the environmental issues that caused by the climate change [1]. In attention to it, new and secure energy supplies emerge. For instance, researchers have focused their investigations to explore plant-based fuels, plant oils, and fats as promising biofuel sources [2]. In this context, biodiesel is a clean non-toxic, and biodegradable fuel. Typically, biodiesel consist of methyl esters of different fatty acid esters. Biodiesel manufacturing at large scale demands an innovative and efficient process. In this sense, reactive distillation (RD) offers interesting and desirable features since RD is a hybrid process where a chemical reaction and a separation take place in the same vessel [3]. However, studies on reactive distillation revel the existence of steady state multiplicities and Hopf bifurcations [4, 5]. The study of the dynamics and control of the biodiesel production has been addressed in various research works [6]. Plant-wide control refers to the control of an entire plant, usually consisting of many interconnected unit operations [7]. Plant-wide control is the most advanced control technique in the literature, and is a very important topic in industrial process control. In general, it involves the selection of controlled and manipulated variables (CVs and MVs), input-output pairing, the definition of the control structure, tuning, etc. The solution to these problems will define (restrict) the future operability degree for the plant under study. In fact, both investment and operating costs can be seriously affected if the plant-wide control problem is not solved properly [8]. There are some approaches for addressing these problems almost systematically and covering the broad spectrum from strategies based on purely heuristics/engineering adjustment to optimization routines and the model-based on control structures design. This approach is a two-stage method starting with the top-down analysis and ending with bottom-up design. In the first stage, a top-down approach is taken to generate a list of manipulated, control, and disturbance variables considering a scalar operation objective and other process constraints. In the second stage, a bottom-up approach is used for designing the control system taking into account the results from the previous stage [9].

In this work, an economical plant-wide control strategy has been designed and implemented for a biodiesel production by transesterification of triglycerides (e.g. soybean oil) with methanol in presence of NaOH as catalyst by reactive distillation, using the Skogestad methodology (both approaches, top-down and bottom-up). This control must be robust to perturbations and set-point changes as well have to provide adequate performance to the biodiesel production process. In order to warrant the robustness and good performance, the loss technique should be used, as well as aspects of robust control such as feedback control.

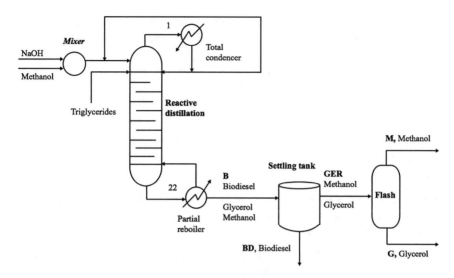

Fig. 1 Flow sheet of biodiesel production by reactive distillation

2 Process Description

The biodiesel production by reactive distillation is carry out by transesterification of the triglycerides (soybean oil) with methanol in presence of NaOH as catalyst. For this case study the kinetic reaction used is reported in [10]. The reactive distillation model for this case study is based on a set of differential algebraic equations such as is given in Eq. (1). This mathematic model is solved considering non-ideal liquid by the UNIQUAC equation.

The reactive distillation has been simulated with the software MATLAB® 2015a and the biodiesel production process (see Fig. 1) has been simulated on Simulink in order to make easier the control implementation. Also, all physicochemical parameters have been taken from ASPEN PLUS® Version 7.2. The dynamic liquid behavior has been considered. Moreover, the Francis equation has been used.

$$\frac{dH_j}{dt} = V_{j+1} + L_{j-1} + F_j + \sum_{m=1}^{r}\sum_{i=1}^{c} v_{i,m}R_{m,j}\varepsilon_j$$

$$\frac{dH_j x_{i,j}}{dt} = V_{j+1}y_{i,j+1} + L_{j-1}x_{i,j-1} + F_j z_{i,j} + \sum_{m=1}^{r}\sum_{i=1}^{c} v_{i,m}R_{m,j}\varepsilon_j$$

$$0 = P - \sum_{i=1}^{c} P_i^{sat} x_{i,j}$$

$$y_{i,j} = K_{i,j}x_{i,j}$$

$$\sum_{i=1}^{c} x_{i,j} = 1; \sum_{i=1}^{c} y_{i,j} = 1$$

$$(1)$$

Nomenclature: H: Hold-up in stage j; F: Feed flow; $v_{i,j}$: Stoichiometric coefficients for the component i in reaction j; R_m: Kinetic reaction; ε: reaction volume; P: Operation pressure; P_i^{sat}: Vapour pressure for component i; K_i: Equilibrium constant; V: Vapor phase; L: Liquid phase; x_i: Molar liquid fraction; y_i: Molar vapour fraction.

3 Systematic Economic Plant-Wide Control System Design

In this section, the plan-wide control design for the biodiesel production by reactive distillation is presented. The control design is based on a Skogestad approach, which has been reduced in seven simple steps [11].

Step 1. Identify the degree of freedom according the valves there are in the system. There are six fixed valves: F, D, B, Ref, V_b, and F_{H_2O}, c

    i. Two valves are used to control liquid level.

    ii. The pressure and feed flow are fixed.

    iii. There are two degree of freedom, so $u = \begin{pmatrix} D \\ B \end{pmatrix}$

Step 2. In Table 1 the functions to be optimized are shown.

Step 3. In Table 2 are present the constraints of the biodiesel production process by reactive distillation.

Step 4. The optimal control structure has been defined in all feasible regions to optimal operation. For this case study there are two active constraints.

    i. The actives constraints must be controlled as the optimal with respect to non-flat variables. Also, it can be possible to move the operation point far enough from the optimal operation point, with an acceptable economic loss (L) in order to have the best control.

Table 1 Functions to be optimized

Name	Function
Objective function	$J_c = J_F + U - G$
Cost feed	$J_F = P_{MeOH}F_{MeOH} + P_{TG}F_{TG}$
Gain products	$G = P_{Biodiesel}F_{Biodiesel} + P_{Glycerol}F_{Glycerol} + P_{MeOH}F_{MeOH}$
Cost utilities	$U = P_{H_2O,c}F_{H_2O,c} + P_{PV}F_{PV}$

J_F Cost feed function; G Products gain function; U Cost utilities function; P_{MEOH} Methanol Price, dollar/ton; P_{TG} Triglyceride price, dollar/ton; $P_{Biodiesel}$ Biodiesel price, dollar/ton; $P_{Glycerol}$ Glycerol price, dollar/ton; $P_{H2O,C}$ Cooling water price, dollar/ton; P_{PV} Pressure steam price, dollar/ton; F_{MeOH} Methanol flow, ton/h; F_{TG} Triglyceride flow, ton/h; $F_{Biodiesel}$ Biodiesel flow, ton/h; F_{MeOH} Glycerol flow, ton/h; F_{MeOH} Distillated methanol flow, ton/h; F_{H2O} Cooling water flow, m³/h; F_{PV} Stream vapor flow, ton/h

Table 2 Constraints applied to the biodiesel production process

Constraint type	Constraint	Number
Fundamentals	$x > 0$	1
	$HR > 0$	2
	$D > 0$	3
	$B > 0$	4
	$Ref > 0$	5
	$T > 0$	6
Decomposition	$T_{eb\_GL} > 423.15$	7
Biodiesel quality	$x_{Metanol}P_{Metanol} < 0.002$	8
	$x_{TG}P_{TG} < 0.002$	9
	$x_{DG}P_{DG} < 0.002$	10
	$x_{MG}P_{MG} < 0.009$	11
	$x_{GL}P_{GL} < 0.002$	12
	$x_{Biodiesel}P_{Biodiesel} < 0.97$	13
Glycerol quality	$x_{C\_GL} > 0.92$	14
Operational	$0.01 < V_b < 8.52$	15

ii. Given the degree of freedom (u) and an important disturbance (d) in the system, it is possible to solve the optimization problem given by Eq. (2). In the case of there is a feasible solution, hence the optimal operational value (u_{opt} (d)) can be obtained as:

$$\min_{\substack{u \\ g(u,d) \leq 0}} J_u(u, d) = J_u\big(u_{opt}(d), d\big) = J_{opt}(d) \tag{2}$$

iii. The current operation value (u) differs from the optimal value (u_{opt} (d)), resulting in a economic loss (L) between current operational cost and optimal operational value. This can be represented by the following equation:

$$L = L_u(u, d) = J_u(u, d) - J_{opt}(d) \\ u = f_c(c_s + d_c, d) \tag{3}$$

where: C_s: is the set-point for the defined disturbances ($d \in D$). The set-point must be the nominal optimum values.

Step 5. The optimal control structure can be chosen for all feasible regions when there are active constraints.

i. Search self-control. For this case study there are two degree of freedom available, which must be used to control the active constraints. Hence, the flow of the vapor stream (V_b) has to be controlled indirectly with the

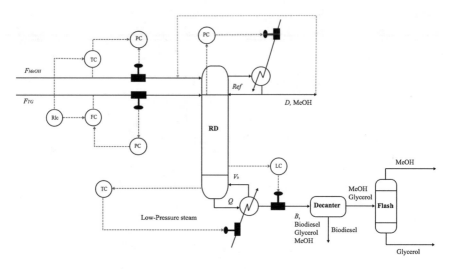

Fig. 2 Control diagram structure of biodiesel production by reactive distillation

input of the heat flow (Q, Heat duty free) as manipulated variable. Similarly, the temperature of the boiler can be computed for supplying low-pressure steam (*FPV*), associating a temperature sensor in the flow rate of the low-pressure steam, as a manipulated variable.

ii. Biodiesel quality can be achieved in terms of the amount of glycerol present in the stream of biodiesel or by hiring a temperature sensor in the distillation column reactor near the desired valuable product. The temperatures chosen for the control loop must be sensitive to inputs.

Step 6. In Fig. 2, the supervisor structure control is presented, using the active constraints. Also, this structure does work in all feasible regions. Therefore, it is an optimal control structure. Simple tuning rules for PI control structures are given in Table 3.

Step 7. For this case study a real time optimizer is not necessary as the supervisor control layer can perfectly control alone the biodiesel production process.

Table 3 Tuning rules to be performed

Tuning rule	K_c	τ_l
Tyreus-Luyben	$k_u/3$	$2P_u$
Skogestad	$1/[k\,(T_u + T_c)]$	$Min\,(T_u, k\,(T_u + T_c))$
Ziegler-Nichols	$0.9T/l$	$0.33\,l$

k_u Last gain; P_u Last period; T_u Last control time constant; T_c Delay time process, l Delay time; T Control time constant

4 Results and Discussions

The prices of the products and utilities to produce biodiesel by reactive distillation have been taking from [12] and the others from the CME group. The results of the optimization of the biodiesel production by reactive distillation are given in Table 4.

Table 5 indicates that all quality constraints are satisfied; giving even an upper than required glycerol purification.

In Fig. 3a displays two different operation regions of the reactive distillation column, which are marked by slope change of the shapes. Also, the cost function diminishes with increments in the feed flow ratio; meaning that optimal operation supports high feed flow ratios. Finally, the cost function diminishes when the flow rates diminish.

The constraints diagram is presented in Fig. 3b. Regions I and IV represent unfeasible regions because region I does not satisfy the norms of biodiesel quality and it region is not possible find a local or global minimum in region IV. Contrarily, region II and III are feasible regions since it is possible to find a local minimum. Finally, in this figure are shows that Regions II and III have the same active constraints numbers. Region II (constraints 12 and 15) corresponds to steeper slope and regions III (7 and 15) corresponds to less steeper slope.

In Table 6, the values of variables process are displayed to reach the optimal operation when the reactive distillation process has been subjected to different disturbances.

Table 4 Parameters values after optimization of the cost function

Parameter	Value	Unit
Cost function, J_c	−946.72	Dollar/min
Vapour, V_b	8.43	Kmol/min
Reflux, Ref	4.51	Kmol/min
Distilled, D	3.79	Kmol/min
Bottom, B	2.10	Kmol/min
Total glycerol	0.12	wt%
Feed flow ratio F_{MeOH}/F_{TG}	333:25	Dimensionless

Table 5 Composition of the purification streams

	BD (wt%)		GER (wt%)		G (wt%)	
	BO	AO	BO	AO	BO	AO
MeOH	0.10	0.09	33.21	27.33	0.56	0.41
TG	1.08	0.04	0.00	0.00	0.00	0.00
DG	1.03	0.09	2.48	0.55	3.22	0.71
MG	0.23	0.05	3.64	1.85	5.23	2.41
GL	2.05	0.02	59.99	69.62	90.12	95.53
Biodiesel	95.59	99.71	0.68	0.65	0.87	0.94

BO Before optimization; *AO* After optimization

Fig. 3 a Operational regions.
b Feasible and unfeasible
regions

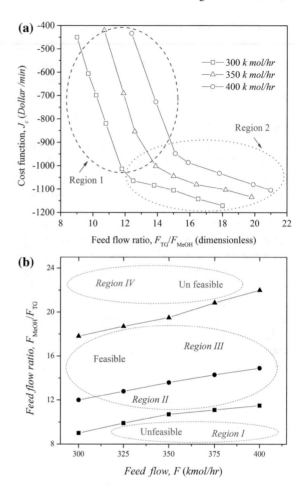

Table 6 Optimal operation conditions for the biodiesel production process

Disturbances	J_c	$X_{Biodiesel}$	$X_{Glycerol}$	F_{MeOH}/F_{TG}	B	D	Ref	V_b
Nominal	−946.721	0.673	0.128	13.32	2.10	3.79	4.510	8.430
P_1: $F = 6.417$ kmol/min	−950.012	0.680	0.124	13.42	2.14	3.75	5.037	8.331
P_2: $Fz_{TG} = 0.917$	−951.310	0.670	0.130	19.99	2.09	3.81	5.027	8.427
P_3: $Fz_{MeOH} = 5.5$ kmol/min	−948.754	0.678	0.130	13.77	2.08	3.82	5.027	8.501
P_4: $q = 1.1308$	−954.217	0.675	0.132	13.35	2.11	3.80	5.017	8.392
P_5: $z_{TG, in} = 0.1375$	−951.136	0.680	0.130	13.40	2.16	3.73	5.037	8.426

Table 7 Loss (Dollar/min) of the biodiesel production process

Disturbances	J_c	$X_{Biodiesel}$	$X_{Glycerol}$	F_{MeOH}/F_{TG}	B	D	Ref	V_b
Nominal	0	0	0	0	0	0	0	0
P_1: $F = 6.417$ kmol/min	3.291	0.002	0.004	0.03	0.04	0.04	0.008	0.009
P_2: $Fz_{TG} = 0.917$	4.589	0.003	0.002	**6.67**	0.01	0.02	**0.517**	0.003
P_3: $Fz_{MeOH} = 5.5$ kmol/min	2.033	**0.005**	0.002	**0.45**	0.02	0.03	**0.517**	**0.071**
P_4: $q = 1.1308$	3.496	0.002	0.004	0.03	0.01	0.01	**0.507**	**0.038**
P_5: $z_{TG, in} = 0.1375$	4.415	**0.007**	0.002	**0.08**	**0.06**	**0.06**	**0.527**	0.004

Table 7 indicates the losses values for the process variables when the reactive distillation process has been subject to different disturbances. The losses are computed by Eq. (3). Unacceptable loss is written in bold (higher than 0.04).

Fig. 4 Plant-wide control performance with the three tuning methods. When it is subject to disturbance of +10 % on feed triglyceride flow at 5 min. Stages (*top* plot) 22 and 11 (*bottom* plot) respectively. **a** Methanol. **b** Biodiesel and **c** Glycerol

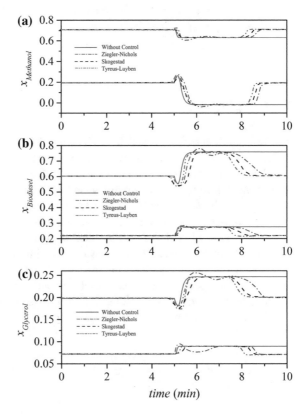

Figure 4 shows the performance of the plant-wide control designed for the biodiesel production by reactive distillation. In this figure, the Ziegler-Nichols's method displays smaller oscillations whereas for Tyreus-Luyben and Skogestad method's there are no notable oscillations. Also, the Tyreus-Luyben present the shortest time to reach the steady state such as is shown with the relationship $\left(\tau^{Tyreus-Luyben} \cong 3.1 \text{ min} < \tau^{Skogestad} \cong 3.4 \text{ min} < \tau^{Ziegler-Nichols} \cong 4.1 \text{ min}\right)$.   Also, the Tyreus-Luybeen's method stabilizes faster the system than the Skogestad and Ziegler-Nichols methods.

Based on the results, the plant-wide control designed is robust to disturbances since it always reaches the set-point. Also, the control objective of the biodiesel production by reactive distillation is largely encountered. The first stage consists of the optimization of the reactive distillation column with the objective function as the cost function (J_c, see Table 1). Finding that optimal operation must be conduced by (1) maximum amount of glycerol present in biodiesel and maximum steam flow and (2) maximum boiler temperature and maximum steam flow, with a cost of 946.72 dollar/min. Finally, second stage consists of two PI control loops structures. (1) the steam flow current and (2) the temperature reboiler, where the manipulated variables are the heat duty and the low-pressure steam. The tuning parameters are $k_{c,1}^{Tyreus-Luyben} = 4.01$; $k_{c,2}^{Tyreus-Luyben} = 2.98$;   , $\tau_{I,1}^{Tyreus-Luyben} = 1.96$; $\tau_{I,2}^{Tyreus-Luyben} = 2.33$.

5   Conclusions

- The operation regions *II* and *III* are feasible since it is always possible to reach a local minimum. Also, there are active constraints for regions *II* and *III*, constraints 12 and 15 (maximum amount of total   glycerol in biodiesel and maximum vapour build-up) and constraints 7 and 15 (maximum temperature in reboiler and   maximum amount of total glycerol in biodiesel), respectively.
- The plant-wide control designed is adequate to control the biodiesel production process by reactive distillation because as the configuration control is stable and robust to perturbations because always reaching the requested set-point.
- Finally, the Tyreus-Luyben tuning method is the best tuning method to synchronize the parameters for biodiesel production process because the stabilization time is lower ($\tau = 186$ s) in comparison with the Skogestad's and Nichols-Ziegler's methods and it has not presented oscillations in in the control performance.

Acknowledgement The authors are grateful to PRODEP (before PROMEP) for the scholarship through PROMEP/103.5/10/5552.

References

1. Benavides, P.T., Diweak, U.: Optimal control of biodiesel production in batch reactor. Part I: Deterministic Control Fuel **94**, 211–217 (2012)
2. Cao, W., Han, H., Zhang, J.: Preparation of biodiesel from soybean oil using supercritical methanol and co-solvent. Fuel **84**, 347–351 (2005)
3. Dimian, A.C., Bildea, C.S., Omota, F., Kiss, A.: Innovative process for fatty acid esters by dual reactive distillation. Comput. Chem. Eng. **33**, 743–750 (2009)
4. Baur, R., Taylor, R., Krishna, R.: Bifurcation analysis for TAME synthesis in reactive distillation column: comparison of pseudo-homogeneous and heterogeneous reaction kinetics models. Chem. Eng. Process. **42**, 211–221 (2003)
5. Chen, F., Huss, R.S., Doherty, M.F., Malone, M.F.: Multiple steady states in reactive distillation: Kinetic effects. Comput. Chem. Eng. **26**, 81–93 (2002)
6. Brásio, A.S.R., Romanenko, A., Leal, J., Santos, L.O.: Nonlinear model predictive control of biodiesel production via transesterification of used vegetables oils. J. Process Control **23**, 1471–1479 (2013)
7. Munir, M.T., Yu, W., Young, B.R.: Plant-wide control: eco-efficiency and control loop configuration. ISA Trans. **52**, 162–169 (2013)
8. Zumoffen, D.A.R.: Oversizing analysis in plant-wide control design for industrial process. Comput. Chem. Eng. **59**, 145–155 (2013)
9. Skogestad, S.: Control structure design for complete chemical plants. Comput. Chem. Eng. **28**, 219–234 (2004)
10. Noureddini, H., Zhu, D.: Kinetics of transesterification of soybean oil. J. Amer. Oil. Chem. Soc. **74**, 1457–1463 (1997)
11. Skogestad, S.: Economic plantwide control. In: Plantwide Control, pp. 229–251. Wiley (2012)
12. Zhang, Y., Dubé, M.A., McLean, D.D., Kates, M.: Biodiesel production from waste cooking oils: 2. economic assessment and sensitivity analysis. Bioresour. Technol. **90**, 229–240 (2003)

State-Space Predictive Control of Two Liquid Tanks System

Lukáš Rušar, Adam Krhovják, Stanislav Talaš and Vladimír Bobál

Abstract This paper presents a process control method called the predictive control used to control a nonlinear process about a selected operating point. The system of the two funnel liquid tanks in series is chosen as an exemplar process. The predictive control is used in its state-space modification for CARIMA mathematical model. This paper describes the linearization process of the nonlinear system at the operating point and a process of a control signal calculation. The designed controller is verified on the process without and with a time-delay.

Keywords State-space · Predictive control · Two liquid tank system · Time-delay · CARIMA model

1 Introduction

Many processes in industry, which need to be control, are complex, nonlinear and also can contain some time-delay. Such processes are very difficult to control, especially when we deal with multi-input multi-output (MIMO) systems. These systems are characteristic by interaction between inputs and outputs which cannot be separated. However, most of the basic control methods are designed only for linear single-input single-output (SISO) systems, so we need to find some more advanced control method. The predictive control based on the state-space CARIMA model is a suitable method to control a complex MIMO system even with the time-delay [1].

The predictive control is a modern method in a control theory. The principle of the predictive control is based on a prediction of the output values on the chosen time horizon. This time horizon should be long enough to cover the step response of the controlled system. The prediction of the output values is based on the

L. Rušar (✉) · A. Krhovják · S. Talaš · V. Bobál
Department of Process Control, Faculty of Applied Informatics,
Tomas Bata University in Zlin, Zlín, Czech Republic
e-mail: rusar@fai.utb.cz

© Springer International Publishing Switzerland 2016
R. Silhavy et al. (eds.), *Automation Control Theory Perspectives
in Intelligent Systems*, Advances in Intelligent Systems and Computing 466,
DOI 10.1007/978-3-319-33389-2_12

119

mathematical model of the controlled system and the control signal is obtained by a minimization of the cost function [2, 3]. The main advantage of this method is its capability to completely eliminate the time-delay of the system [4, 5]. But it is also designed for linear processes.

One additional step has to be done before controlling the nonlinear system. This step is called the linearization of the nonlinear system in a selected operating point. The divergence linear model is the result of the linearization. It means, that the selected operating point is a new origin state for the controller and the input and the output values are a divergence from the equilibrium values [6, 7].

The two funnel liquid tanks in series have been chosen as an exemplar system for the control process. This system is described in the Sect. 2. The predictive control method used to control the chosen system is defined in the Sect. 3 and the results are presented in the Sect. 4. The final section contain the conclusion of this paper.

2 Mathematical Model of the System

The mathematical model of the two liquid tanks system is taken from [8]. The schematic diagram of the controlled process is shown in Fig. 1. The controlled process is formed by two funnel liquid tanks in series. The first tank is filled by the input stream q_{1f}, the second tank is filled by the input stream q_{2f} and the output stream from the first tank q_1.

The mathematical model of this process is obtained from balancing equations [9]. The Eq. (1) stands for the balancing equation of the input and the output of the first tank and the Eq. (2) stands for the balancing equation of the input and the output of the second tank. The balancing equations can be expressed as

$$\pi \frac{D^2}{4H^2} h_1^2 \frac{dh_1}{dt} + q_1 = q_{1f} \tag{1}$$

Fig. 1 Schematic diagram of two funnel liquid tanks

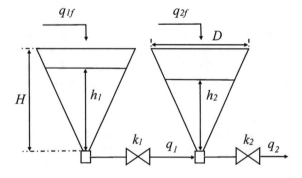

$$\pi \frac{D^2}{4H^2} h_2^2 \frac{dh_2}{dt} - q_1 + q_2 = q_{2f} \tag{2}$$

where D is the maximum diameter, H is the total height and h_1 and h_2 are the liquid levels from the bottom of the tanks. The output liquid streams q_1 and q_2 depend on the liquid levels as

$$q_1 = k_1 \sqrt{|h_1 - h_2|} \tag{3}$$

$$q_2 = k_2 \sqrt{h_2} \tag{4}$$

where k_1 and k_2 are valve constants.

It is obvious that the model of this system is nonlinear. If we want to control this system, we need to linearize Eqs. (1) and (2) about operating points. To do that we choose the input variables as $u_1 = q_{1f}$ and $u_2 = q_{2f}$ and the output (controlled) and the state variables as $y_1 = x_1 = h_1$ and $y_2 = x_2 = h_2$. The balancing equations now can be expressed as a nonlinear state-space model

$$\dot{x} = f(x, u) \tag{5}$$

$$y = g(x) \tag{6}$$

where the particular equations for single states and outputs are

$$\frac{dx_1}{dt} = \frac{4H^2}{\pi D^2 x_1^2} \left(u_1 - k_1 \sqrt{|x_1 - x_2|} \right) \tag{7}$$

$$\frac{dx_2}{dt} = \frac{4H^2}{\pi D^2 x_2^2} \left(u_2 + k_1 \sqrt{|x_1 - x_2|} - k_2 \sqrt{x_2} \right) \tag{8}$$

$$y_1 = x_1 \tag{9}$$

$$y_2 = x_2 \tag{10}$$

We can calculate the operating points as the equilibrium points where are no changes of the liquid levels in the tanks. This equilibrium state can be expressed as

$$0 = \frac{4H^2}{\pi D^2 x_1^2} \left(u_1 - k_1 \sqrt{|x_1 - x_2|} \right) \tag{11}$$

$$0 = \frac{4H^2}{\pi D^2 x_2^2} \left(u_2 + k_1 \sqrt{|x_1 - x_2|} - k_2 \sqrt{x_2} \right) \tag{12}$$

The input and output equilibrium values are changed into the divergence form

$$x_\delta(t) = x(t) - \bar{x} \tag{13}$$

$$u_\delta(t) = u(t) - \bar{u} \tag{14}$$

$$y_\delta(t) = y(t) - \bar{y} \tag{15}$$

where $\bar{x}$ is a vector of the equilibrium state variables, $\bar{u}$ is a vector of the equilibrium input variables, $\bar{y}$ is a vector of the equilibrium output variables, $x_\delta, u_\delta, y_\delta$ are the divergences from the equilibrium values.

The state space model of the linearized system about the operating points is expressed as

$$\dot{x}_\delta = Ax_\delta + Bu_\delta \tag{16}$$

$$y_\delta = Cx_\delta \tag{17}$$

where the matrices A, B are

$$A = \left. \frac{\partial f}{\partial x} \right|_{x=\bar{x}, u=\bar{u}} \tag{18}$$

$$B = \left. \frac{\partial f}{\partial u} \right|_{x=\bar{x}, u=\bar{u}} \tag{19}$$

and C is an identity matrix [6, 7].

The continuous-time input-output linear model of the controlled system has form

$$A(s)y(t) = B(s)u(t)$$

$$\begin{bmatrix} s + a_{01} & a_{02} \\ a_{03} & s + a_{04} \end{bmatrix} \begin{bmatrix} y_1(t) \\ y_2(t) \end{bmatrix} = \begin{bmatrix} b_{01} & 0 \\ 0 & b_{04} \end{bmatrix} \begin{bmatrix} u_1(t) \\ u_2(t) \end{bmatrix} \tag{20}$$

where

$$a_{01}(\boldsymbol{h}) = \frac{4H^2 k_1 \sqrt{h_1 - h_2}}{2D^2 \pi h_1^2 (h_1 - h_2)}, a_{02}(\boldsymbol{h}) = -\frac{4H^2 k_1 \sqrt{h_1 - h_2}}{2D^2 \pi h_1^2 (h_1 - h_2)}$$

$$a_{03}(\boldsymbol{h}) = -\frac{4H^2 k_1 \sqrt{h_1 - h_2}}{2D^2 \pi h_2^2 (h_1 - h_2)}$$

$$a_{04}(\boldsymbol{h}) = \frac{4H^2}{2D^2 \pi h_2^2} \left[\frac{k_1 \sqrt{h_1 - h_2}}{h_1 - h_2} + \frac{k_2 \sqrt{h_2}}{h_2} \right]$$ (21)

$$b_{01}(\boldsymbol{h}) = \frac{4H^2}{D^2 \pi h_1^2}, b_{04}(\boldsymbol{h}) = \frac{4H^2}{D^2 \pi h_2^2}$$

This continuous-time model needs to be transferred into a discrete-time input-output model with the sampling period T_0. The discrete-time model is the CARIMA model

$$\tilde{\boldsymbol{A}}(z^{-1})\boldsymbol{y}(k) = \boldsymbol{B}(z^{-1})z^{-d}\Delta\boldsymbol{u}(k)$$ (22)

where the polynomial matrix $\tilde{\boldsymbol{A}}(z^{-1})$ is

$$\tilde{\boldsymbol{A}}(z^{-1}) = (1 - z^{-1})\boldsymbol{A}(z^{-1})$$ (23)

3   State-Space Predictive Control

The state-space predictive control method used in this paper works with the state-space CARIMA model modified into the form

$$\boldsymbol{x}(k+1) = \tilde{\boldsymbol{A}}\boldsymbol{x}(k) + \boldsymbol{B}\Delta\boldsymbol{u}(k-d)$$
$$\boldsymbol{y}(k) = \boldsymbol{C}\boldsymbol{x}(k)$$ (24)

where the vector of state variables has form

$$\boldsymbol{x}(k) = [\boldsymbol{y}(k), \boldsymbol{y}(k-1), \boldsymbol{y}(k-2), \ldots, \boldsymbol{y}(k-na),$$
$$\Delta\boldsymbol{u}(k-d-1), \Delta\boldsymbol{u}(k-d-2), \ldots, \Delta\boldsymbol{u}(k-d-nb+1)]^T$$ (25)

and the vectors of the outputs variables and the input control increments are

$$\mathbf{y}(k) = [y_1(k) \quad y_2(k) \quad \cdots \quad y_n(k)]^T \tag{26}$$

$$\Delta\mathbf{u}(k - d) = [\Delta u_1(k - d) \quad \Delta u_2(k - d) \quad \cdots \quad \Delta u_m(k - d)]^T \tag{27}$$

where n is a number of outputs and m is a number of inputs [4].

The matrices $\tilde{A}$, B and C can be expressed as

$$\tilde{A} = \begin{bmatrix} -\tilde{A}_1 & -\tilde{A}_2 & \cdots & -\tilde{A}_{na} & -\tilde{A}_{na+1} & B_2 & B_3 & \cdots & B_{nb-1} & B_{nb} \\ I & 0 & \cdots & 0 & 0 & 0 & 0 & \cdots & 0 & 0 \\ 0 & I & \cdots & 0 & 0 & 0 & 0 & \cdots & 0 & 0 \\ \vdots & \vdots & \ddots & \vdots & \vdots & \vdots & \vdots & \ddots & \vdots & \vdots \\ 0 & 0 & \cdots & I & 0 & 0 & 0 & \cdots & 0 & 0 \\ 0 & 0 & \cdots & 0 & 0 & 0 & 0 & \cdots & 0 & 0 \\ 0 & 0 & \cdots & 0 & 0 & I & 0 & \cdots & 0 & 0 \\ 0 & 0 & \cdots & 0 & 0 & 0 & I & \cdots & 0 & 0 \\ \vdots & \vdots & \ddots & \vdots & \vdots & \vdots & \vdots & \ddots & \vdots & \vdots \\ 0 & 0 & \cdots & 0 & 0 & 0 & 0 & \cdots & I & 0 \end{bmatrix} \tag{28}$$

$$B = [B_1 \quad 0 \quad \cdots \quad 0 \quad 0 \quad I \quad 0 \quad \cdots \quad 0 \quad 0]^T \tag{29}$$

$$C = [I \quad 0 \quad \cdots \quad 0 \quad 0] \tag{30}$$

where I is an identity matrix and 0 is a zeros matrix.

The prediction of outputs values can be calculated recursively from the state-space model and it can be expressed in the matrix form

$$\hat{y} = Fx + H_p\Delta u_p + H_f\Delta u_f \tag{31}$$

where $\hat{y}$ is the vector of the predicted output values, Δu_p is the vector of the past control increments and Δu_f is the vector of the future control increments

$$\hat{y} = \begin{bmatrix} \hat{y}(k+d+1) \\ \hat{y}(k+d+2) \\ \vdots \\ \hat{y}(k+d+N) \end{bmatrix}, \Delta u_p = \begin{bmatrix} \Delta u(k-d) \\ \Delta u(k-d+1) \\ \vdots \\ \Delta u(k-1) \end{bmatrix}, \Delta u_f = \begin{bmatrix} \Delta u(k) \\ \Delta u(k+1) \\ \vdots \\ \Delta u(k+N) \end{bmatrix} \tag{32}$$

where d is a number of the time-delay and N is the prediction time horizon. This time horizon should be long enough to cover the step response of the controlled system [4].

The predictive control is based on a minimization of the cost function [10]. The cost function used in this method has a quadratic form

$$J = (w - \hat{y})^T Q_\delta (w - \hat{y}) + \Delta u_f^T Q_\lambda \Delta u_f \tag{33}$$

where w is a vector of the future reference values, $\hat{y}$ is the vector of the predicted outputs values, Q_λ and Q_δ are the diagonal weighting matrices. The vector Δu_f is the unknown vector of the future control increments. This unknown vector is calculated by setting the derivative of the cost function equal zero and then this vector is expressed as

$$\Delta u_f = \left[H_f^T Q_\delta H_f + Q_\lambda \right]^{-1} H_f^T Q_\delta \left(w - \hat{y}_{free} \right) \tag{34}$$

where

$$\hat{y}_{free} = Fx + H_p \Delta u_p \tag{35}$$

is a free response of the controlled system.

4 Results

A functionality of the designed controller was verified by simulations in MATLAB/ Simulink. All of the simulations were executed with the same system parameters shown in Table 1. Three operating points were chosen for the linearization of the system parameters. These operating points values are shown in Table 2 and they are the starting equilibrium points for the simulations. All of the simulations have the same reference values. Figures 2 and 3 show the simulations results without a

Table 1 System parameters

Tank	D [m]	H [m]	K [m³/min]
1	1.5	2.5	0.316
2	1.5	2.5	0.296

Table 2 Operating points values

Operating point	$\bar{h}_1$ [m]	$\bar{h}_2$ [m]	$\bar{q}_{1f}$ [m³/min]	$\bar{q}_{2f}$ [m³/min]
1	2.0	1.6	0.1999	0.1746
2	1.8	1.4	0.1999	0.1504
3	1.5	1.1	0.1999	0.1106

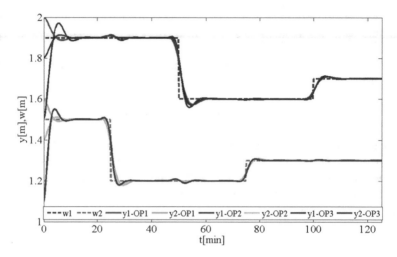

Fig. 2 Simulation outputs without time-delay

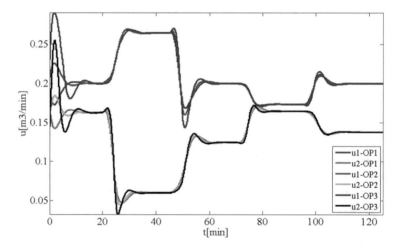

Fig. 3 Simulation inputs without time-delay

time-delay and Figs. 4 and 5 show the simulations results with a time-delay. Table 3 shows the controller parameters used for the simulations.

Following figures show the simulation results:

From the presented figures it can be seen, that the designed predictive controller is able to control the linearized system of two funnel liquid tanks in series about the selected operating point with the properly chosen weighting coefficients. The weighting coefficients influence the speed of control process. The staring equilibrium point has almost no effect on the control process if it is close enough to the

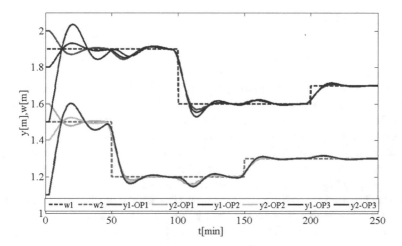

Fig. 4 Simulation outputs with time-delay

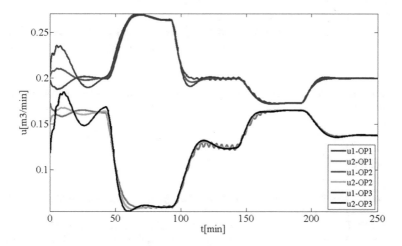

Fig. 5 Simulation inputs with time-delay

Table 3 Controller parameters	D [steps]	T_0 [min]	N [steps]	λ	δ
	0	0.25	20	10	0.1
	10	0.25	20	23	0.01

reference value. The weighting coefficients have a greater impact on the control process then the starting equilibrium point. Wrong choose of these coefficients leads to an unstable simulation. If the time-delay is presented in the controlled system, the weighting coefficients have to be significantly changed to keep the control process stable. The weighting coefficients have been setup experimentally in this case.

5 Conclusion

In this paper, the predictive controller based on the state-space CARIMA model was presented. This method was used to control the nonlinear process about the selected operating point. The problem of controlling the nonlinear process by the predictive control method is that the chosen predictive control method works only with linear processes. So, the linearization of the nonlinear process about the selected operating point was presented after a short description of the chosen nonlinear process. The multi-input multi-output system of the two funnel liquid tanks on series was chosen as the exemplar process. The state-space predictive control showed up to be a suitable method to control such process, if the operating point is close enough to the reference value. This method is also able to control the chosen process with a time-delay. The main advantage of this method is a possibility of control process adjustment by change of the weighting coefficients λ and δ. With proper setting of these coefficients, this method is able to control the chosen system no matter of the starting operating (equilibrium) point. However, a method of the proper setting of the weighting coefficients is subject to a further study. Another advantage of this predictive control method is that it is ready for a control process constraints application directly into the calculation of the control signal. The only difference will be in the way how to obtain the control signal from the cost function.

Acknowledgments This article was created with support of the Ministry of Education of the Czech Republic under grant IGA reg. n. IGA/FAI/2016/006

References

1. Wang, L.: Model Predictive Control: System Design and Implementation Using MATLAB. Springer, London (2009)
2. Camacho, E.F., Bordons, C.: Model Predictive Control. Springer, London (2004)
3. Rossiter J.A.: Model Based Predictive Control: a Practical Approach. CRC Press (2003)
4. Bars, R., Haber, R., Schmitz, U.: Predictive Control in Process Engineering: From the Basics to the Applications. Willey-VCH Verlag, Weinhaim (2011)
5. Camacho, E.F., Normey-Rico, J.E.: Control of Dead-time Processes. Springer, London (2007)
6. Albertos Pérez, P., Sala, A.: Multivariable Control Systems: an Engineering Approach. London. Springer (2004)
7. Hangos, K.M., Bokor, J., Szederkényi, G.: Analysis and Control of Nonlinear Process Systems. Springer, London (2004)
8. Krhovják A., Dostál P., Talaš S., Rušar L.: Multivariable Gain Scheduled Control of Two Funnel Liquid Tanks in Series. In: International Conference on Process Control 2015, pp. 60–65. Štrbské Pleso, Slovakia (2015)
9. Richardson S.M.: Fluid Mechanics. Hemisphere Publishing Corporation, New York (1989)
10. Fikar, M., Mikleš, J.: Process Modelling, Optimisation and Control. Springer, Berlin (2008)

Quantum Evolutionary Cellular Automata Mapping Optimization Technique Targeting Regular Network on Chip

Belkebir Djalila and Boutekkouk Fateh

Abstract This paper presents a novel method for solving the mapping and scheduling problems in network on chip based on quantum evolutionary cellular automata (QECA). The method applies QECA to handle the multimedia application IP placement and scheduling problem. The QECA method is based on the concept and principles of quantum computing, such as quantum bits, quantum gates and superposition of states. Thus, the mechanism of the QECA method can inherently treat the balance between exploration and exploitation where each Q-bit individual can represent and explore all possible states and drive it to exploit a single state. The use of quantum bit representation leads to better population diversity compared with the classical bit representations while the use of quantum gate drive the population towards the best solution. The achieved results are about 0.99 % of the fitness function over 110 generations.

Keywords Quantum genetic algorithm · Cellular automata · Network on chip · Quantum computing · Energy consumption

1 Introduction

Quantum computing has been surpassing classical computers for its ability to solve problems of polynomial-time that classical computers considered it as impossible to be solved unless with million years. Recently, the growing theoretical and practical interest is devoted to researches on quantum computing and quantum computers and due to that many quantum computing algorithms have been developed, such as Shors factorizing algorithm, were explored, quantum search algorithm and Quantum

B. Djalila (✉) · B. Fateh (✉)
Research Laboratory on Computer Science's Complex System (RELA(CS)2),
Oum El-Bouaghi University, Po. Box 358, 04000 Oum El-Bouaghi, Algeria
e-mail: Belkebir.Djalila@univ-oeb.dz

B. Fateh
e-mail: Fateh_boutekkouk@yahoo.fr

© Springer International Publishing Switzerland 2016
R. Silhavy et al. (eds.), *Automation Control Theory Perspectives
in Intelligent Systems*, Advances in Intelligent Systems and Computing 466,
DOI 10.1007/978-3-319-33389-2_13

genetic algorithm (QGA). Quantum evolutionary computing is a branch of study on evolutionary computation and employs the certain principles of quantum mechanics, such as superposition, interference [1]. Based on the concept and principles of quantum computing, Han and Kim [1] developed a quantum-inspired evolutionary algorithm (QEA), which can achieve a better balance between exploration and exploitation of the solution space and also obtain better solutions, even with a small population, compared with the conventional EAs. Moreover, in QGA qubit encoding is used to represent the chromosome, and evolutionary process is implemented by using quantum logic gate operation on the chromosomes. In [2], a comparison is made between a classical genetic algorithm and a quantum inspired method for the travelling salesperson. An improved QGA based on multi-qubit encoding and dynamically adjusting the rotation angle mechanism was presented to separate the blind sources [3].

However, CA presents a highly parallel and distributed system of locally interacting units which are able to produce a global behavior. CA can be considered as a model of naturally existing systems produced by natural evolution. Such systems are capable of producing globally coordinated information processing, unguided by any global criterion or central control. Information processing capabilities of such systems are not explicitly represented in their components but rather in their interconnections. These capabilities are more powerful than ones done by elementary components or their combinations. For these reasons, CA was evolved with GA by Packard and Mitchell et al. where the optimization addressed the density classification problem. Moreover, evolving CAs with GAs also gives us a tractable framework in which an evolutionary process might create complex coordinated behavior in natural decentralized distributed systems and that what encourage its using in embedded system. Evolutionary cellular automata improves many advantage in the field of computer science but the GA search space in CA lattices or the transition rules is increased according to the CA dimension and that leads to increasing the computational time (3D or 4D CA is hard to be updated by GA in short time). Also the classical chromosome are weak in representing the population diversity and as mentioned before, since the use of quantum bit representation leads to better population diversity while the use of quantum gate drive the population towards the best solution. This paper proposes quantum evolutionary cellular automata to improve energy efficiency and timing constrains.

2 Preliminaries and Methods

The NoC main components are links, routers, and network Interfaces. Formally $G(P, L)$ a directed topology graph where P_i represents a node of the network defined by its processor's global voltage or, a directed arc $l_{i,j} = (p_i, p_j)$ represents a physical unidirectional channel connecting two nodes p_i and p_j. The weight of the edge $l_{i,j}$ dented by $bw_{i,j}$ represents the bandwidth capacity for the edge $l_{i,j}$ in flit/time.

The application model is defined by $G(C, E)$: is a synchronous dataflow graph, with each vertex $c_i \in C$ representing a task that is defined by $w(i)$ the task workload. The directed edge $e_{i,j} \in E$ represent the communication between the tasks c_i to c_j. The weight of edge denoted by $v(c_{i,j})$ represents the bandwidth requirement of the communication from c_i to c_j in bits per cycle.

The power model used to estimate the total power dissipated in transferring packets in NoC. It is given by:

$$E_{NoC} = N_{task} \times E_t + M_{channel} \times (n_r \times E_{router} + (n_r - 1) \times E_{link}) \qquad (1)$$

where: N is the number of tasks, M is the number of channels, E_{router}, E_{link} [4, 5] and E_t are the energy consumed during routing the packets in the router, transmitting it from source to destination and executing the task onto the appropriate processor respectively.

$$E_{flit} = n_r \times E_{router} + (n_r - 1) \times E_{link} \qquad (2)$$

E_{flit} is the energy consumed during sending one flit, n_r is the number of hops traversed.

The delay model is defined by the time it takes to transmit, serve and execute all packets (or a flit) is defined by the sum of the execution time of tasks and communication time between those tasks via channels. It is given by:

$$L_{NoC} = \sum_{i=1}^{T} \left(Ex_{ci}^{pj} \right) + \sum_{i=1}^{C} \left((n - 1) \times com_j + N \times R_j \right) \qquad (3)$$

The first part of the equation represents the execution time and the wait time of each task. The execution time Ex_{ci}^{pj} (see Eq. 4) is defined as the number of cycles it takes to execute task c_i on processor p_j.

$$Ex_{Ci}^{Pj} = w(i)/f \qquad (4)$$

The second part from Eq. 3 represents the communication time that is defined as the time it takes for packet to be transmitted from source to destination. It contains the routing decision time and the communication time that is the time it takes to send the packet through the link.

The mapping of the application G(C,E) and NoC G(P,L) is defined by:

Map:C $\rightarrow$ P,s.t.map(c$_i$) = p$_i$, $\forall$c$_i$ $\in$ C,$\exists$p$_i$ $\in$ P

Minimize (E_{NoC}).

Minimize (L_{NoC}).

3   Quantum Evolutionary Cellular Automata

The finite automate is defined by (V, H, f) where V is the number of neighbors per cell, H is number of cell states and f represents the transition rule.

Cellular automata (CA) are dynamic systems whose main characteristics are the number of cell states V and a lattice of N cells where each cell is in one of K possible states at time t. Each cell follows the same rule to update its state. The state "s" of a cell at time $t+1$ depends on its state and the states of a number of neighboring cells at time t. The CA starts with an initial configuration of cells and at each time step the states of all cells in the lattice are updated in a synchronous manner.

A transition function f defines a rule of updating cell i. It is expressed in a table called "look-up table" which lists for each local neighborhood the state's update of the neighborhood's central cell where the neighborhood relation and the transition function are the same for every cell $\{0..n - 1\}$, Formally, the evolution of CA at time $t+1$ is expressed the by the formula: $S_i(t+1) = f(\{S_{j \in H(i)}(t)\})$, where $H(i)$: neighbors list of cell i.

The use of GAs still the subject of intensive study to improve their performance for many problems such as using old instance of populations or the complete population is impossible; general knowledge about previous solutions cannot be used. A new searching process must be started from the beginning by creating an initial random population of potential solutions. Also another problem of using GA are related to crossover and mutation where the chromosome represents one of the mapping solutions so the number of tasks and NoC dimension are demanded before starting because their sizes represent the chromosome size so adding or removing tasks leads to chromosome change, and that will affect the results obtained from GA operators (crossover and mutation). So instead of remapping many times, a fixed chromosome size that will be suitable for any application is needed. Also after many crossovers and mutations, the new individuals will be adjusted to be suitable and accepted as a new mapping solution. Moreover, in the recent year Cellular automata (CA) have been used as a simulation model to solve the problem of embedded system such as allocation and scheduling of multimedia tasks.

In this paper, we motivated by quantum evolutionary cellular automata that will give us a tractable framework in which to study the mechanisms by which an evolutionary process might create complex coordinated behavior in natural decentralized distributed systems and that what encourage its using in embedded system. Evolutionary cellular automata improve many advantage in the field of computer science but the search space (of GA that represent the set of CA transition rules) is increased according to the CA dimension and that leads to increasing the computational time (3D or 4D Cellular automata is hard to be updated by GA in short time). Also the classical chromosome are weak in representing the population diversity and as mentioned before, since the use of quantum bit representation leads to better population diversity while the use of quantum gate drive the population towards the best solution. In following, a detailed mechanism of how using QGA with CA to optimize NoC performances.

3.1 QECA Concepts

(1) Search space: In this work, each CA cell can have only two possible states (0 or 1). The CA dimension is 2-D lattices. The number of neighbors that the transition rules take into account while updating each cell are $H = 5$ (4 neighbors + the cell in the middle). We choose $H = 5$ according to the structure of the 2D mesh NoC topology where each node is connected to four other neighbors' nodes (see Fig. 1). According to this structure, we obtain $card(V)^h$ different neighbors configurations and in our case is $card(2)^5 = 32$, and each cell is updated based on those configurations. CA can be characterized with one transition rule or several where the number of f that can be associated to each CA is the $card(V)^{card(V)^h}$ and in our case it is $card(2)^{card(2)^5} = 2^{32} = 4294967296$ and this number represents the search space.

The next figure is an illustrative example of how CA works. The state of the cell is changed in time according to its previous state and its neighbor's states, so the corresponding decimal value is 28. The new state of the middle is the transition rule cell with position 28.

Given: a multimedia application G(C, E) with C: set of tasks and E: set of channels between them. An NxN 2D mesh based NOC G(P, L) with P: set of cores and L is the set of physical links.

Find: an optimum CA lattice with its optimum set of rules that guarantee optimal energy consumption due to the mapping of an application tasks onto NoC processors.

(2) Problem Representation: In classical genetic algorithm, the chromosome consists of series of genes where each gene is assigned a binary value (1 or 0) with $gene(x, y) = 1$ indicates that $Task(x)$ is allocated to $core(y)$ otherwise $gene(x, y) = 0$, each core can be allocated by several number of tasks whereas each task cannot be assigned to more than one core. In ECA, each transition

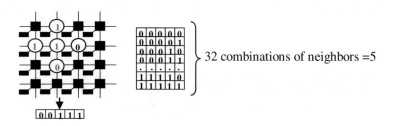

Fig. 1 The combination of neighbors in CA based on NoC neighbors

gene

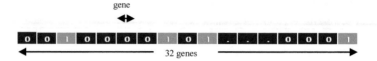

Fig. 2 Structure of the chromosome in ECA [32 bits]

rule represents a chromosome encoded in 32 genes (as illustrated in Fig. 2), the locus represents the $gene(x, y)$ with its 4 neighbors, the corresponding new state of that gene is the state corresponding to the decimal value of the locus situated in the correspondence table.

In QECA, we used the same chromosome representation as ECA but in place of bit, a Q-bit is used. Q-bit is defined as the smallest unit of information. A Q-bit can be represented as:

$$\begin{bmatrix} \alpha \\ \beta \end{bmatrix} \tag{5}$$

where α and β are real numbers whose represent the probabilities that the Q-bit found in the "0" and "1" states, respectively with $|\alpha|^2 + |\beta|^2 = 1$ (Fig. 3).

The state of a Q-bit may be "0", "1", or a linear superposition of the two:

$$|\psi\rangle = \alpha|0\rangle + \beta|1\rangle \tag{6}$$

where $|0\rangle$ and $|1\rangle$ mean the states "0" and "1", respectively. However, a linear superposition of states with m Q-bits can be represented by Q-bit individual (Eq. 7) for that a system with m Q-bits can represent 2^m states at the same time.

$$\begin{bmatrix} \alpha_1 & \alpha_2 & \alpha_3 & \cdots & \alpha_m \\ \beta_1 & \beta_2 & \beta_3 & \cdots & \beta_m \end{bmatrix} \tag{7}$$

with $|\alpha_i|^2 + |\beta_i|^2 = 1$, for $i = 1, 2, \ldots, m$.

(3) Generation of the initial population: The population is the main element of a genetic algorithm, where its size influences the coverage of mapping space. A population that is too large takes time to evolve whereas a population that is

$$|\alpha|^2 + |\beta|^2 = 1.$$

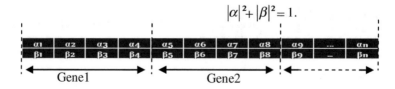

Fig. 3 Quantum chromosome structure

too small lead to a local minimum. We generated an initial population from 2^{32} transition rules, $P = 100$ individuals are chosen randomly and well distributed in the research space.

First, the generation counter is set to $t = 0$, then an initialization of the group of Q-bit individuals $Q(t)$ where $Q(t) = [q_1^t, q_2^t, \ldots, q_n^t]$, n is the total number of Q-bit individuals and q_j^t is the jth Q-bit individual at generation t which is defined as:

$$q_j^t = \begin{bmatrix} \alpha_{j1}^t & \alpha_{j2}^t & \alpha_{j3}^t & \cdots & \alpha_{jm}^t \\ \beta_{j1}^t & \beta_{j2}^t & \beta_{j3}^t & \cdots & \beta_{jm}^t \end{bmatrix} \tag{8}$$

where $j = 1, 2, \ldots, n$ and m is the string length.

(4) Measuring Chromosomes: Generate X(t) by measuring of Q(t):

In this step, each Q-bit is observed and measured from $Q(t)$ in order to extract a classic chromosome $X(t)$ that the evaluation of each quantum chromosome. For that, $X(t)$ is a group of binary solutions where $X(t) = [X_1^t, X_2^t, \ldots, X_n^t]$, and $X_{j1}^t = [x_{j1}^t, x_{j2}^t, \ldots, x_{jn}^t]$, where x_{ji}^t is binary value that is determined as (Fig. 4):

$$x_{ji}^t := \operatorname{get} x_{ji}^t \operatorname{in} [0, 1] ;$$
$$\operatorname{if} (x_{ji}^t < \left| \beta_{ji}^t \right|^2) x_{ji}^t = 1; \operatorname{else} x_{ji}^t = 0;$$

(5) Fitness function: Fitness function represents the desired optimization goal. The evaluation of each classical chromosome $X(t)$ is through its evolution throughout K iterations. A note between [0,1] is assigned to individuals. The fitness function is given as:

$$F_{NOC} = \frac{\lambda \times E_{NOC} + (1 - \lambda) \times L_{NOC}}{F\_min} \times 100\% \tag{9}$$

In the equation, F_min is the minimum energy found and λ is a proportionality coefficient that is used to adjust the proportion of communication power consumption and delay in cost function, and the value range is $0 < \lambda < 1$.

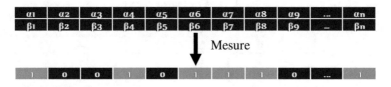

Fig. 4 Measured chromosome

(6) Store the best solution among $X(t)$ into $B(t)$: $B(t)$ is a matrix that stores the best solution in the whole population.

(7) Update $Q(t)$ using Q-gates: Quantum individuals are updated by using Q-gates. The population Q(t) is updated with a quantum gates rotation of qubits constituting individuals. The rotation gate $U(\Delta\theta_i)$ and the update operation are expressed as:

$$U(\Delta\theta_i) = \begin{bmatrix} \cos(\Delta\theta_i) & -\sin(\Delta\theta_i) \\ \sin(\Delta\theta_i) & \cos(\Delta\theta_i) \end{bmatrix} \tag{10}$$

$$\begin{bmatrix} \alpha_{ji}^t \\ \beta_{ji}^t \end{bmatrix} = U(\Delta\theta_i) \begin{bmatrix} \alpha_{ji}^{t-1} \\ \beta_{ji}^{t-1} \end{bmatrix} \tag{11}$$

$\Delta\theta_i$ is a rotation angle which determines the magnitude and direction of rotation. Figure 5 illustrates the polar plot of the rotation gate for Q-bit individuals.

At generation t, the rotation angle $\Delta\theta_i$ is updated according to the criteria summarized in Table 1, where x_{ji}^t and b_i^t are the binary control variables in solution X_j^t and the best solution B^t of $B(t)$, respectively. $f(X_j^t)$ and $f(B^t)$ represent the objective function values of X_j^t and B^t. For example, when x_{ji}^t and b_i^t are 0 and 1, and

Fig. 5 Polar plot of the rotation gate for Q-bit individuals

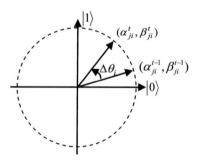

Table 1 Look-up table for quantum gates rotation

x_{ji}^t	b_i^t	$f(X_j^t) > f(B^t)$	$\Delta\theta_i$	$S(\alpha_{ji}^t \times \beta_{ji}^t)$		
				> 0	< 0	$= 0$
0	0	0	$\Delta\theta_2$	−	+	±
0	0	1	$\Delta\theta_2$	−	+	±
0	1	0	$\Delta\theta 1$	−	+	±
0	1	1	$\Delta\theta_2$	−	+	±
1	0	0	$\Delta\theta 1$	+	−	±
1	0	1	$\Delta\theta_2$	+	−	±
1	1	0	$\Delta\theta_2$	+	−	±
1	1	1	$\Delta\theta_2$	+	−	±

$f(X_j^t)$ is larger than $f(B^t)$, the rotation angle $\Delta\theta_i$ is updated according to $S(\alpha_{ji}^t \times \beta_{ji}^t)$ in Table 1 where $S(\alpha_{ji}^t \times \beta_{ji}^t)$ is the sign of $\alpha_{ji}^t \times \beta_{ji}^t$.

In the last step, the best solution among $X(t)$ and $B(t-1)$ is stored to $B(t)$, and terminated if the stopping conditions are met; else generate a new population.

4 Experimental Results

In this section, we define the case of studies experimental used to evaluate QECA and due to that we used benchmark multi-application to test the efficiency of the algorithm and compare it with classical genetic algorithm and evolutionary cellular automata. The benchmark chosen is the multimedia system (MMS) and the automotive/industrial application (auto-indust) for their diversity of both communication patterns and the characteristics of the results produced (Fig. 6).

This paper studies the effectiveness of evolving cellular automata with quantum genetic algorithm where each transition rules represents a chromosome. The network is 5×5 mesh based NoC. The first mapping of the application task onto NoC nodes represents the lattices that are updated according to the CA transition rules. The lattices of each CA are retransformed into mapping and each transition rule evaluated and assigned a fitness value based on the result obtained from each lattice updating. The best transition rule with best fitness function value is saved. The population size is fixed for 150. The termination of GA is set to 500 generation, 300

Fig. 6 Multimedia system task graph

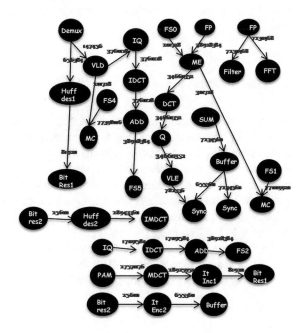

Fig. 7 Automotive/industrial
task graph

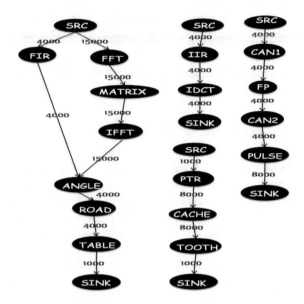

generations for ECA and 150 generation for QECA where $\Delta\theta_2 = 0.001\pi$ and $\Delta\theta_1 = 0.08\pi$.(the same as in [6]) The number of ports per router is fixed to 4 ports and maximum link length is fixed to 4 four virtual channels, the scheduling analysis and the routing algorithm must take the number of ports of the router and the link length as a constraint. For 0.18um technology node (the parameters are obtained from [4, 5, 7 and 8]), the router is operating at voltage range from 1.0 to 0.6 v (Fig. 7).

5 Discussion

Figure 8e and f exhibit our results for the mapping of MMS and auto-indust respectively onto network on chip 5 × 5 mesh using the QECA, the metric chosen is the optimization of the fitness value in each generation whereas in Fig. 8a–d, and under the same assumption used in QECA, we tested the evolution of fitness function in each generation using GA in auto-indust, GA in MMS, ECA in auto-indust and ECA in MMS respectively. A comparison of QECA, ECA and GA finesses function is done in order to demonstrate the efficiency of QECA as it is illustrated in the following figures.

As it is seen in Fig. 8c and d, the fitness function is over 300 generation and an optimization is achieved about 0.98 and 0.97 in both MMS and auto-indust respectively, so the time that the optimization takes is less than the time in GA that

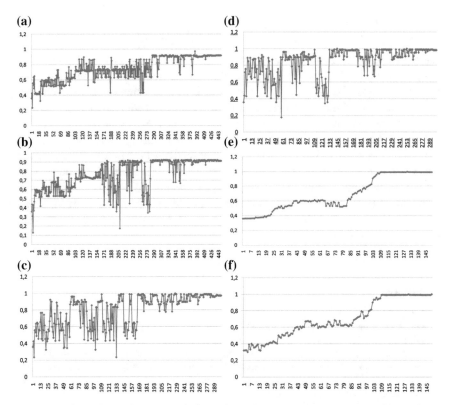

Fig. 8 The average fitness value (%) of each generation. **a** GA in auto-indust. **b** GA in MMS. **c** ECA in auto-indust. **d** ECA in MMS. **e** QECA on MMS. **f** QECA on auto-indust

is about 460 generation where the fitness function optimized about 0.91 on both MMS and auto-indust. So the optimization using ECA is efficient and accurate. Whereas in the following scheme (Fig. 8e and f) and under the same assumption, we can see that the fitness function is achieved about 0.99 on both MMS and auto-indust and only over 110 generation.

Another advantage whose appears in Fig. 8e and f is the stabilization and the incrimination of fitness values of the results obtained compared to GA and ECA, this incrimination is the result of using the quantum gates that have the ability to drive the population toward the best solution. By observing penalty methods, we can then confirm the effectiveness of QEGA's performance over ECA and GA and moreover, quantum algorithms have generally the ability to minimize the complexity of equivalent algorithms that run on classic computers.

6 Conclusion

Energy consumption and latencies are the most critical issues for advanced NoC designs using scaled technologies. In this paper, we have presented a novel approach to optimize the network on chip performance using quantum computing in evolutionary cellular automata in order to guarantee better population diversity while driving the population towards the best solution. We compared classical GA, ECA, and shown that QECA algorithms are a very promising technique stimulating the process of discovering effective rules that leads to best fitness functions with better population diversity that is driven towards the best solution. In future work, we will exhibit our results for the mapping of a multimedia application onto network on chip 2D mesh considering energy and time while comparing the optimization of the fitness value of each GA, ECA and QECA in order to demonstrate the efficiency of QECA.

References

1. Han, K.H., Kim, J.H.: Quantum-inspired evolutionary algorithm for a class of combinatorial optimization. IEEE Trans. Evol. Comput. **6**, 580–593 (2002)
2. Narayanan, A., Moore,M.: Quantum-inspired genetic algorithms. In: Evolutionary Computation, Proceedings of IEEE International Conference on IEEE, 1996, pp. 61–66 (1996)
3. Yang, J., Li, B.: Research of quantum genetic algorithm and its application in blind source separation. J. Electron. **20**(1), 62–68 (2003)
4. Wong, S.C., Winbond TSM: An Extraction Method to Determine Interconnect Parasitic Parameters. Feb. 2000
5. Ho, R.: On-chip wires: scaling and efficiency. On Aug. 2003
6. Laboudi, Z., Chikhi, S.: Comparison of genetic algorithm and quantum genetic algorithm. Int. Arab J. Inf. Technol. **9**, 243 (2012)
7. Bhat, S.: Energy models for network-on-chip components. On Dec 2005
8. Predictive Technology Model (PTM), Arizona State University, Available: http://ptm.asu.edu/, Last accessed: Dec. 2010

Development of a Set of Applications for Intelligent Control System of Compressor Facilities at an Industrial Enterprise

Vadim Kushnikov and Ekaterina Kulakova

Abstract A mathematical model for determining the power of reciprocating compressors used in the production of low pressure compressed air in industrial enterprises has been set up. Simulation experiments have confirmed the adequacy of the developed mathematical support. For the practical application of the research results in intelligent control systems of compression facilities on the basis of the proposed model a set of applications has been created, which allows to minimize the power consumption for production and distribution of compressed air.

Keywords Intelligent control system · Compressor facilities · The AIR complex of application programs

1 Introduction

While solving many problems of optimal operational control of compression facilities at an industrial enterprise on the criteria, which defines electricity consumption for production of compressed air, it is necessary to repeatedly calculate the power of reciprocating compressors for various values of controlling coordinates and environmental variables.

Currently, a large number of methods for calculating the power of compressors are known that allow building the desired dependence for all existing types of units [1–6].

Developed on the basis of these methods, mathematical models can be divided into two large groups. The field of application for the first group, mainly, is the

V. Kushnikov · E. Kulakova (✉)
Department of Applied Information Technologies, Yuri Gagarin State
 Technical University of Saratov, Saratov, Russia
e-mail: kulakovakm@gmail.com

V. Kushnikov
e-mail: kushnikoff@yandex.ru

© Springer International Publishing Switzerland 2016
R. Silhavy et al. (eds.), *Automation Control Theory Perspectives
in Intelligent Systems*, Advances in Intelligent Systems and Computing 466,
DOI 10.1007/978-3-319-33389-2_14

design of compressor units, for the second it is managing the production process of compressed air. The special features of the models of the first group include:

- the need to carry intensive laboratory experiments at an industrial enterprise to determine the characteristics of the compression sections and compressors intercoolers;
- the need for lengthy stoppages and partial dismantling of compressor units in the process of carrying out laboratory experiments;
- increased demands for the accuracy of measurement of the input variables, regardless of their impact on the position of the extremum in the current task;
- significant core storage and disk storage requirements of the supervisory computer control system;
- long duration of calculation and increased demands for speed of the supervisory computer control system;
- a large "redundancy" of applied techniques that allow, along with the required dependency, calculate many other characteristics of the compressor unit which are not used in solving the control problem.

Models of the second, much smaller group of systems are used in operational control systems of the compression facilities to search for extrema of the objective function. The special features of these models are their dependence on the optimality criterion and, as a consequence, a limited scope. Due to this fact, models of the second group, known from the literature, cannot be used unchanged while solving the problems of optimal operational control of compression facilities of an enterprise according to the above criteria.

The purpose of this paper is to construct a mathematical model which allows under operational control to determine the power of the reciprocating compressor within the restrictions of the time of calculation, speed and storage space of computing system connected with real-time environment features.

2   Mathematical Model

Flow diagram of the gas compression process in a three-stage reciprocating compressor is illustrated in Fig. 1.

ρ_0, ϑ_0—the density and temperature of the inlet gas; ρ_1, ρ_2, ρ_3—the density of the gas at each stage; ϑ_w—water temperature in intercoolers.

The pressure in the gas network, which the compressor works on, depends on the customer of the compressed air, i.e. in relation to the compressor it is the independent variable. In determining the power of a reciprocating compressor it is assumed that the gas temperature before and after the check valve remains unchanged. When connecting the reciprocating compressor to the air supply network up until the pressure (ρ_3) to the check valve reaches the pressure in the network (ρ_c), the check valve stays closed and gas will not enter the network. Upon reaching the end compression rate in gas cavities, which is equal to the pressure in

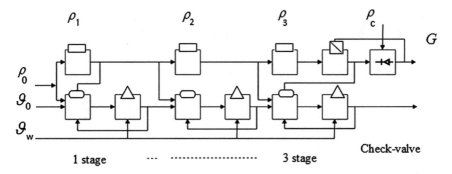

Fig. 1 Gas compression piston compressor

the network, the check valve opens and the compressor output of the gas equals the output of the final stage of compression.

In view of the assumptions made, the mathematical model for determining the power of reciprocating compressors will be as follows:

$$N(t) = G(t)R\frac{k}{k-1}\sum_{j=1}^{2}\left(\vartheta_{H_j}(t) - \vartheta_0\right) + \Delta N, \tag{1}$$

$G(t) = \lambda_3 V_{\Pi_3} n\rho_3(t)$ processor output;

$\Delta N = \dfrac{G(t)RT_1\delta P_1}{\frac{V_{\Pi_1}}{V_{\Pi_2}}P_{atm}} + \dfrac{G(t)RT_2\delta P_2}{\left(\frac{V_{\Pi_1}}{V_{\Pi_2}}P_{atm}-\delta P_1\right)\frac{V_{\Pi_2}}{V_{\Pi_3}}}$ power losses in the intermediate air coolers.

Characteristics of the air cooler of the first stage of the compressor are determined by solving a differential equation:

$$T_{B_1}\frac{d\vartheta_1}{dt} + \vartheta_1 = T_{H_1}\frac{d\vartheta_{H_1}}{dt} + B_{H_1}\vartheta_{H_1} + B_{B_1}\vartheta_B, \tag{2}$$

$T_{B_1} = \dfrac{T'_{B_1}A_1}{A_1 + W_1(A_1-1)}$ index of air cooler time lag normalized to water temperature;

$T_{H_1} = \dfrac{T'_{B_1}}{A_1 + W_1(A_1-1)}$ index of air cooler time lag, normalized to gas temperature;

$B_{H_1} = \dfrac{1 + W_1(A_1-1)}{A_1 + W_1(A_1-1)},$

$B_{B_1} = \dfrac{A_1-1}{A_1 + W_1(A_1-1)}$ nondimensional coefficients;

$A_1 = e^{\frac{K_1F_1}{\overrightarrow{m_1}c_1}}$ gas expenditure function through the air cooler;

$W_1 = \dfrac{\overrightarrow{m_1}c_1}{\overrightarrow{m_{B_1}}c_B}$ water equivalent of the cooled gas;

$T'_{B_1} = \dfrac{m_{B_1}}{\overrightarrow{m_{B_1}}}$ the index of air cooler time lag according to water temperature;

$\vartheta_1 \approx 0.85 \vartheta_{H_1}$ gas temperature after air coolers;

$\overrightarrow{m_1} = \lambda_1 V_{\Pi_1} n \rho_0$ stage gas rate.

The characteristics of air coolers of second and third stages are determined in a similar manner.

$$T_{B_2} \frac{d\vartheta_2}{dt} + \vartheta_2 = T_{H_2} \frac{d\vartheta_{H_2}}{dt} + B_{H_2}\vartheta_{H_2} + B_{B_2}\vartheta_B \tag{3}$$

$$T_{B_3} \frac{d\vartheta_3}{dt} + \vartheta_3 = T_{H_3} \frac{d\vartheta_{H_3}}{dt} + B_{H_3}\vartheta_{H_3} + B_{B_3}\vartheta_B \tag{4}$$

Solving the system of differential Eqs. (2)–(4), the functions $\vartheta_1(t)$, $\vartheta_2(t)$, $\vartheta_3(t)$ are defined, characterizing the temperature of the gas after air coolers on each stage of the reciprocating compressor.

According to a known gas temperature the density of the gas is determined after each stage of compression as well as the power, consumed by the compressor (1):

$$\rho_1(t) = \frac{\rho_0}{0.85} \left(\frac{\vartheta_{H_1}(t)}{\vartheta_0} \right)^{\frac{1}{k-1}}, \; \rho_2(t) = \frac{\rho_1(t)}{0.85} \left(\frac{\vartheta_{H_2}(t)}{\vartheta_{H_1(t)}} \right)^{\frac{1}{k-1}}, \; \rho_3(t) = \frac{\rho_2(t)}{0.85} \left(\frac{\vartheta_{H_3}(t)}{\vartheta_{H_2(t)}} \right)^{\frac{1}{k-1}}$$

$$\tag{5}$$

Developed mathematical model allows to determine the power of the three-stage reciprocating compressor which is widely used in industrial enterprises in the production of low pressure compressed air. Similarly, a model to calculate the power of a reciprocating compressor with a different number of stages of compression can be developed.

3   Adequacy Check of the Mathematical Model

To test the adequacy of the developed mathematical model simulation experiments were conducted, the results of which are presented in graphs in Fig. 2, 3, 4.

Figure 2 shows that about 30s after the unit activation the compressor power reaches its maximum value, which is confirmed by the results of an experiment conducted with units of the Ural Compressor Plant.

Let's construct a graph of the compressor power change based on water temperature and water flow in the intermediate air coolers in the stationary mode. Based on the results of the simulation experiment presented in Fig. 2, after 100s after the compressor activation, the system operation can be considered steady. In view of this we will change the input parameters of the system and calculate the compressor power at t = 100.

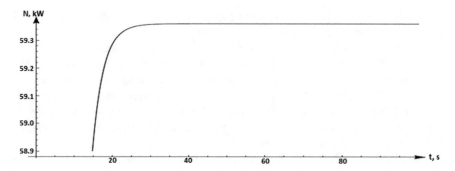

Fig. 2 Changing the power of the compressor when it is switched on

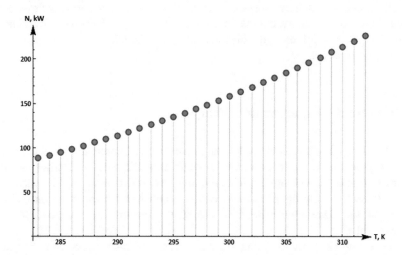

Fig. 3 Dependence of compressor power on water temperature in the intermediate air coolers

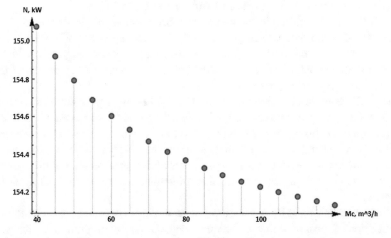

Fig. 4 Dependence of the compressor capacity of the water flow through the intermediate coolers

Graph of dependence of compressor power on water temperature at the inlet of air coolers is shown in Fig. 3.

As we see from this figure, the power of a reciprocating compressor is significantly dependent on the temperature of the water, which is cooling the compressed air in the intermediate air coolers. In particular, when the temperature changes from 283 to 315 K the power consumption increases by more than two times. This effect is known from the specialized literature, and is also confirmed by the results of simulation experiments carried out in [3–6] in the development of a mathematical model of the process of air compressing.

Graph of dependence of compressor power on the water flow through the intermediate air coolers is shown in Fig. 4.

The results of this simulation experiment show that the compressor power to a significantly higher degree depends on the temperature of water flowing through the intermediate air coolers than on the mass flow of the water. This fact is well known from the specialized literature which also confirms the adequacy of the developed mathematical model.

4 Complex of Application Programs

The developed mathematical model has found practical application in the systems of intelligent control of the compression facilities at industrial enterprises as a part of the complex of AIR application programs, the structure of which reflects the basic steps of solving the problem of optimal control of the compression facilities based on the criteria characterizing electricity consumption for production of compressed air.

The complex is built according to a hierarchical approach and consists of a main program MAIN, which organizes the event-driven control of the unit, as well as 22 software units to implement the solution of individual functional problems.

Brief description of the developed software is given in Table 1.

In Table 1, the following notation has been made: KTS—technical means complex; DB—the database of power utilities control system; SW—software.

The structure of the AIR complex of application programs, as part of which the developed mathematical model for calculating the power of reciprocating compressors is implemented, is given in Fig. 5.

The AIR complex of application programs is designed to operate as part of an information support of power utilities intelligent control system of an enterprise and is connected to this system through a common database. The basic software is written in C# language. The software units that implement the procedure of information exchange with the database, interruption processing, coming from the task dispatcher, and a number of other functions are written in ASSEMBLER language.

When creating an AIR software package modern computational technologies, object-oriented programming techniques, capabilities of programming languages

Table 1 Characteristics of the complex of AIR application programs

№	The main functions of AIR complex	Software modules which are used		
		Name	Time slot	Programming language
1	Extrema seeking of objective function	M3, M12	0.5 h	C#
2	Forming set points for controlled variables	M14	24 h period	C#
3	Troubleshooting of KTS and SW	M22	0.5 h–1 year	ASSEMBLER
4	Issuing messages to a supervisor	M2, M13	0.5 h–1 year	C#, ASSEMBLER
5	Database connection, validating the input information	M1, M14, M15	0.5 h	C#, ASSEMBLER
6	Analysis of the given controlling actions	M16	24 h period–1 month	C#
7	Calculation of the economic indicators of the process	M17, M20, M21	1 month–1 year	C#
8	Adjustment of the experimental characteristics of the model	M18, M19	1 quarter	C#
9	Message recording to a journal of emergency and performance situations	M2	0.5 h–24 h period	ASSEMBLER
10	Analysis of performance efficiency of AIR	M20, M21	0.5 h–24 h period	C#

C# and ASSEMBLER, as well as database management system Microsoft SQL Server 2014 were used, which allowed:

- to develop a software product, which is enhancement-open, low-cost, compact and effectively functioning within the constraints of real-time;
- to implement the developed software to the domestic technical means complex, which is widespread among power utilities control system of an enterprise;
- to use technology of open database connectivity, facilitating the integration of the developed software into the existing software and data support of an enterprise;
- to form a friendly graphical user interface, based on the possibilities of visual programming languages;
- to facilitate the operation of the software complex and reduce the time of staff training to work with it;
- to enable the possibility of replicating of the developed software.

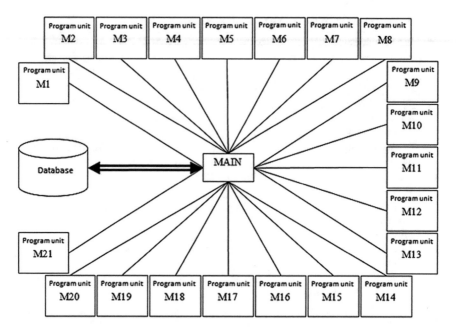

Fig. 5 The structure of the AIR complex of application programs

5 Functionality of the Complex of Application Programs

Let's consider the basic functionality of the software package, as part of which the developed mathematical model is implemented. The AIR complex is designed for use as part of software of power utilities automatic control system of an enterprise. It operates at time slots of 0.5 h, a 24 h period, 1 month, one quarter and 1 year and has five different operation modes.

At the time slot of 0.5 h task dispatcher of the operating system makes a call to the execution of the program MAIN regularly every 20–25 min. The main program will run a program unit M1, which updates the database of the problem and assesses the reliability of the information received from the database.

In case of its invalidation the program stops execution, and a corresponding message is issued to staff and is entered automatically in the journal of emergency and performance situations (program unit M2).

During operation of M3 program unit the dynamics of the main parameters of the mathematical model in between the two problem solvings is estimated. If changes occurred with controlled variables are not significant to the position of the extrema of the objective function, the AIR complex automatically stops the execution and the control is passed to the task dispatcher of the operating system.

In the case of significant changes in the objective function the program units M4–M7 are run, estimating the damage to consumers of compressed air, electric power consumption by compressors, pumps and fans, respectively. While determining the power of reciprocating compressors a mathematical model (1)–(5) is used.

Control of program units M4–M7 is executed from the program unit M8, which initiates calculation of the optimizable objective function as many times as is necessary for the implementation of the decomposition technique for solving the problem of the optimal control of the compression facilities at an industrial enterprise on the criteria defining electricity consumption for production of compressed air. Operational experience of AIR programs complex showed that on average calculation of the objective function is made 105–144 times during each cycle of the problem solution.

As a result of the operation of program units M4–M8 are determined: the total flow of the compressed air at the compressor station reservoir, temperature and mass flow rate of water in the circulation cooling system which ensure a minimum of the optimizable objective function.

In the process of the operation of program units M9–M12 on the known parameters, determined in the operation of program units M4–M8, controlling coordinates of the reasonable operation mode of power utilities are calculated: performance of compressors, pumps and fans; the pressure and flow of compressed air on the pneumatic circuit inputs to the buildings of the enterprise.

Received calculated values of the control parameters are given for the approval of dispatching personnel of the enterprise (program unit M13) and entered into the journal of emergency and performance situations (program unit M2). Upon validation of the results power dispatcher performs the necessary switching of the equipment located at the entrance of the enterprise buildings, changes the composition and performance of compressors, pumps and fans.

When calling the program MAIN at the time slot of a 24 h period, control is transferred to the M14 program unit, which allows to interactively create the settings of process conditions used in the calculation of the mathematical model of the controlled process. The reliability of incoming information is checked by program unit M15, which interrupts the operation of M14 when there is an error in the input data.

Program unit M16 allows to view the list of control actions made over the past 24 h, to identify those that were recommended by the control system, but not implemented by the dispatcher, to build diagrams describing the daily saving of compressed air and electricity, received as a result of solution of AIR programs complex.

When you call the AIR applications complex at the time slot of 1 month (program unit M17), verification of the monthly plan target for saving energy and compressed air, reduction of losses in pneumatic power consumers due to violation of the specified air supply mode is performed. Thus upon user request up to 26 different kinds of graphs can be built illustrating the change in the principal process conditions which characterize the mode of operation of the compression facilities. Calling of AIR applications complex at a given time slot is generally carried out at the initiative of the chief power engineer or specialists of energy companies that control the outcomes of the compression facilities performance for the past month.

At the time slot of 1 quarter the complex of tasks is launched for execution by a system programmer or an expert in the subject area, who is the member of the databank administrator group. When running the program units M18–M19, record database entries are interactively adjusted, which contain information on the experimental performance of compressors, pumps, fans and pneumatic power consumers.

At the time slot of 1 year AIR tasks complex is launched on the initiative of the chief engineer, chief power specialist and energy specialists of an enterprise. Called on this time slot, program units M20–M21 can determine the achieved energy saving, assess reduction of losses in pneumatic energy consumers because of violations of the air supply and calculate the basic technical and economic indicators, characterizing the work of the compression facilities for the past period. Furthermore, these software units allow a user to graphically estimate the change in the bare cost of compressed air for a half-year period, a quarter or certain months of the year.

Program unit M22 diagnoses the performance of AIR software. It is called for execution at any of the time slots in the event of malfunctioning of the technical means complex, the system or application software, preventing the normal functioning of the AIR software (for example, when blocking or absence of the necessary database files, disconnection or malfunctioning of a process interface unit, a sensors defect, unreliability of information provided by power dispatcher, etc.).

6 Conclusion

In this paper, we present a mathematical model for determining the power of reciprocating compressors which allows under operational control to determine the power of the reciprocating compressor within the restrictions of the time of calculation, speed and storage space of computing system connected with real-time environment features. We conducted computational experiments, which confirmed the adequacy of the developed mathematical support. We created a set of applications for intelligent control system of compressor facilities on the basis of the proposed model. Use of the AIR complex of application programs allows to minimize the power consumption for production and distribution of compressed air.

The AIR software package was introduced in intelligent control system of the compression facility enterprise "Transport engineering" (the city of Engels). Its introduction allowed to receive economic benefit by reducing the cost of compressed air (on average by 3.5 %), reducing the damage to consumers from violations of the air supply (on average 1.5 times).

References

1. Hlava, J., Opalka, J., Johansen, T.A.: Model predictive control of power plant superheater—comparison of multi model and nonlinear approaches: advanced materials research. In: 18th International Conference on Methods and Models in Automation and Robotics (MMAR), pp. 311–316. Miedzyzdroje, Poland (2013)
2. Zheng, X., Hu, J.: Air compressor testing system based on the HMI software. In: International Conference on Manufacturing Engineering and Automation (ICMEA2010), pp. 1874–1878. Guangzhou, China (2010)
3. Rezchikov, A.F., Kushnikov, V.A., Lushnikov, I.V.: Optimization problems of intelligent process control systems of air supply on the engineering enterprise: automation and modern technology. Mech. Eng. **1**, 24–26 (1999). Moscow
4. Rezchikov, A.F., Kushnikov, V.A., Evseev, P.L., Kabanov, I.A.: Problems and models for operational control of compressor facilities at industrial enterprises. Mechatron. Autom. Control **3**, 45–53 (2004)
5. Rybolovlev, V.Y., Krasnobaev, A.V., Spirin, N.A., Lavrov, V.V.: Principles of the Development and Introduction of an Automated Process Control System for Blast-Furnace Smelting at the Magnitogorsk Metallurgical Combine. Metallurgist. pp. 1–6 (2015)
6. Kushnikov, V.A., Fedorov, A.V.: Mathematical modeling of energy consumption of a centrifugal compressor. Actual Probl. Humanit. Nat. Sci. **6**, 17–20 (2010)

Simulation Approach for Optimal Maintenance Intervals Estimation of Electronic Devices

Alexander Lyubchenko, Pedro A. Castillo, Antonio M. Mora, Pablo García-Sánchez and Maribel G. Arenas

Abstract Simulation is a powerful and flexible technique for imitation of variety of stochastic processes and it has attractive advantages in comparison to analytical routine solutions. In this paper, the Monte Carlo simulation technique is used for imitation of operational process of electronic devices which is formalized by the model of Semi Markov process. The model considers sudden, gradual, latent and fictitious failures, human factor of service staff and time parameters of preventive maintenance. Simulation approach permits to obtain necessary data for estimation of recommended value of maintenance interval according to suggested optimality criterion. Moreover, it could be easily used for investigation and analyzing of the process with different combinations of input parameters.

Keywords Preventive maintenance (PM) · Optimization · Simulation · Monte carlo · Semi markov process

A. Lyubchenko (✉)
Omsk State Transport University, Omsk, Russia
e-mail: allyubchenko@gmail.com

P.A. Castillo · A.M. Mora · P. García-Sánchez · M.G. Arenas
University of Granada, Granada, Spain
e-mail: pacv@ugr.es

A.M. Mora
e-mail: amorag@ugr.es

P. García-Sánchez
e-mail: pablogarcia@ugr.es

M.G. Arenas
e-mail: mgarenas@ugr.es

© Springer International Publishing Switzerland 2016   153
R. Silhavy et al. (eds.), *Automation Control Theory Perspectives*
in Intelligent Systems, Advances in Intelligent Systems and Computing 466,
DOI 10.1007/978-3-319-33389-2_15

1 Introduction

Exploitation of up-to-date electronic devices and equipments are basically organized according to preventive maintenance (PM) procedures which are a base for reliable operation [1]. As a rule, recommended periodicity is specified in technical documentation of the corresponding system and preventive operations could be performed regarding this information. Determination of recommended intervals is a traditional part of electronic means design and calculations are implemented according to the states of reliability as PM is a method for supporting of required reliability level. Under the circumstances, the important objective is to obtain an optimal value of periodicity which could ensure possible maximum of reliability. In order to solve an optimization problem, it is necessary to have a mathematical model describing the operational process of corresponding system and taking into account the main factors which could significantly influence on system's failure rate. Obviously, on the one hand considering multiple of exploitation factors could lead to analytical model's complexity increase and routine operations for changing parameters, on the other hand application of simple models could provide inadequate solutions. Simulation-based approach could ensure good effectiveness and the possibility to make model's corrections without complicated developments of formulas. Thus, the simulation technique was used to construct a model of operational process of electronic devices considering the following factors: appearance of sudden, gradual, latent and fictitious failures, human factor of service staff and time parameters of PM.

State of the art analysis showed that research works exist [2, 3] devoted to the development of a model with the abovementioned factors, but generally the authors used analytical approaches. The most recent work [4] with similar way relates to estimation of maintenance intervals of fiber-optic lines.

2 Methodology

In order to make a basis for the development of simulation models that could be used in optimization procedures of maintenance intervals the following steps were implemented: suggestion of optimality criterion, conceptual description of the researched process and mathematical formalization of the previous explanation. These stages are disclosed in the sections below.

2.1 Optimality Criterion

Modern electronic devices are generally continuously working complex systems. Integrated reliability indexes such as coefficient of operation efficiency (K_{OE}) and availability coefficient (K_A) could serve as estimations of the systems' reliability. Thus, optimality criterion based on these coefficients is presented by the following expression:

$$\begin{cases} K_{OE} = f(T_{int}) \rightarrow \max \\ K_A \geq K_{A.A.} \end{cases} \tag{1}$$

where T_{int} is the maintenance interval and $K_{A.A.}$ is the allowable value of availability.

The formula of objective function $K_{OE}(T_{int})$ can be written as:

$$K_{OE}(T_{int}) = \frac{T_{OS}(T_{int})}{T_{OS}(T_{int}) + T_{RS}(T_{int}) + T_{MS}(T_{int})} \tag{2}$$

where $T_{OS}(T_{int})$, $T_{RS}(T_{int})$ and $T_{MS}(T_{int})$ are the mean times of operable, repair and maintenance states, respectively.

The expression of availability function $K_A(T_{int})$ can be written as:

$$K_A(T_{int}) = \frac{T_{OS}(T_{int})}{T_{OS}(T_{int}) + T_{RS}(T_{int})} \tag{3}$$

Typical graphs of dependences $K_{OE}(T_{int})$ and $K_A(T_{int})$ are depicted in Fig. 1. According to the suggested criterion, it is possible to estimate optimal T_{opt} and allowable T_{all} maintenance intervals using the following expressions:

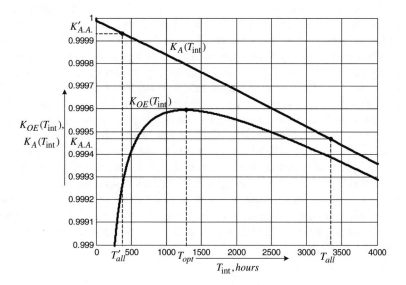

Fig. 1 Typical graphs of dependences $K_{OE}(T_{int})$ and $K_A(T_{int})$

$$T_{opt} = \arg \max K_{OE}(T_{int})$$
$$T_{all} = K_A^{-1}(K_{A.A.}) \tag{4}$$

Besides, if there are high requirements for devices' reliability, e.g. large value of $K_{A.A.}$, then it could be a situation when maximally achieved level of $K_{OE}(T_{opt})$ will be less than allowable coefficient $K_{A.A.}$ as it is shown in Fig. 1 for $K'_{A.A.}$. Furthermore, it is suggested to use a notion of rational maintenance interval which is defined as an advisable interval between preventive procedures T_{rat} on the basics of expression:

$$T_{opt} \leq T_{rat} \leq T_{all}, \ if \ T_{opt} < T_{all}$$
$$T_{opt} \geq T_{rat} \geq T_{all}, \ if \ T_{opt} > T_{all} \tag{5}$$

Hence, the rational maintenance interval could be determined as:

$$T_{rat} = \min\{T_{opt}, T_{all}\} \tag{6}$$

2.2   Conceptual Content

Conceptual description of researched process is one of the steps for development of any mathematical model and includes all distinguished information about its peculiarities [5]. The paper is confined with important knowledge in order to provide necessary comprehension.

The functioning process of electronic devices consists of two stages: normal operation and active ageing. The service staff implements PM at the point of normal operation in order to keep workable condition and required reliability level. PM is performed with periodicity T_{int} and intended to prevent partially sudden failures, but generally gradual failures. This kind of failures is preceded by a misalignment of system's parts in irreversible physicochemical processes of components, in other works, due to ageing. Gradual failures of components could amount 30–80 % of the total quantity of failures. Consequently, accounting of misalignment state of repairable systems is necessary to analyze its reliability. Time parameters of preventive maintenance are defined according to branch documentation and usually are presented by the following quantities: testing time (t_t), time for system tuning and configuration (t_a), fault search time (t_s) and emergency repair time (t_r). Control and diagnosis of system characteristics could be implemented by different measurement devices and automated diagnosis apparatuses. Validity of diagnostic devices information could be estimated by the probability of erroneous determination of real system state.

The following assumptions were made for the development of the conceptual model:

1. The electronic equipment is subjected to misalignments during its exploitation. Misalignments' rate and failures of detuned electronics follow on exponential law;
2. Misalignment of equipment is not detected during its exploitation;
3. Failures of diagnostic devices are not taken into account;
4. Recovery time and duration of maintenance are constants;
5. Repair starts immediately after a failure occurs or its determination by diagnostic devices;
6. Service staff can make errors during preventive maintenance which could result to failures;
7. Probability of service staff errors vanishes while the maintenance intervals tend to zero.

Devices could stay during its operational process in the following states:

1. Operable state (S1);
2. Misalignment state (S2);
3. Non-operable state (S3);
4. Preventive maintenance of operative system (S4);
5. Maintenance of system with misalignment (S5);
6. Latent failure (S6);
7. Maintenance of system being in latent failure (S7);
8. Fictitious failure (S8).

Sojourn time of state S1 has stochastic character and is determined by a distribution function, failure rate and misalignment rate of system. System moves to state S2 if one of its parameters exceeds its assumed value and to state S3 if there is a failure. The system remains in S3 during the time needed for repair which includes the testing time t_t, fault search time t_s and emergency repair time t_r.

The system moves to states S4, S5 and S7 at the scheduled time, according the maintenance system for testing and adjusting of equipment if the last is necessary. Testing is an operation for confirmation of operable system's state or detection of misalignment or identification of latent failure state, and it is characterized by the duration of testing time t_t. Adjusting is a set of operations for recovery of normative values of system's parameters. Testing and tuning are performed only in state S5.

Maintenance operations are carried out by service staff. Therefore, during testing (t_t) and adjusting times (t_a) an error could be made leading to equipment failure and its transition to state S3. The distribution function of probability of service staff errors $F_p(T_{int})$ is offered to compute this factor.

Control of system condition is performed by external and embedded diagnostic devices (DD). Built-in DD implement periodic diagnostic and fix failure with delivery of beep and/or fault light. External DD are exploited during the implementation of PM procedures. The main DD's parameters are I type (α) and II type (β) errors.

Fig. 2 State diagram of
functioning process of
repairable and maintainable
electronic devices

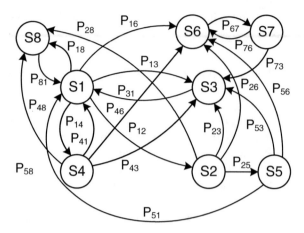

Diagnosis error α is a probability to declare the operable device as out of order and error β is a probability to declare the inoperable device as faultless. Diagnostic devices which are used for maintenance have a higher grade of accuracy than embedded means, therefore, $\alpha_1 > \alpha_2$, $\beta_1 > \beta_2$. For the stochastic process under study the diagnosis errors are supposed not to equal to zeros, consequently, there are states of latent failure and fictitious failure. In state S6 the system stands until maintenance procedures are done, and during it, operative staff can detect the malfunction. The system moves to state S8 when embedded DD makes an error giving a signal about a fictitious failure.

Generally, next step of mathematical model development is the interpretation of conceptual description with illustrative graphical model. Graphical interpretation is presented in Fig. 2 using a graph with 8 states and transitions between states implementing with accordance to the values of one-step transition probabilities P_{ij}, $i \in [1, 8]$, $j \in [1, 8]$.

2.3 Mathematical Formulation

According to conceptual description and graphical model, the observable operational process could be presented as a random sequence of transitions from a current state S_i to the following state S_j which could be mathematically formalized using theory of Markov and Semi Markov [6, 7]. Then the sequence of stochastic transitions could be described by Markov chain with continuous time and discreet state space. Furthermore, Markov property is obligatory to be hold true. This property inherent in exponential law, however, sojourn time complies with this law only in state S1 and S2. In the rest of states the duration of stay is a constant. Therefore, it was suggested to use the model of the model of Embedded Markov chain of Semi Markov process [6, 7]. Application of simulation technique gives the opportunity to take into account different character of input parameters without difficult routine analytical operations [5].

Model's parameters are presented as:

- vector of initial states:

$$P_0 = \{P_1^0, P_2^0, P_3^0, P_4^0, P_5^0, P_6^0, P_7^0, P_8^0\} \qquad (7)$$

- square matrix of transition probabilities:

$$P = \begin{vmatrix} 0 & (1-F_{13})\cdot F_{12} & (1-\beta_1)\cdot F_{13} & \begin{matrix}(1-\alpha_1)\cdot \\ (1-F_{13})\cdot \\ (1-F_{12})\end{matrix} & 0 & \beta_1 F_{13} & 0 & \begin{matrix}\alpha_1\cdot \\ (1-F_{13})\cdot \\ (1-F_{12})\end{matrix} \\ 0 & 0 & (1-\beta_1)\cdot F_{23} & 0 & (1-\alpha_1)\cdot (1-F_{23}) & \beta_1 F_{23} & 0 & \begin{matrix}\alpha_1\cdot \\ (1-F_{23})\end{matrix} \\ 1 & 0 & 0 & 0 & 0 & 0 & 0 & 0 \\ \begin{matrix}(1-\alpha_2)\cdot \\ (1-F_p)\end{matrix} & 0 & (1-\beta_2)\cdot F_p & 0 & 0 & \beta_2 F_p & 0 & \begin{matrix}\alpha_2\cdot \\ (1-F_p)\end{matrix} \\ \begin{matrix}(1-\alpha_2)\cdot \\ (1-F_p)\end{matrix} & 0 & (1-\beta_2)\cdot F_p & 0 & 0 & \beta_2 F_p & 0 & \begin{matrix}\alpha_2\cdot \\ (1-F_p)\end{matrix} \\ 0 & 0 & 0 & 0 & 0 & 0 & 1 & 0 \\ 0 & 0 & 1-\beta_2 & 0 & 0 & \beta_2 & 0 & 0 \\ 1 & 0 & 0 & 0 & 0 & 0 & 0 & 0 \end{vmatrix} \qquad (8)$$

where F_{12} is the exponential distribution function of transition probability from S1 to S2 with periodicity equal to T_{int}; F_{13} is the exponential distribution function of transition probability from S1 to S3 with periodicity equal to T_{int}; F_{23} is the exponential distribution function of transition probability from S2 to S3 with periodicity equal to T_{int} and $F_p(T_{int})$ is the distribution function of service staff errors probability.

- vector of density functions:

$$f = \left\{(\lambda_{13}+\lambda_{12})\cdot e^{-(\lambda_{13}+\lambda_{12})\cdot T_1}, \lambda_{23}\cdot e^{-\lambda_{23}\cdot T_2}, t_t+t_s+t_r, t_t, t_t+t_a, T_{int}, t_t, t_t\right\} \qquad (9)$$

where λ_{12}, λ_{13} are the rates of misalignments and sudden failures, respectively, hours^{-1}; λ_{23} is the rate of sudden failures of misaligned system, hours^{-1}; t_t is testing time, hours; t_a is the time for system tuning and configuration, hours; t_s is the fault search time, hours; t_r is the emergency repair time, hours.

Thus, multiple simulation of transitions sequence from state S_i to S_j in accordance with the abovementioned model, gives opportunity to get statistical data about operational process, and to estimate values of integrated reliability indexes K_{OE} and K_A depending on different values of T_{int}. Consequently, on the basis of the obtained data it is possible to solve the optimization task with offered optimality criterion.

2.4 Monte Carlo Simulation Approach

Simulation algorithm was developed applying a discreet-event approach [8]. According to this method, the imitation of the process is implemented only in its significant states and transitions between them are conducted by means of a specially organized statistical sampling procedure, which is based on Monte Carlo method [7, 9]. The method includes three main steps:

1. Determination of the first state of the process described by a Semi Markov model according to the initial states' vector (7);
2. Calculation system's sojourn time in current state using the vector (9) before moving to another state. In order to perform this operation with specified density function the inverse transforms method could be used [8];
3. Definition of the next state S_j according to the square matrix of transition probabilities (8) when the following inequality is correct:

$$\sum_{j=1}^{f-1} P_{ij} < u \leq \sum_{j=1}^{f} P_{ij}, i = const, j = \overline{1,8}, \tag{10}$$

where u is a uniformly distributed number in the range [0, 1].

After some corrections of the expression (10), the target value of j-index is equal to variable f. In order to generate the random values, the L'Ecuyer algorithm was used, considering period of reiteration of pseudorandom numbers of 10^{18}, as it is recommended in [10]. The properties of the pseudorandom generator with the L'Ecuyer algorithm were analyzed in [11].

Thus, if we simulate the investigated process using the described algorithm during a time T_k for each value of maintenance periodicity $T_{int} \in [0, T_k]$ with step ΔT; then, it is possible to calculate sojourn time for each process's state S_i, $i \in [1, 8]$. On the basis of collected statistical data, the estimations of integrated reliability indexes $\overline{K_{OE}}(T_{int})$ and $\overline{K_A}(T_{int})$ are implemented with a specified accuracy ε and a confidence probability Q:

$$\overline{K_{OE}^m}(T_{int}^m) = \frac{1}{N_m} \sum_{j=1}^{N_m} \frac{T_1^j(T_{int}^m) + T_2^j(T_{int}^m) + T_8^j(T_{int}^m)}{\sum_{i=1}^{8} T_i^j(T_{int}^m)}, m = \overline{1,M} \tag{11}$$

$$\overline{K_A^m}(T_{int}^m) = \frac{1}{N_m} \sum_{j=1}^{N_m} \frac{T_1^j(T_{int}^m) + T_2^j(T_{int}^m) + T_8^j(T_{int}^m)}{\sum_{i=1}^{3} T_i^j(T_{int}^m) + T_6^j(T_{int}^m) + T_8^j(T_{int}^m)}, m = \overline{1,M} \qquad (12)$$

where T_i^j is the total stay time in S_i for j replication of the model; M is the quantity of points for calculation determined as $T_k/\Delta T$; N_m is the quantity of replications of simulation model for each estimated point of the dependences $\overline{K_{OE}}(T_{int})$ and $\overline{K_A}(T_{int})$, since the calculation is performed using a different amount of sampling of simulation experiment which is organized by automatic stop principle with achievement of specified accuracy ε [12].

3 Numerical Example

In order to perform experimental tests, the following parameters were specified: $\lambda_{12} = 50 \times 10^{-6} \text{ h}^{-1}$, $\lambda_{13} = 5 \times 10^{-6} \text{ h}^{-1}$, $\lambda_{23} = 0.5 \times \lambda_{12}$, $\alpha_1 = 0.01$, $\alpha_2 = 0.005$, $\beta_1 = 0.01$, $\beta_2 = 0.005$, $t_t = 2 \text{ h}$, $t_a = 1 \text{ h}$, $t_s = 3 \text{ h}$, $t_r = 3 \text{ h}$, $F_p(T_{int}) = 0$, $T_k = 10^6 \text{ h}$ and $\Delta T = 600 \text{ h}$.

The first experiment was devoted to demonstration of the dependency $\overline{K_{OE}}(T_{int})$ with different accuracy ε and with increased failure rate λ_{13}. The results are shown in Fig. 3.

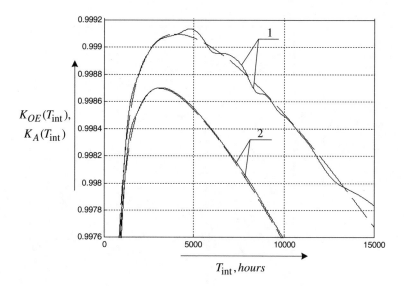

Fig. 3 Results of the first experiment

The illustration contains two types of dependences for the same specified input data: analytical results (dashed line) and calculations of simulation experiment (solid line). The analytical model was taken from the research work [2] and gives opportunity to verify the simulation model. The lines marked with digit "1" were obtained for above mentioned input parameters and for simulation estimations the accuracy was specified as $\varepsilon = 1 \times 10^{-2}$. The lines marked with digit "2" were acquired with three times increased failure rate $\lambda_{13} = 15 \times 10^{-6}\,h^{-1}$ and the accuracy $\varepsilon = 1 \times 10^{-5}$.

The second experiment was implemented only with simulation models for two different values of service staff errors probability $F_p(T_{int}) = 0$ and $F'_p(T_{int}) = 0.3$, the accuracy was kept the same $\varepsilon = 1 \times 10^{-5}$. Obtained dependencies $\overline{K_{OE}}(T_{int})$ and $\overline{K_A}(T_{int})$ are presented in Fig. 4, bold line shows results for increased probability of staff errors.

Results analysis showed that for zero staff errors probability $T_{opt} = 4226$ h and $T_{all} = 14015$ h, for probability specified as 0.3 the results are $T'_{opt} = 3314$ h and $T'_{all} = 9617$ h. It could be concluded that allowable intervals T_{all} reduces greater than optimal ones T_{opt}, it means that allowable periodicity is more sensitive to changing of staff errors probability.

Besides, according the formula (6) rational maintenance intervals T_{rat} equal optimal periodicity T_{opt} as in both cases allowable level of availability was chosen 0.998. If $K_{A.A.}$ was specified more than 0.9995 it would be possible to get results when rational values equal allowable values and offered criterion takes into account such a variant.

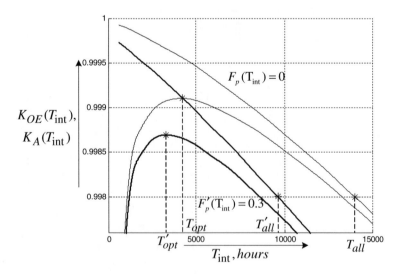

Fig. 4 Results of the second experiment

4   Conclusions

In this paper, a methodology for estimation of recommended PM intervals of electronic devices was proposed. The approach is based on two main issues: optimality criterion and simulation model. The criterion includes reliability coefficients such as availability and operation efficiency for calculation of allowable and optimal intervals, respectively. Such technique permits of estimating of recommended rational interval as minimum between allowable and optimal values.

The simulation model was developed for analyzing of operational process of electronic devices and collecting of required statistical data. The model is based on Monte Carlo method. The model of Embedded Markov chain for semi Markov process was chosen for mathematical formalization of the process. Common usage of a simulation model and optimality criterion provides an opportunity to examine one of the most important parameter of PM system, interval of preventive procedures, depending on the following factors: appearance of sudden, gradual, latent and fictitious failures, human factor of service staff and time parameters of preventive maintenance. Application of simulation approach is more flexible and allows the easy analysis of the model's input parameters influence on PM intervals with comparison to analytical routine manipulations.

Acknowledgments This work has been supported in part by projects ERANET-Plus (European Commission), TIN2014-56494-C4-3-P (Spanish Ministry of Economy and Competitiveness), PROY-PP2015-06 (Plan Propio 2015 UGR), and UMNIK Program (Russian Foundation for Assistance to Small Innovative Enterprises in Science and Technology).

References

1. Gertsbakh, I.: Reliability Theory: With Applications to Preventive Maintenance. Springer Science and Business Media (2000)
2. Lutchenko, S.S.: Optimization of control and preventive maintenance of radio communication devices. Ph.D. Thesis, Omsk State Technical University (2000)
3. Filenkov, V.V.: Improvement of methods for safety control and estimation of automation and radio communication facilities. Ph.D. Thesis, Omsk State Technical University (2004)
4. Lutchenko, S.S., Bogachkov, I.V., Kopytov, E.Y.: The technique of determination of fiber-optical lines availability and maintenance intervals. In: 2015 International Siberian Conference on Control and Communications (SIBCON 2015), pp. 184–187. IEEE Press, Omsk (2015). doi:10.1109/SIBCON.2015.7147004
5. Sirota, A.: Computational Modeling and Effectiveness Estimation of Complex Systems. Technosfera, Moscow (2006)
6. Barbu, V.S., Limnios, N.: Semi-Markov Chains and Hidden Semi-Markov Models Toward Applications: Their Use in Reliability and DNA Analysis. Springer Science and Business Media (2009)
7. Pardoux, E.: Markov Processes and Applications: Algorithms, Networks, Genome and Finance. Wiley (2008)
8. Kelton, W.D., Law, A.M.: Simulation Modeling and Analysis. McGraw Hill, Boston (2000)

9. Asmussen, S., Glynn, P.W.: Stochastic Simulation: Algorithms and Analysis. Springer Science and Business Media (2007)
10. Press, W.H.: Numerical recipes 3rd edn: The Art of Scientific Computing. Cambridge University Press (2007)
11. Bartosh, S.V., Kogut, A.T., Lyubchenko, A.A.: The analysis of properties and selection of pseudorandom number generators for simulation. Vestnik Kompyuternyih i Informatsionnyih Tehnologiy, vol. 2 (116), pp. 52–57. Moscow (2014)
12. Sokolowski, J.A., Banks, C.M.: Principles of Modeling and Simulation: A Multidisciplinary Approach. Wiley (2011)

Modeling of Consumption Data for Forecasting in Automated Metering Infrastructure (AMI) Systems

A. Jayanth Balaji, D.S. Harish Ram and Binoy B. Nair

Abstract The Smart Grid is a new paradigm that aims at improving the efficiency, reliability and economy of the power grid by integrating ICT infrastructure into the legacy grid networks at the generation, transmission and distribution levels. Automatic Metering Infrastructure (AMI) systems comprise the entire gamut of resources from smart meters to heterogeneous communication networks that facilitate two-way dissemination of energy consumption information and commands between the utilities and consumers. AMI is integral to the implementation of smart grid distribution services such as Demand Response (DR) and Distribution Automation (DA). The reliability of these services is heavily dependent on the integrity of the AMI data. This paper investigates the modeling of AMI data using machine learning approaches with the objective of load forecasting of individual consumers. The model can also be extended for detection of anomalies in consumption patterns introduced by false data injection attacks, electrical events and unauthorized load additions or usage modes.

Keywords Automated metering infrastructure · Smart grid · Load forecasting · Distribution side management · Soft computing · Artificial intelligence

The CER Electricity Dataset used in this work was provided by Irish Social Science Data Archive (ISSDA).

A. Jayanth Balaji (✉) · D.S. Harish Ram · B.B. Nair
Computing, Hardware Systems and Architectures Group, Department of Electronics and Communication Engineering, Amrita School of Engineering, Amrita Vishwa Vidyapeetham University, Coimbatore, India
e-mail: a_jayanthbalaji@cb.amrita.edu

D.S. Harish Ram
e-mail: ds_harishram@cb.amrita.edu

B.B. Nair
e-mail: b_binoy@cb.amrita.edu

1   Introduction and Motivation

AMI has evolved from smart meter (SM) and automated metering reading [1] (AMR) technologies which were initially introduced with the objective of faster collection of energy consumption data with minimal human intervention. However unlike AMR, AMI envisages a galaxy of distribution side services aimed at improving power quality and reliability, increase energy savings, facilitate distributed generation, reduce energy costs for the consumers and reduce generation/distribution costs for the utilities. The data from the smart meter front-end is processed by the *Metering Data Management Side* (MDMS) that forms the intelligent backbone enabling a host of AMI Demand Side Management (DSM) services at the consumer end. The block diagram depicting AMI components is shown in Fig. 1. The MDMS provides the database and analytics for the metering data acquired from the AMI network and interacts with other DSM services such as *Demand Response (DR), Distribution Automation (DA)* and *Outage Management Services (OMS)* among others. *Demand Response* is the modification of consumer behavior patterns triggered by pricing changes, incentives or imminent events affecting power system reliability [2, 3].

The presence of extensive computing and communication infrastructure exposes AMI systems to severe cyber security vulnerabilities which will compromise the fundamental requirements of *availability, integrity* and *confidentiality* of any system. *Confidentiality* is addressed by encryption schemes [4, 5]. *Data Integrity* will be affected by *False Data Injection* and *Replay attacks* [6] which will impact DR and Dynamic Pricing services. *Integrity* is primarily addressed using encryption and authentication. When all these measures are compromised by attackers, an additional layer of intelligence has to be provided to identify anomalous data patterns introduced by the attackers. This is accomplished by developing intelligent models for the AMI data for non-intrusive analytics. In addition to security, intelligent analytics has to be built into the MDMS for services such as DR and load management. The AMI models have to be capable of extracting behavioral trends and

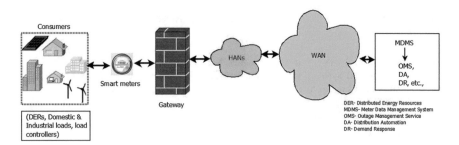

Fig. 1 Block diagram of AMI

cyclic patterns in energy consumption data. Several attempts have been made at modeling consumption based on smart meter data [7–13]. The following points are evident from the reported work.

- There is a pressing need for reliable models for AMI data for devising schemes to ensure data integrity. Modeling schemes reported in the literature do not provide accurate insights into anomalies in consumer behavioral patterns. Also accurate load forecasting on the consumer side and profiling of consumption data shall enable better demand side management
- The impact of cyber security attacks on DA services at the LV stages has not been analyzed in the reported literature. Hence these vulnerabilities have to be modeled with the intent of developing mechanisms for detection of false command injection and command disruption. The models will serve to assist in authenticating the commands. In the event of command disruption, the system can recover by a predictive mechanism using the model.

With the advent of AMI, energy consumption data is available at typically hourly or half-hourly resolutions. The high resolution obtained from AMI makes consumption data amenable to state of the art machine learning approaches. The time series comprised of the AMI data of a single consumer is resolved into the cyclic and trend components. This segregation leads to a more robust model which can be tuned to the application that is envisaged. This is detailed in the next section. The rest of the paper is organized as follows. Section 2 outlines the methodology proposed in the paper. Section 3 presents the results of the proposed models on the CER dataset and analyzes the outcomes. Section 4 concludes the paper.

2 Methodology

The model was developed for the CER smart meter dataset consisting of consumption data for households as well as small businesses in Ireland (a total of 6445 consumers). The dataset consists of readings for half-hourly intervals for duration of 74 weeks for three types of customers: 1. Small and Medium Enterprises (SME) with 485 consumers and 2. Residential, with 4225 consumers and 3. Other with 1735 consumers. In the present study, an attempt has been made to model and forecast the energy consumption of SMEs.

Initially, the energy consumption data from each of the SME consumers was tested for unit-root stationarity to identify the possibility of using Auto Regressive Moving Average (ARMA) models for modeling the data. Kwiatkowski-Phillips-Schmidt-Shin (KPSS) test was used for the purpose. Results indicated that ARMA models would be unsuitable for the modeling/forecasting purpose. Since machine learning based techniques have been widely used for modeling non-stationary data, in the present study, Artificial Neural Network (ANN) (Single hidden layer feed-forward ANN) and Ensemble Regression Trees (ERT) have been used for modeling and forecasting the consumption.

The procedure for generating the consumption forecast is as follows:

Step 1: Normalize the consumption data for each consumer.

Step 2: Segregate the consumption data based on seasons into summer, winter, autumn and spring. This was done due to the fact that energy consumption patterns are dependent on the seasons. The months and the seasons to which the consumption data is grouped are: Summer: May-July, Autumn: August-October, Winter: November-January and Spring: February-April. Forecasting without segregation of data into seasons was also carried out to identify if segregating the data improves the forecast or not.

Step 3: Separate the trend and cyclic components in the data for each user using the *Hodrick-Prescott* (*HP*) [14] filter. HP filter has been used widely used in macroeconomic theory and more recently, has been successfully employed for design of stock trading recommender systems [15] and for forecasting financial time series [16, 17]. However, it is observed by the authors that the utility of HP filter in modeling and forecasting of consumption data has not been explored. Hence, this study is also unique in that aspect. Using the HP filter, the energy consumption time series is split into two components (trend and cyclic) i.e. $y(t) = y_{trend}(t) + y_{cyclic}(t)$, where $y(t)$ is the consumption at time t; $y_{trend}(t)$ is the trend component of $y(t)$ and $y_{cyclic}(t)$ is the cyclic component of $y(t)$. In the context of AMI data, the cyclical component can be used for identifying events at the consumer end. This includes events caused by peak surges in load that are part of routine behavioural patterns. Such information can be valuable in modifying consumption behaviour in services such as demand response and improving reliability on the distribution side. Moreover, the cyclic component may be further analysed for detection of potentially anomalous behaviour from metering information having potential cyber security implications. The trend component can predict long term behaviour for applications such as distribution management. The HP filter used in the present study uses a smoothing parameter (Λ) set to 100, selected using trial and error. To illustrate the functioning of the HP filter, an example is presented in Fig. 2. The user ID 1023 from the AMI dataset is considered with 1000 data points from June 12, 2010 morning 08.30 am onwards. The user and the time frame are selected at random. As can be seen, the HP filter is able to separate the trend and cyclic components quite effectively.

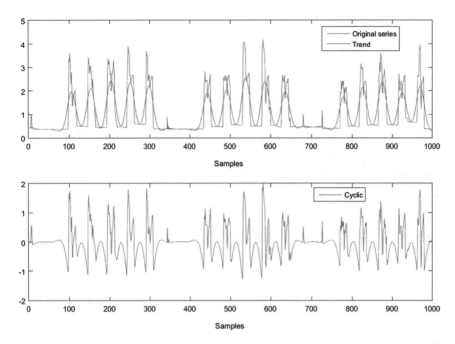

Fig. 2 HP filter used to separate the energy consumption time series into its trend and cyclic components

Step 4: Forecast the trend and cyclic components using ANNs and ERTs. Four different combinations are attempted:

    i. Both trend and cyclic components forecast separately using ANN.
    ii. Both trend and cyclic components forecast separately using ERT.
    iii. Trend component forecast using ANN and cyclic using ERT.
    iv. Trend component forecast using ERT and cyclic using ANN.

Step 5: Combine the two forecasts to generate the overall consumption forecast. Two more forecasting systems using ANN and ERT respectively, without using the HP filter (i.e. considering the time series as it is, without splitting it into trend and cyclic components) are also evaluated. The procedure is illustrated in detail in Fig. 3.

Single hidden layer feed forward ANN trained using the *Levenberg-Marquardt* (*LM*) algorithm was used in all the cases. ERTs used in the present study employed bagging and 5 trees in all the cases.

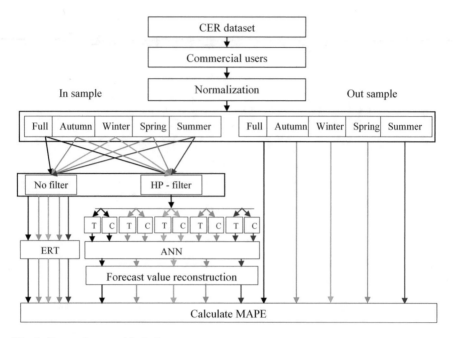

Fig. 3 Proposed system block diagram

3 Results and Analysis

All the six systems considered in the present study are trained and used to generate 2-h-ahead, 4-h-ahead, 6-h ahead and 8-h-ahead consumption forecasts. Sliding window technique was used for generating the forecasts. The performance measure employed in the current study is Mean Absolute Percentage Error (MAPE), as suggested in [18]. It was observed that 2-h-ahead consumption forecasts were consistently more accurate when compared to forecasts made over longer horizons of 4 and 8 h. It was also observed that of all the systems considered, forecasting system that employed HP filter for separation of trend and cyclic components and trained using ANNs generated better forecasts on a consistent basis. Forecasting system trained using ERTs without splitting the time series using HP filter, also generated good results, although slightly worse than the HP filter-ANN based system discussed previously. The increase in the MAPE as the forecasting horizon increases from 2 h to 4, 6 and 8 h for both these systems is presented in Fig. 4. As can be observed from the figure, the MAPE worsens as the forecast horizon increases. This was found to be true for all forecasting systems.

The results for the above two forecasting systems are presented in the form of histograms in Figs. 4 and 5. It must be noted that the histograms in Figs. 4 and 5 represent histograms for the Median MAPEs generated by SME consumers. The x-axis in each histogram plot represents the Median MAPE and the y-axis, the

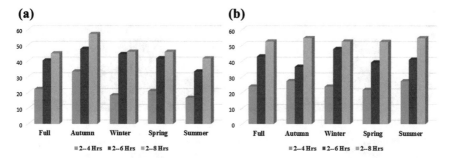

Fig. 4 Percentage increase in MAPE between different forecast periods for (**a**) forecasting system employing HP filter for separation of trend and cyclic components and trained using ANNs and (**b**) forecasting system trained using ERTs without splitting the time series using HP filter

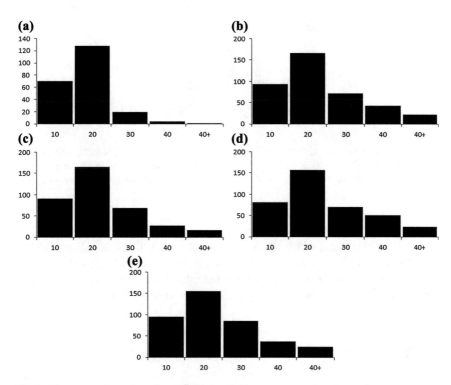

Fig. 5 Histogram plots of median MAPE for SME consumers using forecasting system using ANNs. **a** ANN based forecasting results using HP filter and without segregating data into seasons. **b–e** Results for HP filter based separation of trend and cyclic components and ANNs for **b** autumn, **c** winter, **d** spring and **e** summer seasons

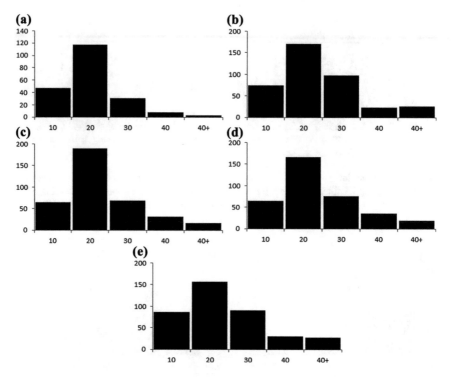

Fig. 6 Histogram plots of median MAPE for SME consumers using forecasting system using ERTs. **a** ERT based forecasting results without using HP filter and without segregating data into seasons. **b–e** Results for data segregated into **b** autumn, **c** winter, **d** spring and **e** summer seasons

number of users for which the consumption forecasts produced Median MAPE in the specified bin. Equal width binning was considered with the bin size of 10 (i.e. 10 % Median MAPE). Some of the users where found to exhibit MAPE in excess of 100 %. The results pertaining to these users have not been taken into account while plotting the histograms since they are likely to be outliers (Fig. 6).

4   Conclusions

From the results, it can be concluded that ANNs and ERTs can be successfully employed for forecasting electricity consumption for SME consumers. It was observed that the forecasting accuracy deteriorated as the forecast horizon was increased from 2 h-ahead to 4,6 and 8 h-ahead. It is also observed that the use of HP filter to separate the trend and cyclic components in the consumption time series, forecasting them separately and then combining the trend and cyclic component forecasts to generate the overall consumption forecast tends to improve the prediction accuracy with the best performance being demonstrated by HP filter-ANN based forecasting system. The study can be further extended to forecasting of energy consumption by residential consumers for further validation of the proposed system and develop a generalized consumption forecast model.

References

1. Meters, S.: Smart meter systems: a metering industry perspective. An Edison Electric Institute-Association of Edison Illuminating Companies-Utilities Telecom Council White Paper, A Joint Project of the EEI and AEIC Meter Committees, Edison Electric Institute (2011)
2. Siano, P.: Demand response and smart grids. A survey. Renew. Sustain. Energy Rev. **30**, 461–478 (2014)
3. Balakrishna, P. et al.: Analysis on AMI system requirements for effective convergence of distribution automation and AMI systems. In: 2014 6th IEEE Power India International Conference (PIICON) (2014)
4. Deng, P., Yang, L.: A secure and privacy-preserving communication scheme for advanced metering infrastructure. In: 2012 IEEE PES Innovative Smart Grid Technologies (ISGT) (2012)
5. Chen, J. et al.: A key management scheme for secure communications of advanced metering infrastructure. In: Communications in Computer and Information Science Applied Informatics and Communication, pp. 430–438 (2011)
6. Wang, W., Lu, Z.: Cyber security in the Smart Grid: Survey and challenges. Comput. Netw. **57**(5), 1344–1371 (2013)
7. Tasdighi, M. et al.: Residential microgrid scheduling based on smart meters data and temperature dependent thermal load modeling. IEEE Trans. Smart Grid. **5**(1), 349–357 (2014)
8. Rahman, M.A. et al.: A noninvasive threat analyzer for advanced metering infrastructure in smart grid. IEEE Trans. Smart Grid **4**(1), 273–287 (2013)
9. Guruprasad, S. et al.: A learning approach for identification of refrigerator load from aggregate load signal. In: 2014 International Conference on Advances in Computing, Communications and Informatics (ICACCI) (2014)
10. Chan, S. et al.: Load/price forecasting and managing demand response for smart grids: methodologies and challenges. IEEE Signal Process. Mag. **29**(5), 68–85 (2012)
11. Hong, T. et al.: Long term probabilistic load forecasting and normalization with hourly information. IEEE Trans. Smart Grid **5**(1), 456–462 (2014)
12. Kwac, J. et al.: Household energy consumption segmentation using hourly data. IEEE Trans. Smart Grid **5**(1), 420–430 (2014)
13. Krishna, V.B. et al.: PCA-based method for detecting integrity attacks on advanced metering infrastructure. In: Quantitative Evaluation of Systems Lecture Notes in Computer Science, pp. 70–85 (2015)
14. Hodrick, R., Prescott, E.: Postwar U.S. Business Cycles. In: Real Business Cycles A Reader, pp. 593–608 (1998)
15. Nair, B.B., Mohandas, V.: An intelligent recommender system for stock trading. Intell. Decis. Technol. **9**(3), 243–269 (2015)
16. Nair, B.B., Mohandas, V.: Artificial intelligence applications in financial forecastinga survey and some empirical results. Intell. Decis. Technol. **9**(2), 99–140 (2015)
17. Nair, B.B., et al.: A stock trading recommender system based on temporal association rule mining. SAGE Open **5**, 2 (2015)
18. Gooijer, J.G.D., Hyndman, R.J.: 25 years of time series forecasting. Int. J. Forecast. **22**(3), 443–473 (2006)

Scanning System for Ballistic Analysis

Tomáš Martínek, Josef Kudělka, Milan Navrátil
and Vojtěch Křesálek

Abstract This paper presents hardware and software solution of the scanning system designed for ballistic analysis of cartridge cases which were deformed by firing pin after shooting for identification process of firing weapon. The system is based on resistivity measurement between the scanning tip and a sample and three axis servomotor positioning controlled by application created in MATLAB programming environment. Output of the application is matrix of coordinates for 3D visualization of deformed cartridges for firearm identification process. Measured example results are included.

Keywords Ballistic analysis · Forensic ballistic · Metrology · Scanning system · Programming · MATLAB

1 Introduction

Identification of firearms according to the cartridges found on the crime scene is one of the most common methods of criminological forensic ballistics [1]. Marks left on the cartridges or bullets after firing can be examined by different methods. Optical comparison microscopes [2] are one of the most common techniques for this. Another possible method is scanning system. Scanning systems have the advantage of providing 3D model, where another marks like depth or direction of deformation can be found.

Basic idea of scanning system for ballistic analysis came from scanning probe microscopy [3].

Smaller countries are unable to buy and maintain commercial Integrated Ballistic Identification Systems (IBIS) [4] for automated analysis and database comparison but this can be done with minimal financial expenses as this article presents.

T. Martínek (✉) · J. Kudělka · M. Navrátil · V. Křesálek
Faculty of Applied Informatics, Tomas Bata University in Zlín,
Nad Stráněmi 4511, 76005 Zlín, Czech Republic
e-mail: tmartinek@fai.utb.cz

© Springer International Publishing Switzerland 2016
R. Silhavy et al. (eds.), *Automation Control Theory Perspectives in Intelligent Systems*, Advances in Intelligent Systems and Computing 466,
DOI 10.1007/978-3-319-33389-2_17

2 Used Instruments

2.1 Mercury M-110 1DG

Mercury M-110 1DG is a very high motoric resolution servomotor with linear motion in the maximum range of 5 mm. Minimal guaranteed resolution by manufacturer is 0.5 μm [5]. Velocity and acceleration of this servomotor can be precisely controlled. They can be attached together to create multi axis positioning system. These servomotors are controlled by Mercury C-862. Count unit is used in this servomotor for movement.

2.2 Mercury C-862

Mercury C-862 is a networkable single-axis DC-motor controller unit using quadratic encoder signal to determine the actual position. The controller itself is connected with computer via RS-232 as well as with controlled motor and other controllers. In the first step the individual controller is addressed and activated and in the second step is send command for setting of parameters of the motor or for its movement.

2.3 Hewlett Packard 34401A

HP 34401A is digital multimeter with maximum resolution of 6½ digits. It is connected with computer via GPIB. The most important parameter in this case was measurement speed—up to 1000 measurement per second depending on the number of digits used. This multimeter was used for resistance measuring with relative measurement uncertainty 0.0015 % for DC values [6].

2.4 Scanning Tip

Scanning tip, especially its point radius, determine maximum resolution of scanning process. In this case tip point was in size of 35 μm.

However, during a control measurement, sometimes the physical contact does not lead to correct evaluation. This was the reason to put the tip into the housing to allow it to move freely in Z axis to prevent the tip and a sample from a damage. The tip then moves back in the housing to the default position by its weight.

3 System Setup and Software

Figure 1 presents the connection scheme of the scanning system. Each servomotor is connected with its own controller unit and it is capable of movement in one of the axis—X, Y and Z. On this configuration of servomotors is placed movable stage with scanned sample.

HP 34401A is measuring resistivity between the scanning tip and moving stage with the sample. Figure 2a shows the electric scheme for measuring the resistivity, where the switch is representing scanning tip and the sample. When the scanning tip is in contact with the sample, the resistivity of the circuit changes from $R = R1 + R2$ to $R = R2$, where $R1 = 1 M\Omega$, $R2 = 100 \Omega$.

Figure 2b presents the feedback loop for evaluating the contact between the scanning tip and the sample. If the resistivity is higher the R2, the movement command is send to the controller unit and then to the servomotor to move another step in Z axis. This is done for the each point of the scanning raster as shown in Fig. 3.

When the contact is evaluated, the coordinates are saved to the file, sample moves to its default Z-axis position and then to the next point of the scanning raster. Simplified flowchart can be seen at Fig. 4.

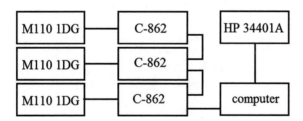

Fig. 1 Connection of the used instruments

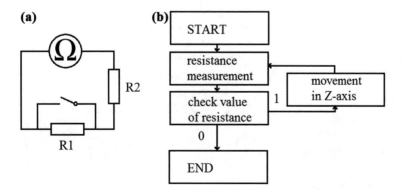

Fig. 2 a Electric circuit scheme. **b** Flowchart of the feedback loop

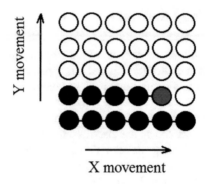

Fig. 3 Scanning raster

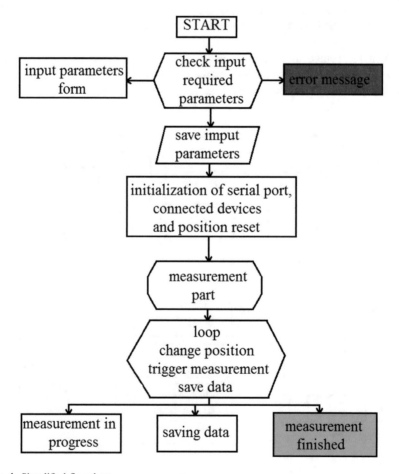

Fig. 4 Simplified flowchart

4 Results and Discussion

The 2 CZK coin was selected for the test measurement as can be seen in Fig. 5. Test measurement was used to verify the hardware and software solution of the scanning system. Parameters were set to 45 lines, 90 point per line, distance between points was set to 70 μm and approach step in Z axis was set to 2 μm.

Test measurement takes about 20 h to scan the selected area. This process is very time consuming but the main advantage of scanning system compared to comparative microscope is ability to determine another marks like depth or direction of the deformation on the cartridge (Fig. 6).

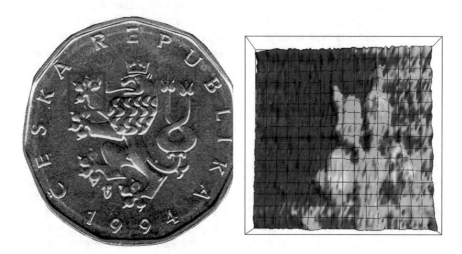

Fig. 5 Test measurement on the coin

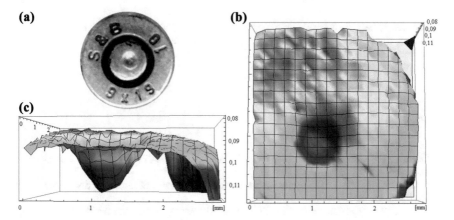

Fig. 6 a Cartridge for example measurement. **b** Top bottom view of 3D visualization. **c** Side view of 3D visualization

Figure 5a shows 9×19 mm cartridge selected as the example measurement. 30×30 points raster size was set with 100 μm distance between points and 2 μm approach step. Figure 5b presents top bottom view and Fig. 5c show side view at the 3D visualization of coordinates. The radius of firing pin can be determined from this visualization as well as depth of deformation, its shape and direction. The example measurement takes about 10 h.

Coordinates are saved into the file by defined format—in this case: $\{x_1, y_1, z\}$, $\{x_2, y_1, z\}, \ldots$ allowing to use these matrix of coordinates also in other programming environments.

Resolution of this system is based mainly on scanning tip radius—30 μm in this case. Resolution can be improved by reducing tip radius and decreasing distance between adjacent points. However distance between two points should be at least twice as big as the tip radius. Increasing of the resolution as mentioned would lead into smaller scanning area so more points are needed to keep scanning area in the same size. However this would lead into longer processing time. Balance between resolution and time consumption must be determined for each specific application.

5   Conclusion

This article presents system for ballistics analysis created from three servomotors, attached controllers, digital multimeter and personal computer with GPIB and RS-232 interface. The control software was made in MATLAB environment however it is possible to use some other software packages. Whole scanning process is quite time consuming but after initial settings is fully automated. This system is very cheap compared to commercial IBIS and can be created with common laboratory equipment and with basic programming skills. Main advantage of scanning systems is 3D model of scanned sample providing additional marks for firearm identification process.

Acknowledgments This work was performed with financial support by the Ministry of Education, Youth and Sports of the Czech Republic within the National Sustainability Programme project No. LO1303 (MSMT-7778/2014) and with support by the Internal Grant Agency of Tomas Bata University under the project IGA/CebiaTech/2016/003.

References

1. Dimaio, V.: Gunshot Wounds: Practical Aspects of Firearms, Ballistics, and Forensic Techniques. CRC Press, Boca Raton (1998)
2. Hawkes, P., Spence, J.: Science of Microscopy, vol. 2, p. xviii, 1265, I26. Springer, New York (2007). ISBN 03-872-5296-7
3. Tsukruk, V., Singamaneni, S.: Scanning Probe Microscopy of Soft Matter: Fundamentals and Practices. Wiley-VCH, Weinheim (2012). ISBN 978-3-527-32743-0

4. Thompson, R.M.: Automated firearms evidence comparison using the integrated ballistic identification system (IBIS). In: Proceedings of SPIE—The International Society for Optical Engineering, vol. 3576, pp. 94–103 (1999)
5. MS 74E User Manual. Physik Instrumente. Mercury C-862 Networkable Single-Axis DC-Motor Controller. Release 8.41. Karlsruhe (2004)
6. Service Guide. Agilent Technologies. Agilent 34401A 6½ Digit Multimeter, 7th edn. Santa Clara.   http://www.home.agilent.com/agilent/product.jspx?cc=CZ&lc=end&nid=-536902435. 536880933&pageMode=PL (2007)

Predictive-Based Stochastic Modelling of Power Transmission System for Leveraging Fault Tolerance

G. Raghavendra and Manjunath Ramachandra

Abstract With the dynamicity in the load requirements of the user, it is essential that power transmission lines should be highly resilient against any possibilities of error or outage. The existing techniques have witnessed some good solutions towards ensuring fault tolerance, but it was never analyzed from the viewpoint of smart grid system using predictive principles. Hence, this paper presents a system that performs stochastic modelling of the distributed power generation system considering dual power states i.e. available and outage. The study also implements a cost effective algorithm for random modelling the distributed generation system of power to reduce the outage probability and increase the capacity probability considering all the possible cases of susceptible errors in transmission lines with islanding mechanism. The study outcome shows superior analysis of load capacity with respect to the existing system.

Keywords Power transmission system · Stochastic modelling · Fault tolerance · Outage · Prediction

1 Introduction

In the present scenario, there is a massive revolution in the electrical power dissemination techniques in the form of infrastructure, techniques, delivery mechanism etc. With the radical change of conventional power transmission lines to smart-grid system, it is believed that electrical systems will introduce a new era in energy for users [1]. The upcoming trends of power transmission lines calls for adoption of numerous distributed power resources, managing dynamic loads, storage units etc.

G. Raghavendra (✉) · M. Ramachandra
Jain University, Bangalore, India
e-mail: researchraghavendra@gmail.com

© Springer International Publishing Switzerland 2016
R. Silhavy et al. (eds.), *Automation Control Theory Perspectives in Intelligent Systems*, Advances in Intelligent Systems and Computing 466,
DOI 10.1007/978-3-319-33389-2_18

However, there are various forms of the sophisticated components in electrical transmission system that is highly susceptible to serious faults degrading the quality of power transmission. Therefore, in order to leverage an upcoming concept of the power grids, it is essential that present day electrical system must be resilient against various potential faults which may finally result in outage. Therefore, it is necessary to identify the position of the source point of fault and address it effectively. A robust detection mechanism of fault must ensure that identification of fault should be instantly detected with highly reduced rate of error. Although, in digital mechanism of smart-grid system makes the distribution system not only efficient but also easy to monitor, but there are various unpredictable situation in the transmission lines leading to outage. Hence, such problems are quite difficult to solve and calls for using extensive testing [2]. Therefore, we use a computational model which plays a significant role in carrying out optimization process using state-based modelling technique. The prime contribution of the study is its design of predictive principle considering the possible cases of islanding mechanism using stochastic mechanism. The paper is organized as follows Sect. 2 highlights prior research work. Section 3 describes problems explored after reviewing the literatures. Discussion of the proposed system using stochastic modelling approach is presented in Sect. 4 followed by research methodology to accomplish the stated goal of study in Sect. 5. Algorithm description is discussed in Sect. 6 followed by Sect. 7 that demonstrates the outcomes of the proposed study followed by summary of paper in Sect. 8.

2 Related Work

This section discusses about the studies being carried out by the significant authors pertaining to the error analysis or fault tolerance aspects in distributed power generation system.

Bazrafshan and Gatsis [3] have performed optimization of distributed power system using stochastic approach over rural distribution circuits. The design principle of the technique also uses regular update transmission in closed formed of circuits. Behrens et al. [4] have presented a technique of estimating errors in the power distribution system and then proposes a mechanism to segregate the error. Usage of sophisticated signal processing methods is highly encouraged by Bracale et al. [5]. Most recently, study toward detecting fault in power distribution system is identified in the work of Jamil et al. [6]. The authors have also used evolutionary technique to perform categorization of the faults in electrical systems. Maa et al. [7] have presented a mechanism for evaluating the amount of voltage being used over smart grid system. The study was carried out over real time circuits. From the usage of optimization viewpoint, Babu et al. [8] have used fuzzy logic for overcoming the issues in controlling strategies over power transmission system. Hafez [9] has also emphasized on power generation system taking the case study of electric aircraft. According to the study, power distribution in aircraft is entirely different from frequently studied conventional and hypothetical designs.

Das [10] have also carried out the similar line of research emphasizing on the usage of the fuzzy logic for identification of the fault. Similar line of research work have been also focused by Faig et al. [11] who attempted to excavate the location of the faults for single-phase circuits in power transmission system. Li et al. [12] have presented a middleware-based solution for strengthening the power distribution system. All the studies discussed above have assisted the upcoming researchers with its significant techniques however, very fewer techniques are found to cost effective. The problems pertaining to existing studies are discussed briefly in next section.

3 Problem Description

After reviewing the existing techniques of distributed power generation system, it was seen that although majority of the studies attempts to enhance the availability of the power loads in distributed traffic but there is very few of them which performs a predictive analysis of it from the viewpoint of demands of the users. The demands of the present customer base have become highly dynamic which doesn't cope up with the existing method of power transmission system. Although there are studies calculating fault tolerance of the system but very few studies have been witnessed to reliably forecast the extent of retention in fault tolerance with the random and stochastic environment in distributed power generation system. The biggest trade-off is being found is that existing studies are more inclination towards circuitry-based approach and less on computational modelling which doesn't seem to be much aligned with the present electrical distribution system which is highly digital and less analogue. Hence, the next section discusses about proposed system to overcome this issues.

4 Proposed System

The prime purpose of the proposed study is to develop a computational model that can be used to perform predictive analysis of the power distribution generation system. Although, there has been various studies towards distributed power generation system, but it has been seen that such studies were usually focused with an aid of circuitry principle and less towards cost-effective optimization process.

From our prior studies, we find that error-prone distribution network as well as intermittent connectivity is still an unsolved problem in this area. Hence, our contribution/novelty laid in this research manuscript are as follows:

- **Cost-Effective Computational Model**: The proposed study presents a novel algorithm that is mainly meant for carrying out computational optimization process by considering the stochastic properties of distributed power generation system.
- **Predictive Analysis**: The algorithm in the proposed system is designed considering two different states of power load availability and outage considering the need of carrying out islanding mechanism.
- **Unfailing Distribution System**: The algorithm and the mechanism of proposed system considers evaluating errors in every iterations of load and chooses to minimizes the probability of error on multiple epochs leading to highly fault tolerant outcomes in predictive analysis of distribution system.

Hence, the proposed system takes the input of randomness of power distribution system and applies the algorithm to accomplish the probability of its capacity and minimization of the errors in the prediction process. The next section discusses about the research methodology adopted for designing the proposed system.

5   Research Methodology

The proposed study is developed using analytical research methodology with a goal of evolving up with a computational model for distributed power system for furnishing efficient and optimal power supply to its end user. The prime design aspect was considered for designing the parameters of distributed power generation system considering parameters e.g. error-prone factors, resource availability, outage, etc. The study also develops a novel stochastic framework for distributed power system.

- **Designing a Simulation Test-bed**: In order to study the considered issues as well as challenges of quality loss while attempting to design distributed power system, an analytical approach is presented in the proposed study. Various parameters like resource and load supply, states of power, power output, probability of capacity, resource availability, outage etc. is considered while preparing the simulation test bed. Based on the review of literatures, conventional simulation parameters are considered in this phase of study. New simulation parameters are also formulated based on the need of the design level. The prime purpose of this phase is to design a system model for evaluating the effectiveness of distributed energy system in most challenging environment.

Once the simulation test bed is design, the proposed computational model can be designed on the top of it for performing simulation and extract the outcomes of the study. Accomplishment of this phase study will achieve second objective of the research work.

- **Designing Stochastic Modelling**: This phase of the study will include designing a stochastic modelling of distributed power system. After studying the literature, it can be inferred that non-conventional resources of energy are stochastic in nature and hence the study prefers to perform stochastic modelling. Another reason for performing stochastic modelling is because nowadays the substitution of conventional for non-conventional resources in the energy sector is among the most important environmental and economic issues. Therefore, this part of the study will focus on non-conventional distributed power generation along with the traditional approaches. The stochastic modelling will be designed considering a novel design of load demand provisioning process for enhancing the distributed power system for overcoming the issues of outage in both critical and non-critical areas of supply. Continuous monitoring of the networked distributed system will be performed throughout the computation process to extract the sufficiency transition rate from the output data along with evaluating the demand probability. This information are computed and used in measuring the scalability of the proposed stochastic model to ensure consistency in power supply to its end user. Accomplishment of this phase study will achieve study objective of the research work.

Therefore, the proposed study focuses on performing predictive analysis of the fault tolerance in the form of capacity of load available and outage using probability theory as well as stochastic approach. The next section discusses about the algorithm implementation for the proposed study to meet the study objectives.

6 Algorithm Implementation

The design of the proposed system is carried out in Matlab and uses a stochastic approach to compute. In order to design the proposed algorithm, we more focus to be laid on the mechanism to identify the availability as well as unavailability in the forms of outage in power distribution system. Therefore, we use stochastic approach to understand the impact of various power factors on its usage as well as its utility. The algorithm for stochastic modelling of distributed power generation system is as follows:

Algorithm for stochastic modelling of distributed power generation

Input: α_1 (power system breakdown), α_2 (power system restore), β_p (presence of power system) and β_o (outage of power system), c (number of components),

Output: T (cumulative hours), C_{prob} (probability of capacity)

Start

1. init α_1, α_2, c

2. Estimating β_p and β_o

$$.\beta_p = \alpha_1 . \arg_{\max}\left(\frac{\alpha_2}{\alpha_1 + \alpha_2}\right)$$

$$.\beta_o = \alpha_1 . \arg_{\min}\left(\frac{\alpha_1}{\alpha_1 + \alpha_2}\right)$$

3. DG=rand[1, 365*24]

4. Develop State-Based Power Matrix (SPM)

5. Sort→desc(DG)

6. For i=1:c

7.   For j=DG

8.   If DG(j)<p_{level}(i) && DG(j)>p_{level}(i+1)

9.   time=count++

10.   End

11.   End

12. End

13. T=time / DG //Stochastic model

14. For i=1:c

15.   For j=1:DG

16.   If (DG(j)<p_{level}(i) && DG(j)>p_{level}(i+1))

17. t_i=count++

18.   End

19.   End

20.   Estimate outage factor

$$O_{cap} = 100.\left[1 - \frac{1}{N}\right]$$

21.   If (i= = c) //Condition for Estimating Probability

22. C_{prob}(i)= β_{p^*}(t_i/DG+ β_o)

23.   End

24. End

End

The algorithm takes the input of α_1 (power system breakdown) and α_2 (power system restore), which is user-defined. In order to possess a good challenge, we initialize the value of power system breakdown as well as power system restore as a probability factor of 0.3 and 0.5 respectively. The algorithm than estimates other two probability-based power factors e.g. β_p (presence of power system) and β_o (outage of power system). We also apply stochastic modelling considering distributed generation system to be a random factor for one year with 24 h of observation time. The distribution of data points for distribution generation (DG) will be arranged in descending order. The algorithm also consider c = 10 initialized number of components to be observed in one cycle from the distribution curve. Hence, analyzing the distribution curve (Line-7) for all components (Line-6), the system checks for the value of DG to be lying within certain range of points as specified in Line-7. The count is increased to get the time. Hence, this is a unique strategy to check of the value of resultant power comes within the value of initialized power components (c = 10). We also calculate cumulative power in this process (line-9). Hence, it is completely a time-based evaluation process that generates total time for representing power availability or outage. The same process is again repeated back from line-14–19 in order to check the outage factor. Finally for a stated condition, we calculate probability of the distribution C_{prob} as an outcome of the proposed study using Line-22. The execution of the algorithm is ultimately stored in a matrix called as State-Based Power Matrix (SPM), which follows a binary state of power system availability and outage. For power system availability, the output power will be computed to be C_{prob}, while that of outage will be zero. Similarly, for state of power availability, the probability factor will be corresponding to β_p (presence of power system) whereas in case of outage, the probability factor will be $(1 - \beta_p)$.

Further, the proposed system also uses a unique islanding mechanism in the availability of the distributed generated loads for enhancing the better prediction. Normally, the existing system fails in prediction as it is not able to identify the appropriate situation when the supply from distributed generation system is cut-off completely. Hence, the system may lead to false positive if islanding mechanism is not considered to be assessed. Hence, the system continues to supply for the load even if the supply point is cut off as the proposed system can identify the point of probability when islanding is required. The proposed system considers non-conventional source of energy which can be used to minimize the operational cost of existing electricity. Such sources are used to carry out islanding operation. We also consider a probability of outage when the load as well as power output couldn't follow a similar pattern leading to disconnection of source point. We carry out the analysis of islanding by combining the stochastic trend of distributed generated output of power with the time-series distribution of the load for evaluating the probability that the system will not cut-off its power supply for any reason. This calculations, although simple, but yet can affect the higher degree of accuracy in the analytical outcome of prediction.

7 Result Discussion

This section discusses about the result analysis of the proposed as well as existing system. The power distribution system is quite dynamic as well as quite unpredictable and hence it is highly stochastic in nature. The prime motive of the

Fig. 1 Individual outcome of
capacity outage

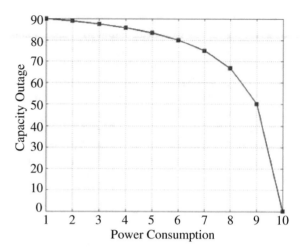

implementation discussed in prior section is to ensure the accuracy in prediction. We also used a Markov modelling for developing states of the power load capacity and outage factor. The individual outcome of the capacity outage is shown in Fig. 1.

The outcome shows a superior form of gradient descent nature of the curve for the proposed system in Fig. 1. We accomplish this type of result as the proposed modelling offers higher degree of decision making parameters for a system to identify the breakpoints, reduce errors in increasing iterations of the loads and the most important it can compute two essential probability-factors e.g. β_p (presence of power system) and β_o (outage of power system). Hence, if the system has more information in increasing time-series about the output power than it can render the distributed generation system to be available whenever there is no incident of power system breakdown. From cost-effectiveness viewpoint, the system constructs a state-based power matrix that holds all the computed predictive values based on two different forms of state (when power is available and when there is an outage). The interesting point of the proposed system is that the system is free from any training-based or predefined data-based prediction.

We find that study using fuzzy logic is more prevalent in power electronics as well as in the study of distribution system. Hence, we consider the signature work carried out by Das [10] and Babu [8]. Das [10] have developed a fuzzy technique for analyzing the faults in radial distribution system. Similarly, faster response time is achieved by Babu [8] using similar fuzzy logic technique. In order to make similar forms of test-bed, we do specific modification in the existing system by extracting only the time-series based prediction using fuzzy logic. The outcomes of the study are showcased in this section as follows:

i. Probability of Capacity:
   The outcome of the study in Fig. 1 shows that proposed system offer higher probability of capacity as compared to fuzzy-based existing system. This

Fig. 2 Comparative analysis
of probability of capacity

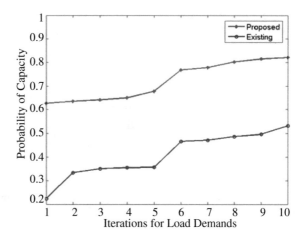

parameter calculates the trend of capacity of making predictions for capacity in increasing iterations of load demands. We apply the stochastic modelling in order to design the power distribution system based upon its randomness. Hence, the system make use of more number of parameters e.g. power system breakdown, power system restore, presence of power system, and outage of power system, number of components etc. for performing decision thereby enhancing the decision of prediction with higher accuracy. Hence, probability trend is found to be enhanced compared to fuzzy-based existing system. A significant pitfall of fuzzy-based existing system is also its dependency of rules and membership functions (Fig. 2).

ii. Error-Per Epoch:

Error-per-epoch is a parameter used for assessing the effectiveness of the prediction for power distribution in time-series. The distribution system is highly complex and requires faster decision making algorithm. Application using fuzzy-based processor is completely based on rule and hence is not suitable for any dynamic cases of outage probability study that doesn't comply with the rules. Hence, it may shoot in massive errors in the distribution system which can later adversely affect the mechanical aspects of the devices. Figure 3 shows that although the start point of both the techniques have similar value of error but proposed system has better performance of error minimization as compared to existing system. The prime reason behind this fact is that proposed system also makes use of islanding scheme and asses it to multiple iterations in time-series to take better decision. For this purpose, we also calculate antici-pated new probability factor of the power outage, which is just a power function of the load that is greater than capacity of the power distribution generation system. Hence, the prediction modelling considers this mechanism for performing much better forms of the prediction, which is missing from the existing system and therefore, the error per epoch is evident with better per-formance as compared to existing fuzzy-based approach.

Fig. 3 Comparative analysis of error per Epoch

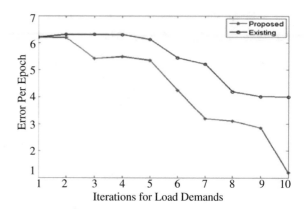

Iterations for Load Demands

8   Conclusion

This paper has presented a cost effective mechanism of identifying the fault in the power transmission line based on smart grid system. Our algorithm design is purely stochastic in nature and emphasizes mainly time factor to perform prediction of the probability. At this stage, we have emphasized on the calculating the probability of capacity and error per epoch as these are the major parameters that affects the outcome in predictive analysis of power distribution system. Compared with the existing system, we find that proposed system offer better predictive analysis as compared to fuzzy-based predictive analysis in power generation system.

References

1. Bush, S.F.: Smart Grid: Communication-Enabled Intelligence for the Electric Power Grid. John Wiley & Sons, Technology & Engineering (2014)
2. Huang, Q., Jing, S., Yi, J., Zhen, W.: Innovative Testing and Measurement Solutions for Smart Grid. John Wiley & Sons, Technology & Engineering (2015)
3. Bazrafshan, M., Gatsis, N.: Decentralized stochastic programming for real and reactive power management. In: IEEE International Conference on Smart Grid Communications in Distribution Systems (2014)
4. Behrens, D., Serafini, M., Arnautov, S.: Scalable error isolation for distributed systems. In: Proceedings of the 12th USENIX Symposium on Networked Systems Design and Implementation (2015)
5. Bracale, A., Barros, J., Cacciapuoti, A.S.: Guest editorial introduction to the special issue on advanced signal processing techniques and telecommunications network infrastructures for smart grid analysis, monitoring, and management. Springer—EURASIP J. Adv. Signal Process. **48** (2015)
6. Jamil, M., Sharma, S.K., Singh, R.: Fault detection and classification in electrical power transmission system using artificial neural network. SpringerPlus **4**(334) (2015)

7. Maaß, H., Cakmak, H.K., Bach, F.: Data processing of high-rate low-voltage distribution grid recordings for smart grid monitoring and analysis. Springer—EURASIP J. Adv. Signal Process. **14** (2015)
8. Babu, P.C., Dash, S.S., Bayindir, R.: A new control strategy with fuzzy logic technique in distribution system for power quality issues. In: IEEE-International Power Electronics and Motion Control Conference and Exposition (2014)
9. Hafez, A.A.: Power generation and distribution system for a more electric aircraft—A review. In: IntechOpen (2012)
10. Das, B.: Fuzzy logic-based fault-type identification in unbalanced radial power distribution system. IEEE Trans. Power Deliv. **21**(1) (2006)
11. Faig, J., Melendez, J., Herraiz, S., Sánchez, J.: Analysis of faults in power distribution systems with distributed generation. In: International Conference on Renewable Energies and Power Quality (2010)
12. Li, Q., Zhou, M.: P.R. China, Research on dependable distributed systems for smart grid. J. Softw. **7**(6) (2012)

A Matlab Program for Analysis of Robust Stability Under Parametric Uncertainty

Radek Matušů and Diego Piñeiro Prego

Abstract The main aim of this contribution is to present a Matlab program for robust stability analysis of families of polynomials with parametric uncertainty. The created software is applicable for basic uncertainty structures such as single parameter uncertainty (including quasi-polynomials), independent (interval) uncertainty structure, affine linear, multilinear, polynomial or general uncertainty structure. Moreover, the discrete-time interval polynomials can be analyzed as well. From the viewpoint of available tools, the program incorporates the Root Locus, the Bialas Eigenvalue Criterion, the Kharitonov Theorem, the Tsypkin-Polyak Theorem, the Edge Theorem and the Value Set Concept combined with the Zero Exclusion Condition. The use of the toolbox is briefly outlined by means of the simple example.

Keywords Robust stability analysis · Parametric uncertainty · Matlab

1 Introduction

Parametric uncertainty is commonly used tool for description of real systems because it allows using natural and relatively easily understandable mathematical models for plants which true behaviour can be much more complicated. The structure of the models with parametric uncertainty is considered to be fixed, but its parameters can lie within given bounds. Frequently, these bounds are assumed in the shape of a box, which practically means that the uncertain parameters are bounded by the intervals. Alternatively, other shapes of bounds such as spheres or diamonds are also available [1, 2].

R. Matušů (✉) · D.P. Prego
Information and Advanced Technologies (CEBIA—Tech),
Faculty of Applied Informatics, Tomas Bata University in Zlín, Nám. T. G.
Masaryka 5555, 760 01 Zlín, Czech Republic
e-mail: rmatusu@fai.utb.cz

© Springer International Publishing Switzerland 2016
R. Silhavy et al. (eds.), *Automation Control Theory Perspectives
in Intelligent Systems*, Advances in Intelligent Systems and Computing 466,
DOI 10.1007/978-3-319-33389-2_19

Since the most critical feature of control systems is their stability, investigation of robust stability under conditions of parametric uncertainty has represented an attractive research goal for decades. The direct calculation of roots of polynomial families could be highly computationally demanding and so the more sophisticated and efficient techniques had to be developed [2, 3].

This contribution is aimed to present a Matlab program, developed under student project of Prego [4] in Faculty of Applied Informatics, Tomas Bata University in Zlín, which allows a user-friendly investigation of robust stability for families of polynomials with parametric uncertainty. The toolbox is usable for all basic uncertainty structures, i.e. single parameter uncertainty (including quasi-polynomials), independent (interval) uncertainty structure, affine linear uncertainty structure, multilinear uncertainty structure, polynomial uncertainty structure and general uncertainty structure. On top of that, the discrete-time interval polynomials can also be analyzed via the program. Altogether, the software offers implementation of the following tools: the Root Locus, the Bialas Eigenvalue Criterion, the Kharitonov Theorem, the Tsypkin-Polyak Theorem, the Edge Theorem and the Value Set Concept combined with the Zero Exclusion Condition. The possible application is briefly outlined through the simple illustrative example.

The program is available for free on demand at the contact e-mail.

2   Robust Stability of Systems with Parametric Uncertainty

The systems with parametric uncertainty can be described though a vector of real uncertain parameters (or just uncertainty) q. This uncertainty enters into the coefficients of the model, for example into the transfer function:

$$G(s,q) = \frac{b(s,q)}{a(s,q)} = \frac{\sum_{i=0}^{m} \beta_i(q)s^i}{\sum_{i=0}^{n} \alpha_i(q)s^i} \tag{1}$$

or into a generally written uncertain polynomial (typically a closed-loop characteristic polynomial):

$$p(s,q) = \sum_{i=0}^{n} \rho_i(q)s^i \tag{2}$$

where β_i, α_i and ρ_i are coefficient functions.

Then, so-called family of polynomials combines together the structure of uncertain polynomial given by (2) with the uncertainty bounding set Q. Such family of polynomials can be denoted as:

$$P = \{p(\cdot, q) : q \in Q\} \tag{3}$$

The uncertainty bounding set Q is supposed as a ball in an appropriate norm. The most frequently used case employs L_∞ norm:

$$\|q\|_\infty = \max_i |q_i| \tag{4}$$

The family of polynomials (3) is robustly stable if and only if $p(s, q)$ is stable for all $q \in Q$. As it had been already stated, the direct calculation of roots is not convenient way due to potentially extreme computational requirements. The choice of suitable method depends mainly on the complexity of the uncertainty structure. It can increase from the independent (interval) uncertainty structure, through the affine linear uncertainty structure, multilinear uncertainty structure, and polynomial uncertainty structure up to the general uncertainty structure. Besides, the special case is represented by the single parameter uncertainty. On top of that, the developed program is capable to analyze the robust stability also for a quasi-polynomial with single parametric uncertainty and for the discrete-time interval polynomials.

The potential tools for robust stability analysis of specific uncertainty structures (i.e. the Root Locus, the Bialas Eigenvalue Criterion [5], the Kharitonov Theorem [6], the Tsypkin-Polyak Theorem [7], the Edge Theorem [8] and the Value Set Concept in combination with the Zero Exclusion Condition [2, 9–11]) are well-known and their description as well as some application possibilities can be found in related literature (e.g. [2, 3, 5–11]).

3 Program Description

The main window of the program is shown in Fig. 1. As can be seen, it consists of three main parts, namely "Main Parameters", "Polynomial", and "Theorems".

The first part, called "Main Parameters", serves for adjustment of basic information such as uncertainty structure and dimension, input form of the family of polynomials, frequency range or step of uncertain parameters. The list of supported uncertainty structures can be seen in Fig. 2 and the list of potential input forms of the polynomial families is shown in Fig. 3. Naturally, only appropriate forms are available for the specific selected uncertainty structure.

The middle part of the main screen, called "Polynomial", is intended for entering information about the polynomial family (i.e. the detailed structure and uncertain parameters). The specific way of putting the data depends on the selected uncertainty structure and selected "Form of the polynomial" (if available).

Finally, the applicable analysis tools can be run via the third part called "Theorems". Obviously, the list of offered methods is also structure dependent. The "Value Set Concept" as the most general tool is available for all possible uncertainty structures. On top of that, the "Root Locus" and "Bialas Eigenvalue

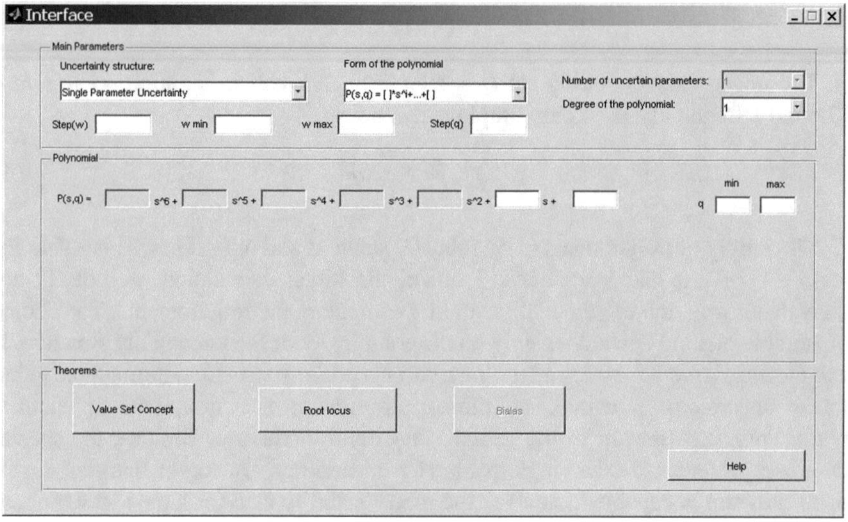

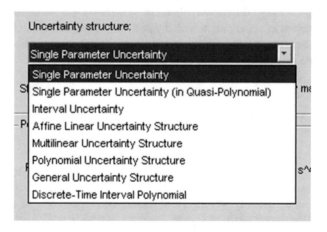

Fig. 1 Main screen of the program

Fig. 2 List of available uncertainty structures

Fig. 3 List of available input forms of polynomial families

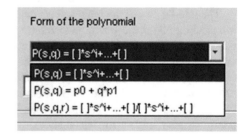

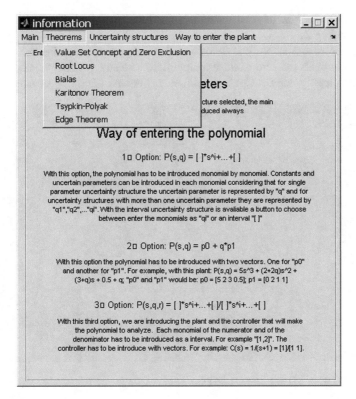

Fig. 4 Example of "Help"

Criterion" are available for "Single Parameter Uncertainty", then "Kharitonov Theorem" and "Tsypkin-Polyak Function" are available for "Interval Uncertainty", and "Edge Theorem" is available for "Affine Linear Uncertainty Structure". Moreover, the simple user's guide can be called by pushing the "Help" button (see the example in Fig. 4).

4   Example of Use

This section is intended to present a simple example of application of the developed software.

Suppose the family of polynomials with single parameter uncertainty (and formally affine linear uncertainty structure) [4]:

$$p(s, q) = 5s^3 + (2 + 2q)s^2 + (3 + q)s + 1.5 + q$$
$$q \in [0, 1]$$

(5)

The aim is to analyze its robust stability. The filled main window for this case can be seen in Fig. 5.

First, the robust stability is investigated by means of the value set concept in combination with the zero exclusion condition. The straight-line value sets plotted by pushing the "Value Set Concept" button are shown in Fig. 6. As can be seen, the

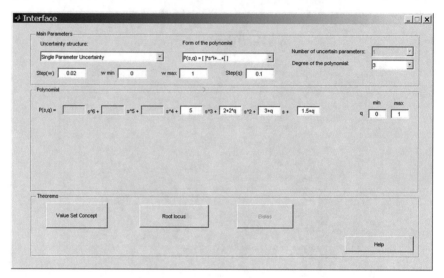

Fig. 5 Filled main window for the example

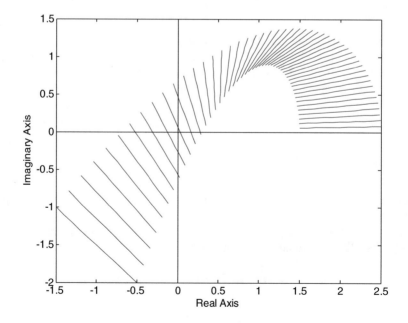

Fig. 6 The value sets of the family (5) for the frequencies from 0 to 1 with step 0.02

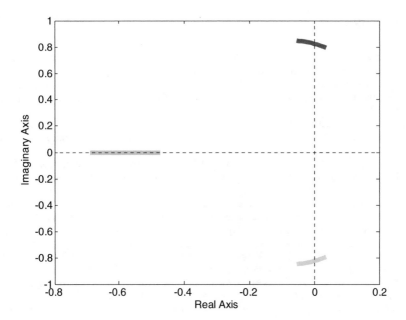

Fig. 7 Root locus for the family (5)

origin of the complex plane (zero point) is included in the value sets and thus the family (5) is not robustly stable [2, 9, 10].

Alternatively, the family can be analyzed by a simple test through "Root Locus" button. Figure 7 depicts the obtained plot. The roots reaching also the right half-plane clearly confirms the fact of robust instability of the tested family (5).

5   Conclusion

This contribution has been focused on brief presentation of the Matlab program for user-friendly testing of robust stability for families of polynomials affected by parametric uncertainty. The toolbox, which was developed under student project of Prego [4] in Faculty of Applied Informatics, Tomas Bata University in Zlín, is applicable for various uncertainty structures since it incorporates a number of analysis tools. The use of the software is shortly illustrated via the simple example. The program is available for free on demand at the contact e-mail.

Acknowledgments The work was supported by the Ministry of Education, Youth and Sports of the Czech Republic within the National Sustainability Programme project No. LO1303 (MSMT-7778/2014). This assistance is very gratefully acknowledged.

References

1. Matušů, R., Prokop, R.: Robust stability analysis for families of spherical polynomials. In: Advances in Intelligent Systems and Computing, vol. 348. In: Proceedings of the 4th Computer Science On-line Conference 2015 (CSOC2015), vol. 2: Intelligent Systems in Cybernetics and Automation Theory, pp. 57–65. Springer International Publishing, Cham (2015)
2. Barmish, B.R.: New Tools for Robustness of Linear Systems. Macmillan, New York (1994)
3. Bhattacharyya, S.P., Chapellat, H., Keel, L.H.: Robust control: The Parametric Approach. Prentice Hall, Englewood Cliffs (1995)
4. Prego, D.P.: A software program for graphical robust stability analysis. Student Project Report, Faculty of Applied Informatics, Tomas Bata University in Zlín, Czech Republic (2015)
5. Bialas, S.: A necessary and sufficient condition for the stability of convex combinations of stable polynomials or matrices. Bull. Polish Acad. Sci. Tech. Sci. **33**, 473–480 (1985)
6. Kharitonov, V.L.: Asymptotic stability of an equilibrium position of a family of systems of linear differential equations. Differentsial'nye Uravneniya **14**, 2086–2088 (1978)
7. Tsypkin, Y.Z., Polyak, B.T.: Frequency domain criteria for l^p-robust stability of continuous linear systems. IEEE Trans. Autom. Control **36**(12), 1464–1469 (1991)
8. Bartlett, A.C., Hollot, C.V., Huang, L.: Root locations of an entire polytope of polynomials: it suffices to check the edges. Math. Control Signals Syst. **1**, 61–71 (1988)
9. Matušů, R., Prokop, R.: Graphical analysis of robust stability for systems with parametric uncertainty: an overview. Trans. Inst. Meas. Control **33**(2), 274–290 (2011)
10. Matušů, R., Prokop, R.: Robust stability analysis for systems with real parametric uncertainty: implementation of graphical tests in Matlab. Int. J. Circuits Syst. Signal Process. **7**(1), 26–33 (2013)
11. Matušů, R.: Robust stability analysis of discrete-time systems with parametric uncertainty: a graphical approach. Int. J. Math. Models Methods Appl. Sci. **8**, 95–102 (2014)

FPGA Based Self-tuning PI Controller Using IFT Technique

Grayson Himunzowa and Farouck Smith

Abstract In this paper the FPGA based self tuning Proportional Integral (PI) controller using Iterative Feedback Tuning (IFT) Technique is proposed. This is accomplished by carrying out an overview of IFT technique and simulation, using Matlab/Simulink platform. The matlab m-file for the Self Tuning PI Controller is used to develop the VHDL code that describes the Self Tuning PI Controller hardware. Fixed point number representation is used in preference to floating point numbers for good performance in respect of reduced power consumption and enhanced speed in program execution. Finite state machine with Datapath (FSMD) model is preferred to Finite state machine plus Datapath (FSM + D) model for ease of implementation. Finally, the Self Tuning PI Controller VHDL code is simulated before implementing it on Altera EP4CE6E22C8 N FPGA.

Keywords Self tuning PI controller · Iterative feedback tuning · FPGA

1 Introduction

The implementation of Self tuning PI controller using IFT technique is an on-going research in an effort to develop a novel hardware that tunes PID controller parameters on-line. The self tuning PI controller using IFT technique was successfully implemented on to the Motorola DSP56F807C microcontroller at University of Cape Town, South Africa in 2006. But computing time in DSPs, especially when dealing with complex algorithms is increased, since software programs are serial in nature and hence causing delays in the loop. Delays in the loop have potential to significantly affect bandwidth of a closed loop control.

G. Himunzowa (✉) · F. Smith
Faculty of Engineering, the Built Environment and Information Technology,
Department of Mechatronics, Nelson Mandela Metropolitan University,
Port Elizabeth, South Africa
e-mail: ghimunzowa@fastmail.fm

© Springer International Publishing Switzerland 2016   203
R. Silhavy et al. (eds.), *Automation Control Theory Perspectives*
in Intelligent Systems, Advances in Intelligent Systems and Computing 466,
DOI 10.1007/978-3-319-33389-2_20

Finally, the quantization inherent in DSPs due to analog to digital converters (ADCs) introduces nonlinearities that can significantly affect performance of digital controllers [1]. These are the main reasons that invoked the design of FPGA based self tuning PI controller.

1.1 Motivation

Self tuning controllers are a dream of industrial control since modeled and external uncertainty problems always deteriorate the control performance of controllers widely used in the industry (such as e.g. proportional-differential-controllers (PID) controller). In the absence of self tuning, the fixed controller parameters can hardly adapt to uncertainty or time varying systems [2]. In addition to this is a particular case of industrial interest in which tuning of PI or PID controller need to be adaptive since classical approaches contain a number of fundamental problems, such as:

- The amount of offline tuning required,
- The assumption on the plant structure,
- The issue of system stability, and
- The difficulties in dealing with nonlinear, large time delay and time variant plant.

Due to the rapid advancement in VSLI technology, the FPGA has become an important resource in the development of specialized digital hardware: with advantages of fast time-to market, shorter design cycle, lower power consumption and higher density for implementing digital control systems. Hence, an FPGA can be used to implement a complex system in a compact, low cost and high performance manner [3, 4].

Research indicates that there is currently no stand alone self tuning PI controller hardware using IFT technique or commercial product developed to the level of the chip even when IFT is proven to be versatile and robust technique as illustrated by a vast number of papers currently published [5–12]. This is one of the motivations we have, to come up with the self tuning PI controller hardware, at the level of an integrated circuit (IC) as this can lead to commercialization of IFT Technique or development of an industrial product. Literature clearly shows the applications of the method since its inception in 1994 by Hjalmarsson [13–18].

To the best of the knowledge of authors, FPGA based self tuning PI controller using IFT technique has never been published in literature to date.

This paper is structured as follows: Sect. 2 introduces an overview of IFT technique's basic theory, and detailed interpretation of the technique's equations; in Sect. 3 the IFT technique is simulated using Matlab/Simulink software; in Sect. 4 matlab m-file is used to develop VHDL code for self tuning PI controller which then is simulated using ISim Xilinx simulator; in Sect. 5 the VHDL code is tested by running it in the FPGA; and in Sect. 6 the conclusion is presented.

2 Iterative Feedback Tuning Technique Overview

Since self tuning PI controller is based on IFT technique, an overview of its basic theory is paramount and is considered in this section.

The IFT methodology can be outlined as follows: the controller structure is given in advance, thereafter the performance specification indices in terms of a Linear Quadratic Gaussian (LQG) criterion are pointed out, after which closed loop experiments are run to evaluate the gradient of the performance indices by summation of the generated signals from closed loop experiments and finally, minimization of the performance criterion is carried out by use of Robbins and Monro stochastic approximation algorithm. A minimum of the criterion LQG can be searched by either of the gradient based local minimization techniques such as steepest descent, or Gauss-Newton method.

In this section we formulate a deterministic version of Iterative Feedback Tuning for the case of simple Proportional Integral (PI) controller, so that investigations of some issues relevant to practical implementation are simplified. The PI controller is chosen because it is a very common control law and hence a reasonable starting point, and the one to consider because of its importance to industrial application [19].

We investigate a single input, single output control system of one-degree-of-freedom (1DOF) control law as shown in Fig. 1.

r(t), e(t), d(t), u(t), v(t) and y(t) represent reference signals, error, disturbance, control signal, output noise, and controller output signal respectively. G(s) is a linear time invariant plant (LTI) and C(s) is the PI controller. The internal closed loop signals can be described via sensitivity functions (considering d(t) = 0) as follows (omitting the variable s for notation simplicity):

$$u(t) = C * S * r(t) - C * S * v(t) \tag{1}$$

$$y(t) = G * C * S * r(t) + S * v(t) \tag{2}$$

where

$$S = \frac{1}{1 + G * C} \tag{3}$$

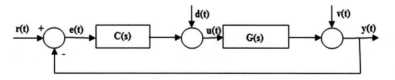

Fig. 1 Closed loop system

and

$$G * C * S = T = \frac{G * C}{1 + G * C} \tag{4}$$

so that Eq. (2) can be rewritten as

$$y(t) = T * r(t) + S * v(t) \tag{5}$$

We will consider here v(t) as zero mean weakly stationary random noise.

The IFT procedure for 1DOF controller operates using two experiments. The first experiment measures and stores, the signals, e(t), u(t) and y(t) for N length of time and the second experiment optimizes the PI controller (as explained in Sect. 2.1) using the signals obtained in the first experiment.

2.1 Controller Optimization

Controller optimization require that a control objective is defined, usually by expressing it in terms of a criterion function, J and given as

$$J = \frac{1}{2N} \sum_{k=0}^{N} [(L_e * e(\partial))^2 + \lambda * (L_u * u(\partial))^2] \tag{6}$$

where, rho (ρ) is the controller parameter vector. The first term in J is the closed loop tracking error, and the second term is the penalty on the control effort [20]. Le and Lu in the first and second term are filters. Minimization of criterion function, J, requires the calculation of the gradient of the criterion function J with respect to the controller parameters ρ from (6) and the gradient of J is given by

$$\frac{\partial J}{\partial \rho}(\rho) = \frac{1}{N} \sum_{k=0}^{N} [e(\rho) \frac{\partial e}{\partial \rho}(\rho) + \lambda * u(\rho) \frac{\partial u}{\partial \rho}(\rho) \tag{7}$$

where, for simplicity reasons we assume

$$L_e = L_u = 1$$

With this criterion function, the desired model(that resemble the process) is a scalar (desired model, M(z) = 1), implying that the cost function that optimises the controller parameters is the result of the summation of the absolute error obtained from the comparison of the reference signal, r(t) and the plant response, y(t). The advantage of this criterion function is that it presents a good balance between overshoot and settling time [21] and it simplifies the hardware design since the desired model is reduced to unit and also the task of making an assumption for the plant structure becomes insignificant since the desired model is reduced to a scalar. However, the main drawback of this criterion function, as compared to other minimisation criterion functions is the difficulty of obtaining an analytical solution to the controller design [21].

If the gradient $\frac{\partial J}{\partial \rho}$ could be computed, then the solution of (7) would be obtained by stochastic approximation of Robins and Monro of 1951, given as

$$\rho_{i+1} = \rho_i - \gamma_i * R_i^{-1} \frac{\partial J}{\partial \rho}(\rho_i) \tag{8}$$

Here γ_i is a positive real scalar that determines the step size and R_i is some appropriate positive definite matrix. R_i determines the amplitude of the steps or the step sizes in the direction of each parameter provided by either steepest descent or Gauss Newton method. The choice of R_i as an identity matrix renders a steepest descent gradient which is normally negative and also slow to converge. The choice of R_i generated by the Gauss Newton method (as shown in (9)) yields quick convergence [21, 22].

$$R_i = \frac{1}{N} \sum_{k=0}^{N} \left(\left[\frac{\partial e}{\partial \rho}(\rho)\right] * \left[\frac{\partial e}{\partial \rho}(\rho)\right]^T + \lambda * \left[\frac{\partial u}{\partial \rho}(\rho)\right] * \left[\frac{\partial u}{\partial \rho}(\rho)\right]^T \right) \tag{9}$$

All the signals required for optimization of the criterion function are generated by the IFT experiment 1 and 2.

2.2 Calculation of the Gradient

Noting that, $\frac{\partial e}{\partial \rho}(\rho) = \frac{\partial y}{\partial \rho}(\rho)$ since it is the plant that causes variation in error, e(t). Hence,

$$\frac{\partial y}{\partial \rho}(\rho) = \frac{(1 + G * C) * G * \frac{\partial C}{\partial \rho} * r - G * \frac{\partial C}{\partial \rho} * r * G * C}{(1 + G * C)^2}$$

$$= \frac{1}{C} * \frac{\partial C}{\partial \rho} * T * S * r \tag{10}$$

Therefore, the gradient is given as

$$grad = \frac{1}{C} * \frac{\partial C}{\partial \rho} \tag{11}$$

Equation (11) is also known as a filter. This filter can be used to generate signals given as

$$\frac{\partial e}{\partial \rho}(\rho) = -grad * T(\rho) * e(\rho) \tag{12}$$

and

$$\frac{\partial u}{\partial \rho}(\rho) = grad * S(\rho) * e(\rho) \tag{13}$$

2.3 PI Controller

The PI controller is given by

$$C(z) = \frac{\rho_1 z + \rho_0}{z - 1} \tag{14}$$

Hence, the expression in (11) is a vector of filters as shown in (15)

$$grad = \frac{1}{C} * \frac{\partial u}{\partial \rho} = \begin{bmatrix} \frac{1}{\rho_1 * z + \rho_0} \\ \frac{z}{\rho_1 * z + \rho_0} \end{bmatrix} \tag{15}$$

2.4 Self Tuning PI Controller Architecture

Having studied the IFT technique basic theory, the self tuning PI controller architecture is formulated in this subsection. The block diagram illustrating the architecture of the self tuning PI controller for imbedding into the FPGA is given in Fig. 2.

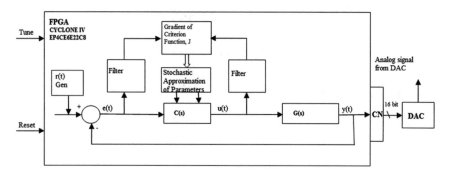

Fig. 2 Self tuning PI controller architecture

The r(t) generator generates a series of step signals to represent the reference signal, r(t); C(s) is a PI controller; G(s) is a plant or a process; the filter is the gradient, $\frac{1}{C} * \frac{\partial C}{\partial \rho}$ used to generate the derivative of the error signal, e(t) and control signal, u(t) in order to calculate the gradient $(\frac{\partial J}{\partial \rho})$ of the criterion function, J. After calculation of the gradient of the criterion function J, stochastic approximation of Robins and Monro of 1951 is applied to update or tune the controller parameters. External signals (tune and reset in Fig. 2) serve the purpose of tuning the controller and resetting the hardware respectively.

3 Simulation of IFT Technique Using Matlab/Simulink Platform

Following an overview of the IFT technique basic theory carried out and also the formulation of the self tuning PI controller hardware architecture in Sect. 2, simulation on matlab/simlink platform is considered in this section. This is done in order to validate the IFT's viability of realising it on FPGA. The matlab code is developed from the self tuning PI controller architecture formulated in Sect. 2.

The plant in simulation is expressed in the Laplace variable domain as shown below:

$$\frac{1.01}{2s+1} \tag{16}$$

The results of the simulation for the output response of a linear plant shown in Eq. (16) are depicted in Figs. 3, 4, and 5. Figure 3 illustrates initial response, while Fig. 4 illustrates medium response and Fig. 5 illustrates the final response of the IFT technique.

Fig. 3 Initial IFT response

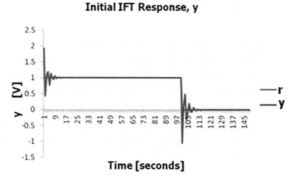

Fig. 4 Medium IFT response

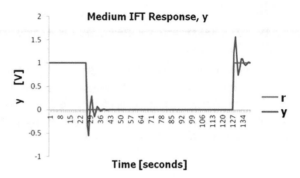

Fig. 5 Final IFT response

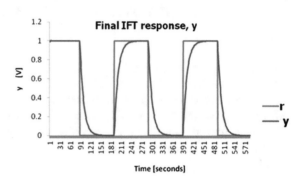

Notice that the initial response has an overshoot of approximately 1 V and as the parameters begin to update the amplitude of the initial overshoots reduces to 0.5 V and as tuning progresses overshoots are eliminated completely and convergence is achieved at 0.02 for rho0 and rho1. The results of simulation compare well with the results in [23].

4   Development of VHDL Code for Self Tuning PI Controller

Following the simulation of self tuning PI controller on Matlab/Simlink platform, the VHDL code is developed and simulated in this section and the results are illustrated in Fig. 6. Matlab m-file used for simulation of self tuning PI controller was ported to a VHDL platform and a behavioural model VHDL code that describes self tuning PI controller hardware was developed with slight modification of matlab m-file. This modelling is also known as finite state machine with datapath (FSMD) and was preferred to finite state machine plus datapath (FSM + D) for ease of implementation.

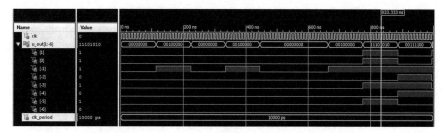

Fig. 6 Simulation results for the IFT VHDL code

The IFT VHDL code is composed of two processes: a process for clock division and a process for the PI controller and the IFT technique. The IFT technique tunes the parameters of the PI controller as stated already in Sect. 2.

In the development of self tuning PI controller VHDL code, the matlab m-file is used. We demonstrate here how the VHDL code for the PI controller is formulated by using a chunk of the code for the PI controller from the m-file. To transform the m-file into the VHDL code, a word length of 16 bit was chosen simply because word length of 16 bit or higher is more accurate than word length below 16 bit. Fixed point number representation is denoted by $Q(m, -n)$ format. Where Q is a fixed point number variable, m is an integer part of a fixed point number, and n is a fraction part of a fixed point number. For 16 bit word length and a 1 V maximum for input/output signal, the format is given as $Q(1, -14)$, meaning that the fixed point number has 2 bits for integer part and 14 bits for the fractional part. Hence, using the specified Q format, the m-file expressions are expressed in terms of fixed point representations then the modified instruction is converted to VHDL expression. An example is demonstrated here: Taking the instruction, $e = r - y$ from the m-file, we write it in form of fixed point number representation as $e(2, -14) = r(1, -14) - y(1, -14)$. Notice that, the result of subtraction register is higher than 1 bit than the operand registers. This is also known as an overflow and since in this design, the reference signal varies between 0 V and 1 V, meaning that, the result is upper bounded by 1 V and if an overflow occurs to raise the voltage above 1 V then the most significant bit (msb) is dropped without any loss of accuracy. This is carried out by a way of scaling or shifting. Hence, register $e(2, -14)$ is shifted to the left by 1 bit yielding $e(1, -15)$ which can further be truncated to $e(1, -14)$ to maintain the 16 bit format. The expression is now converted to VHDL language as follows:

```
Variable e: ufixed(2 downto -14); Variable r: ufixed(1 downto
-14); Variable y: ufixed(1 downto -14);
e := r - y;
```

Taking the next instruction, $u = rho1 * e + rho0 * e + u(-1)$ from the m-file we write it in form of fixed point number representation as

```
u(4, -29) = rho1(1, -14) * e(1, -15) + rho0(1, -14) * e(1,
-15) + u - 1(1, -14).
```

Notice that the result of the control register, u has 4 bits for the integer part and 29 bits for the fractional part. This is scaled by shifting 3 times to the left yielding u $(1, -32)$. Finally, the process plant expression, y = gain * y + u is written in the form of fixed number representation as $y(2, -30) = gain(1, -14) * y(1, -14) + u(1, -14)$. Notice the result of the output, y has 2 bits for the integer part and 30 bits for the fractional part and need to shifted once to the left to obtain $y(1, -31)$. This is further resized or truncated to allow it fit on the output port, u_out which is 16 bit. Then the modified expressions are converted to VHDL language as follows:

```
Variable e: ufixed(2 downto -14); Variable u: ufixed(4 downto -
29); Variable rho1: ufixed(1 downto -14);
Variable rho0: ufixed(1 downto -14); Variable gain: ufixed(1
downto -14); Variable y: ufixed(1 downto -14);
u: = rho1 * e + rho0 * e + u; y: = gain * y + u;
```

A skeleton of Self Tuning PI VHDL code is indicated below:

```
Entity s_pi_controller is
  Port( clk : in std_logic;
    u_out : ufixed(1 downto -14));
End s_pi_controller;
Architecture behavioral of s_pi_controller is
Begin
Process(clk)
begin
If rising_edge(clk) then
--clock divider code here
Process(toggle, reset)
e = r - y; u = [rho1*e] sll 2 + [rho0*e] sll 2 + re-
size(u, 1, -14);
y = [gain*y] sll 1 + u;
      Optimisation code needs to be inserted here;
end process;
end behavioural;
```

Behavioral modelling was applied in this research since the IFT technique is sequential in operation, meaning that experiment 1 runs before experiment 2 begins.

Speed of execution, power consumption, and optimised hardware utilisation are not investigated in this research but the table showing the usage of resource is illustrated in Table 1.

Table 1 Device utilisation summary

	Used	Available	Utilization (%)	Comment
Number of slice flip flops	130	9,312	1	Ok
Number of 4 input LUTs	4,603	9,312	49	Ok
Number of occupied slices	2,430	4,656	52	Ok

5 Results of Self Tuning PI Controller Using IFT Technique and Imbedded into the Altera EP4CE6E22C8 FPGA

The VHDL code developed in Sect. 4, was tested in the Altera, EP4CE6E 22C8 FPGA and a 16 bit output, u_out(u_out is the output port of the hardware implemented on the FPGA) of the FPGA was fed via a 16 bit digital to analogue converter (DAC) to a National Instruments USB-6009 multifunction interface device in order to display the output signal, y. The NI USB-6009 device, analogue input, interfaces with labview driver (DAQ Assistant). The DAQ Assistant processes the analogue signal from the NI USB-6009 in order to display it on labview graph. The VHDL code that describes self tuning PI controller hardware was imbedded into the FPGA. The results obtained after running the code imbedded into the FPGA were compared to those of simulation with matlab.

The obtained results are depicted in Fig. 7 showing initial IFT response, Fig. 8 showing medium IFT response, Fig. 9 showing final IFT response. These results demonstrate asymmetrical response, mainly because of feedbacks present in the developed hardware. The problem can be resolved by driving the plant externally and not internally as is the case in this research.

Results obtained from running the self tuning PI controller VHDL code loaded on to the FPGA does show some similarities but suffer from bias and asymmetrical errors. This is attributed to fixed point implementation problems such as coefficient quantization, overflows, and truncation/rounding in arithmetic operations. This is

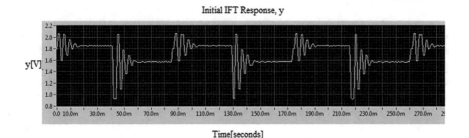

Fig. 7 Initial IFT response-simulation in the FPGA

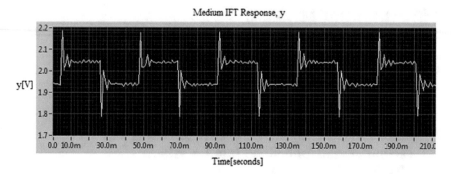

Fig. 8 Initial IFT response-simulation in the FPGA

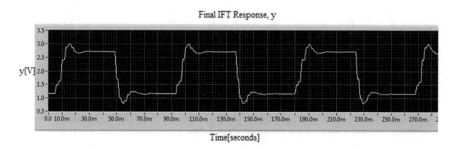

Fig. 9 Initial IFT response-simulation in the FPGA

one disadvantage with fixed point number representation as compared to floating point number representation. But due to tight speed requirements (that can only be achieved by fixed point numbers) in hardware design, this will still be a relevant problem for research [24].

6 Conclusion

In this paper the self tuning PI controller is implemented into the FPGA by describing it using VHDL language. Fixed point number representation is opted for good performance and FSMD modelling is applied for ease of implementation. This has laid foundation for future implementation of many other forms of self tuning PI controller on a novel hardware platform.

References

1. Monmasson, E., Idkhajine, L., Naou, M.W.: FPGA-based controllers. IEEE Ind. Electron. Mag. (2011)
2. Kung, Y.S., Wang, M.S., Chuang, T.Y.: FPGA-based self-tuning PID controller using RFB neural network and its application in X-Y table. In: IEEE International Symposium on Industrial Electronics, 5–8 July 2009
3. Monmasson, E., Cirstea, M.N.: FPGA design methodology for industrial control systems—a review. IEEE Trans. Ind. Electron. **54**(4), 1824–1842 (2007)
4. Naouar, M.W., Monmasson, E., Naassani, A.A., Slama-Belkhodja, I., Patin, N.: FPGA-based current controllers for AC machine drives—a review. IEEE Trans. Ind. Electron. **54**(4), 1907–1925 (2007)
5. Hjalmarsson, H.: Iterative feedback tuning—an overview. Adapt. Control Signal Process. **16**(5), 373–395 (2002)
6. Huusom, J.K., Poulsen, N.K., Jørgensen, S.B.: Improving convergence of iterative feedback tuning using optimal external perturbations. In: IEEE Conference on Decision and Control, Cancun, Mexico, no. WeB04.3, pp. 2618–2623 (2008)
7. De Bruyne, F., Kammer, L.C.: Iterative feedback tuning with guaranteed stability. In: Proceedings of the American Control Conference, San Diego, California, June 1999, pp. 7–11 (1999)
8. Hjalmarsson, O.L.H., Gevers, M., Gunnarsson, S.: Iterative feedback tuning: theory and applications. IEEE Control Syst. Digit. Stock. **1996**, 26–41 (1998)
9. Proch, H., Gevers, M., Anderson, B.D.O., Ferrera, C.: Iterative feedback tuning for robust controller design and optimization. In: Proceedings of the 44th IEEE Conference on Decision and Control, and the European Control Conference, TuC04.5, 12–15 Dec 2005 pp. 3602–3607 (2005)
10. Bitmeadt, R.: Iterative Feedback tuning via minimization of the absolute error. In: Proceedings of the 38th IEEE Conference on Decision and Control (1999)
11. Sharifi, A.: The design of frequency filters of iterative feedback tuning using particle swarm optimization. Adv. Electr. Eng. **2014**, 1–5 (2014)
12. Hjalmarsson, H., Gunnarsson, S., Gevers, M.: A convergent iterative restricted complexity control design scheme. In: Proceedings of the 44th IEEE Conference on Decision and Control, December 1994, pp. 1735–1740 (1994)
13. Weihong, W., Zhongsheng, H., Shangtai, J.: Overview of iterative feedback tuning. In: Proceedings of the 26th Chinese Control Conference, 26–31 July 2007
14. Radac, M.B.: Iterative techniques for controller tuning. PhD thesis (2011)
15. Hjalmarsson, H., Gunnarsson, S., Gevers, M.: A convergent iterative restricted complexity control design scheme. In: Proceedings of 33rd IEEE Conference on Decision and Control (CDC), vol. 2, pp. 1735–1740, Orlando, FL, USA (1994)
16. Hjalmarsson, H.: Control of nonlinear systems using Iterative Feedback Tuning. In: Proceedings of the American Control Conference, vol. 4, pp. 2083–2087, Philadelphia, PA, USA (1998)
17. Hjalmarsson, H., Gevers, M., Gunnarsson, S., Lequin, O.: Iterative feedback tuning: theory and applications. IEEE Control Syst. Mag. **18**, 26–41 (1998)
18. Weihong, W., Zhongsheng, H., Shangtai, J.: Overview of the iterative feedback tuning. In: 2007 Chinese Control Conference, no. 2, July 2006, pp. 14–18 (2006)
19. Hjalmarsson, O.L.H., Gevers, M., Gunnarsson, S.: Iterative feedback tuning: theory and applications. IEEE Control Syst. Digit. Stock. **1996**, 26–41 (1998)
20. Bitmeadt, R.R.: Iterative feedback tuning via minimization of the absolute error. In: Proceedings of the 38th IEEE Conference on Decision and Control (1999)

21. Hjalmarsson, H., Gunnarsson, S., Gevers, M.: A convergent iterative restricted complexity control design scheme. In: Proceedings of the 44th IEEE Conference on Decision and Control, Dec 1994, pp. 1735–1740 (1994)
22. Hjalmarsson, H.: Iterative feedback tuning—an overview. Adapt. Control Signal Process. **16** (5), 373–395 (2002)
23. Himunzowa, G.: Investigations into implementation of IFT technique into microcontroller. Thesis for Degree of Masters, University of Cape Town (2008)
24. Aoki, T.: Implementation of fixed-point control techniques based on the modified delta operator and form for intelligent systems. J. Adv. Comput. Intell. **11**(6), 5–6 (2007)

Design and Implementation of an Integrated System with Secure Encrypted Data Transmission

Adam Hanacek and Martin Sysel

Abstract The aim of this work is to describe the developed system which provides broad integration and better protection against a sabotage than commonly used systems. The communication bus which is used for message transmission is named Controller Area Network. The security of the CAN bus communication is ensured by the implementation of 3DES encryption method with regular changing of the cipher key. One of the main parts describes a designed database, a method of key assignment, a data encryption, a method of device addressing and a collision protection. Finally, the protocol testing is depicted.

Keywords Protocol · 3DES · Addressing · Collision · Encryption

1 Introduction

This article solve many problems in the three main areas—intrusion and hold-up alarm system, fire alarm system and systems that collect data from secured object. The detectors in the first area are divided into analog, digital or wireless. Unfortunately, none of these transmission methods provide sufficient protection against sabotage. The main problem lies in European standards that should be changed and efforts of manufacturers to keep financial development costs as low as possible. As a result, people worldwide use the systems that can be overpowered and the ways of overcoming depends on the selected type of transmission path

A. Hanacek (✉) · M. Sysel
Faculty of Applied Informatics, Tomas Bata University, Nad Stranemi 4511,
760 05 Zlin, Czech Republic
e-mail: ahanacek@fai.utb.cz; hanacekadam@seznam.cz
URL: http://www.fai.utb.cz

M. Sysel
e-mail: sysel@fai.utb.cz

© Springer International Publishing Switzerland 2016
R. Silhavy et al. (eds.), *Automation Control Theory Perspectives
in Intelligent Systems*, Advances in Intelligent Systems and Computing 466,
DOI 10.1007/978-3-319-33389-2_21

217

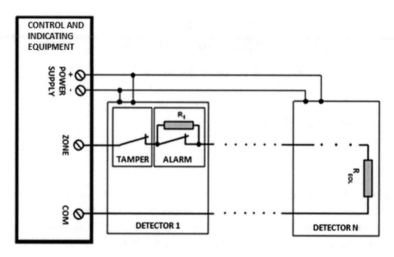

Fig. 1 Connecting of analog detectors to the I&HAS [1]

between a central and detectors. Feasible ways how to overcome security systems are narrowly illustrated in [1, 2]. Evolved system is primarily designed to connect analog detectors and its purpose lies not only in accomplishing the European standard EN 50131 but it is designed also in order to offer better protection against sabotage by digitization with 3DES encryption and by regular changing of cipher key. Almost 90 % of current developers from the field of intrusion alarm system do not encrypt the communication and none of them change cipher key during a communication. The connection of analog detectors to standard intrusion alarm system is depicted in the Fig. 1. *The principle of analog I&HAS consists in measuring the resistance of each circuit. Each detector has one alarm and one tamper contact connected to the circuit* [1].

The protocol is also designed to implement analog devices from the field of fire alarm system and systems that collect data from object. The advantage lies in the use of a secured encrypted communication. In addition, the current systems require a control unit for each type of area, but the developed system was designed to accomplish the standards from all areas; therefore, it is possible to use one control unit and save money. This paper is separated into four main parts called "Entire View on The Evolved System", "Encryption, Addressing and Collision Protection and Method of Assignation" and the last part is named "Testing and Price Comparison". The communication between connected devices is ensured by the bus named Controller Area Network. This bus grants up to join 110 devices without any repeater and allows to utilize the message priority. The CAN bus was basically designed for usage in automotive industry, but currently it is often implemented in a lots of development kits. CAN is based on the CSMA/CR mechanism to prevent frame collisions during transmission messages on the CAN bus [3, 4].

It is a high-integrity serial data communication bus for real-time applications and more cost effective than any other serial bus systems. Other advantages are written below [5, 6].

CAN protocol supports four types of messages [7].

- Data frame
- Remote frame
- Error frame
- Overload frame

The thorough description of layers, message frames and communication speed is explained in [8, 9]. Two wires used for transmission a message across the CAN bus are named CAN_L and CAN_H. The described bus has two states named dominant and recessive [10–12].

2 Entire View of the Evolved System

The purpose of this section is to provide general view of the system which can be seen in the Fig. 1. The main parts of this section are composed of the main control unit, of the CAN subnetwork and of the terminated equipments. Mentioned control unit is represented by Ixora board with the processor Apalis T30 clocked at 1,4 Ghz. Moreover, Angstrom Linux was installed into the Ixora board. Supported buses are PCIe, USB, I2C, SPI, UART, Ethernet, Serial ata, CAN, HDMI, parallel camera interface and serial camera interface. The firmware in the control unit was evolved by Eclipse environment and it was divided into two parts. The first part guarantees a message assembling in the correct order and the second part is used to assign connected devices and for monitoring of the whole system. Further, the data was sent via the CAN bus with the speed set to 250 Kb/s; therefore, the max length is set to 250 m. It is evident that connecting to the network requires another PCB with CAN bus as a communication interface between a subnetwork and a device. The processor lpc11c24 was chosen for this purpose because of the very low cost and the high frequency clocked at 50 MHz. The disadvantage of lpc11c24 lies in a small 32kB flash memory; however, mentioned size is sufficient for this purpose. The PCB is recommended to be placed inside of each element. The system affords to join any analog devices such as an analog gas sensor, a temperature sensor, smoke sensor, a PIR detector, an ultrasonic detector, a microwave detector and other analog input devices. Moreover, it was possible to implement security keyboards for arming of assigned area and the protocol also allows to change the state of output devices in case of alarm message declaration in the CAN bus (Fig. 2).

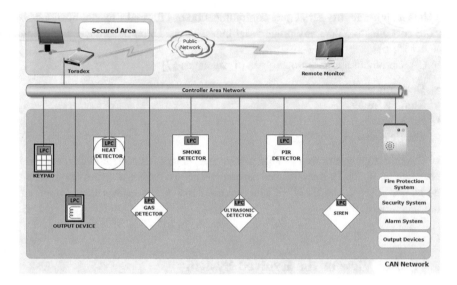

Fig. 2 Entire view on the evolved system

2.1 Created Database to Store the History and Device Setting

Postgres database was implemented in the Ixora board to store the history and devices setting; however, another benefit lies in the fact that the system can be monitored remotely by ethernet connection with the public IP address. The database is composed of the following tables:

- Serial_Number
- Devices
- Sector_Assignment
- Sectors
- Inputs
- Device_Type

The table named Serial_Number contains information of the serial number which is obtained from the processor lpc11c24; therefore, the number serves not only for processor identification but also for basic device identification. The name of the most important table is Devices and provides information of device network id which is used for network identification, information of the device type that specifies its purpose and detailed specification of this device. Other two tables called Sectors and Sector_Assignment represent type M:N. The purpose of this tables consists in the separation of the system into 256 sectors. Finally, the table Inputs affords information of input history. The Fig. 3 shows each part of the database and its relationships.

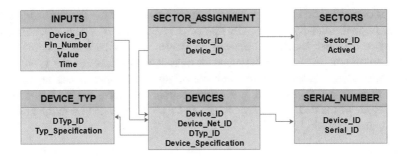

Fig. 3 Created database

3 Encryption, Addressing and Collision Protection and Method of Assignation

One of the most important parts is named network assignment which includes following steps:

- Connecting of all devices to the CAN bus
- Sending a request for SID for all devices
- Saving to the database
- Network assignment
- Cipher key sending and saving to a flash memory

It is recommended to carry out all this steps in the secured area before placing the PCB at the final position. The purpose of this step is to prevent sending the key on the whole CAN bus. The next Fig. 4 illustrates the division of a CAN frame in the system. Extended frame contains 29 bits identifier and 64 bits data field. The identifier is divided into 11b and 18b by the protocol. The first 10 bits of identifier are used to part assignment and next 18 bits are used to device identification in the network. The first 8 bits of the data field contains information of the command and the last 56 bits of data field contains the data of the message. The reason of placing the command into the data field lies in the protection against commands reading by an intruder. The entire data field is encrypted by 3DES and the cipher key is changed every frame which is sent to the CAN bus. The length of cipher key is set to 192 bits.

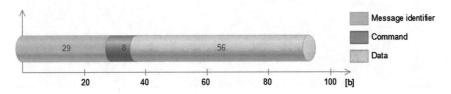

Fig. 4 Message frame

The communication between components is provided by the protocol addressing that is implemented by the last 18 bits of message identifier and represents the address of a sender and the address of a receiver. The value 0 is assigned to the main control unit, the value 510 represents unassigned devices and the value 511 is used for sending a message to all connected elements. Further, it was necessary to avoid a message collision during communication; therefore, the answer delay of each device was added. This delay consists of a hash generated by unique serial id of each processor.

4   Hardware Design

In the Fig. 5 there is illustrated the hardware developed as a terminated PCB for placing into the detectors. The name of the processor used in the PCB is LPC11C24. This processor was chosen for the high speed (50 MHz) and for the

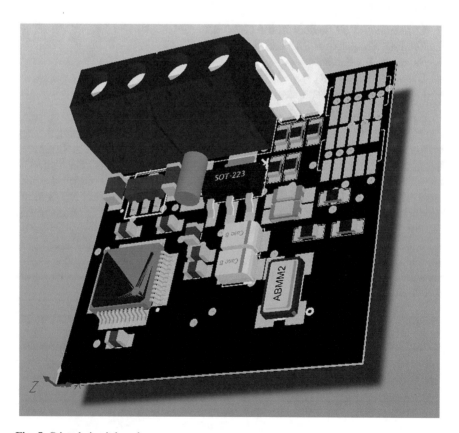

Fig. 5 Printed circuit board

low price. It was necessary to implement two voltage regulators. First regulator is named "MCP1702T-5002E/MB" and provides 5 V output voltage. The name of the second regulator is "AP1117E18G-13 IC" and its purpose lies in supplying the processor by 3 V. Further, the PCB also includes a connector for J-Tag, I2C, SPI, two I/O pins, one A/D converter and two led diodes.

5 Testing and Price Comparison

The testing of the evolved system was realized by 14 kits with the processor lpc11c24, 14 detectors, one keyboard and one Toradex kit. The system was armed by the keyboard connected to the CAN BUS via one of the kits. While an alarm was induced by a detector, the alarm message is transmitted via CAN bus and the Toradex kit sends output activation requirement via CAN bus. Finally, the development kits from the same sector activated their outputs. The usage of the system in practice requires inserting of the PCB into the interior area of the detector or preventing an access to the processor by the box with sabotage protection. The system allows future extension like connecting of cameras to the main control unit, an integration with intelligent building or a substitution of the Toradex kit by cheaper main control unit connected to the CAN bus. Then, the Toradex kit just keep system settings, provides a remote access and saves an alarm messages to the database. In addition, mentioned control unit may create subnetworks what causes separating the CAN bus and extending the number of connected devices without repeaters. From the view of price, the system is cheaper then digital systems currently in use because of low processor price but more expensive than analog systems; however, analog system cannot be recommended for the possibility to be overcome.

6 Conclusion

Illustrated system was developed in order to ensure higher protection against sabotage than commonly used digital systems, to provide high integration, high number of connected devices and in order to lower the price of digital integrated systems. The system was designed to accomplish the standards from the field of an intrusion and hold-up alarm system, fire alarm system and from the field of access control system. Moreover, the protocol corresponds with the highest security level of the intrusion European standard EN 50-131. In addition, another advantage lies in data encryption by 3DES and changing the cipher key every frame. The protocol was tested in laboratory environment of Tomas Bata University. Future extension of the presented protocol includes a connection of cameras to the Toradex kit, the implementation of AES encryption and changing not just a cipher key but also an encryption methods during the communication. Further, the extension also includes development of a hardware for placement inside detectors and the implementation of a control units for placing into the CAN bus.

References

1. Hanacek, A., Sysel, M.: the methods of testing and possibility to overcome the protection against sabotage of analog intrusion alarm systems. Intell. Syst. Cybern. Autom. Theory 119–128 (2015)
2. Hanacek, A., Sysel, M.: Universal system developed for usage in analog intrusion alarm systems. Intell. Syst. Cybern. Autom. Theory 129–138 (2015)
3. Xiao, H., Lu C.: Modeling and simulation analysis of CAN-bus on bus body. In: 2010 International Conference on Computer Application and System Modeling (ICCASM), vol. 12, pp. V12-205, V12-208, 22–24 Oct 2010. doi:10.1109/ICCASM.2010.5622238
4. Jiang, Y., Liang, B.R., Ren, R.: Design and implementation of CAN-bus experimental system. In: 2011 6th International Forum on Strategic Technology (IFOST), vol. 2, pp. 655, 659, 22–24 Aug 2011 doi:10.1109/IFOST.2011.6021111
5. VOSS, Wilfried: A Comprehensible Guide to Controller Area Network. 2. vyd. Canada: Copperhill Media Corporation (2005). ISBN 0976511606
6. Li, R., Wu, J., Wang, H., Li, G.: Design method of CAN BUS network communication structure for electric vehicle. In: 2010 International Forum on Strategic Technology (IFOST), pp. 326, 329, 13–15 Oct 2010
7. Vehicle Test Data Visualization and Processing. Prague, 2011. Diploma thesis. CZECH TECHNICAL UNIVERSITY IN PRAGUE
8. SPURN, Frantiek. Controller Area Network. Automatizace. 41, 7, s. 397–400 (1998)
9. Device for transmitting messages on CANbus. Plzen, 2012. Diploma thesis. University of West Bohemia
10. Implementation of Controller Area Network (CAN) Bus. Springer, Berlin, Heidelberg (2011). ISBN 978-3-642-18439-0
11. Understanding and Using the Controller Area Network Communication Protocol. Springer Science Business Media, London (2012). ISBN 978-1-4614-0313-5
12. Measurement system with CAN interface. Plzen, 2013. Diploma thesis. University of West Bohemia

Web Application for Simple System Identification from Experimental Data

Frantisek Gazdos and Petr Micola

Abstract This paper presents a project for simple web-based system identification from experimental data. A user can upload input/output data from a process to be identified in three common file formats, the data can be filtered simply using two approaches and a structure of the identified model is user-controlled. The resultant discrete-time ARX model is obtained in a common form together with the possibility to assess quality of identification and to obtain also a continuous-time model. The developed application is built on the interconnection of the latest web technologies with the MATLAB computing system where the obtained results can be easily exported. The paper explains motivation for development of this site and gives also detailed description of the whole process including the Web—MATLAB interconnection. The results are presented using selected screen-shots of the application and discussed further.

Keywords Web-application · MATLAB · System identification

1 Introduction

As the capabilities of web-applications improve together with the speed, quality and accessibility of the Internet connection there is a growing trend of software products and applications being provided on-line, using standard web browsers rather than installing on workstations. These include e.g. various office tools and organizers, maps, media players, translators, programming frameworks, photoeditors, etc. Complex computing systems such as The MathWorks's MATLAB [1, 2] and Wolfram Research's Mathematica [3, 4] are also following this trend, as seen in their cloud solutions MATLAB Online [5] and Wolfram Mathematica Online [6]. However these powerful tools are paid services, require a registered user with a

F. Gazdos (✉) · P. Micola
Faculty of Applied Informatics, Tomas Bata University in Zlin,
Nam. T.G. Masaryka 5555, 760 01 Zlin, Czech Republic
e-mail: gazdos@fai.utb.cz

© Springer International Publishing Switzerland 2016   225
R. Silhavy et al. (eds.), *Automation Control Theory Perspectives
in Intelligent Systems*, Advances in Intelligent Systems and Computing 466,
DOI 10.1007/978-3-319-33389-2_22

certain valid license and are partly limited compared to their workstation versions. Consequently there are many free applications on-line enabling different simple or even more complex calculations useful not only for students of various levels, but also for teaching staff and even researchers, e.g. interactive Math Apps and calculators from the Möbius project of the Maplesoft [7]. For control engineering students, teachers and researchers, within the scope of the project [8], a free web-based application for controller design and tuning from experimental data had been developed [9], followed by an application for easy linear time-invariant systems analysis [10]. Both applications were developed to support pedagogical activities at the Faculty of Applied Informatics, Tomas Bata University of Zlin but due to the web-based solution they can be used freely by broader community of students, teachers and researchers. This paper presents another similar web-application developed at our faculty focused on other aspects of control system design, namely the process identification. It is also built on the interconnection of the MATLAB system with latest web technologies and it enables to identify a process simply using its input/output data. It supports three common data-file formats, enables basic filtration of the data and returns the identified model in both discrete-time and continuous-time form with the possibility to export the results into the MATLAB system for further work. The presented application was designed in a bilingual (English/Czech) version within the scope of the Master's thesis [11] and further tuned by the authors.

This paper is structured as follows: after this introductory part, main goals are outlined in the next section, followed by the employed hardware, software and Internet tools. Next part is devoted to the process of connecting the MATLAB system with web-applications and a brief description of the main implemented functions for the process of identification follows. Further, the web-application user interface is presented and described in detail, followed by discussion of the results and conclusion at the end of the paper.

2 Main Goals

For the web-application development, main goals were formulated as follows: to develop a web-application which enables to identify a given system using uploaded input/output data from the process, with the following requirements:

- Bilingual (English/Czech) version;
- Identification using a linear discrete-time ARX model;
- User-controlled structure of the model;
- Possibility to use following data file-formats: MAT, TXT and XLS;
- Possibility to filter the uploaded data simply;
- Possibility to transform the resultant model into its continuous-time counterpart;
- Possibility to export the results into the MATLAB system;
- Display selected statistical characteristics related to quality of the identification.

Besides this, the application has to be user-friendly and provide help information where needed.

For the systems identification, standard functions of the MATLAB SW are employed, mainly from the popular Control System Toolbox [12]. More information about the used functions are provided further in this contribution.

The designed web-application is intended to support pedagogical and research activities and one of the main advantages can be seen in the open access to the application without the need to install any kind of software on a user computer or to have a particular kind of operating systems. Apart from this, it can be administered and updated easily. The application is accessible directly at the URL: http:// matserver.utb.cz/ExpIdent [13].

3 Methodology

This section further explains used software and hardware tools, the process of MATLAB ↔ Web interconnection and main implemented functions for the systems identification and related tasks.

3.1 Hardware and Software Tools

Web-applications are generally operating on web-servers—computers with a suitable operating system and a special application, web service enabling operation of the web-applications. The developed application runs on the PC Fujitsu Siemens Esprimo P5625 with the AMD Athlon 64 X2 5600+ processor with 4 GB of RAM and two 500 GB hard-drives in the RAID 1 configuration. It hosts the Microsoft Windows Web Server operating system with the web service IIS.

Web-applications have usually several tiers and most common is the 3-tiered architecture with a presentation, application and storage tier.

In the developed application, the first, presentation tier accessible by a web browser was created using the HTML, CSS and AJAX technologies.

The middle application tier (logic tier) processing requests and data from the presentation tier and generating user-friendly interface dynamically was designed using the ASP.NET technology, a part of the Microsoft .NET Framework. For algorithms implementation, C# programming language and the Microsoft Visual Web Developer Express Edition software was used. Complex computations needed for the system identification were realized with the help of the MATLAB system—partly using its implemented functions and partly programmed in matlab scripts and functions. The following MATLAB components were employed for the required functions, corresponding data-processing and deployment to the Web:

- MATLAB;
- Control System Toolbox;
- MATLAB Builder NE [for .NET];
- MATLAB Compiler.

The last, third web-application tier is the storage one, also known as the data tier, which is accessed through the middle application tier and enables data-retrieving, storage and update. In the developed application, it is represented by a file-system of the server operating system.

3.2   Web ↔ MATLAB Interconnection

Up to the MATLAB version R2006b it had been possible to connect web-pages with the MATLAB functions simply using the MATLAB Web Server component, where the functions were implemented directly in the form of MATLAB source codes (m-files). Further versions of the MATLAB system do not include nor support the Web Server component and the connection has to be realized in a different way. MATLAB functions can be implemented (deployed) in the form of so-called components—dynamic-link libraries (DLL-files) in the Microsoft Windows operating system. Then the development of a web-application connected to the MATLAB system can be divided into two independent parts: preparation of the source m-files with required MATLAB functions and components generation in the first step, and development of the web-application and implementation (deployment) of the generated components (DLL's) in the second step. Web-applications connected to the MATLAB system can be developed using various technologies such as Microsoft .NET Framework or JAVA. In this work the former framework has been used.

An example of a source MATLAB m-file used for a web-application component is presented in Fig. 1. This function simply plots a graph from input data.

```
mPlot.m  ×
1      function [] = mPlot(fileName, axis_X, axis_Y,axis_Xlbl,axis_Ylbl)
2 -        fhandle = figure('Visible','off'); %hide figure
3 -        plot(axis_X, axis_Y,'LineWidth',2)
4 -        xlabel(axis_Xlbl);
5 -        ylabel(axis_Ylbl);
6 -        grid on;
7 -        print((fhandle), '-dpng', fileName); %print in png format
8
9 -     end
```

Fig. 1 Deployed m-file example

Fig. 2 Deployment tool

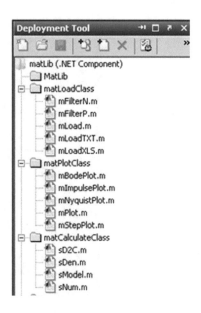

The resultant components for the web-application are then created from the m-files (and relevant toolboxes) using the MATLAB Compiler toolbox. This toolbox uses other supporting toolboxes for the compilation, depending on the chosen technology of web-application development. For the applications developed in the Microsoft .NET environment, the MATLAB Builder NE [for .NET] is needed which creates .NET components containing Microsoft .NET Framework classes. An example of a .NET component with the source m-file above is displayed in Fig. 2. Components (DLL's files) created using the MATLAB Builder NE can be further used in the standard common way as other components of the Microsoft .NET Framework technology. It is only necessary to make a reference to the components in the web-application development environment, e.g. in the Microsoft Visual Studio as illustrated in Fig. 3.

Then it is possible to use all the classes and functions from the source MATLAB m-files during the application development. An example of a simple class definition used for plotting according to specified parameters is given in Fig. 4.

3.3 Main Implemented Functions

Presented application fruitfully exploits capabilities of the MATLAB system and used toolboxes. Besides their standard functions implemented within it was necessary to programme several m-scripts for the desired functions of the application. These include the following main ones (as seen also in Fig. 2):

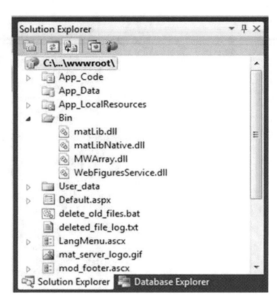

Fig. 3 Reference to a .NET component

```
matPlotClass mPloter = new matPlotClass();
mPloter.mPlot(sFileLocationYT, mwnaT, mwnaY,xlabel,ylabel);
```

Fig. 4 Class definition from the imported library

- *mfilterN.m, mFilterP.m* for simple filtration of uploaded data using standard functions *filter* (MATLAB) and *lsim* (Control System Tbx);
- *mLoad.m, mLoadTXT.m, mLoadXLS.m* for import of user-uploaded data using standard functions *load*, *textscan* and *xlsread* (MATLAB);
- *mPlot.m, mPlotMulti.m* for display of results in the form of graph(s) and graphical file generation using the standard functions *plot* and *print* (MATLAB);
- *sD2C.m, sDen.m, sNum.m* to transform the resultant discrete-time model into its continuous-time counterpart and corresponding manipulations with numerator and denominator coefficients—uses functions *tf* and *d2c* from the Control System Tbx;
- *sSim.m* for simulation of the identified model response to a given input data— uses function *lsim* from the Control System Tbx;
- *sModel.m* for the identification of a user-defined discrete-time (ARX) model from the uploaded input-output process data and corresponding statistical information; here, the classical least squares method, e.g. [14, 15], was implemented using the standard MATLAB routines. Tests performed using this function indicate that it provides comparable results to the standard function *arx* from the System Identification Toolbox, popular MATLAB extension for system identification (this function, as other model estimation commands from this

toolbox, cannot be compiled/deployed unfortunately, due to the restrictions concerning some of the functions from this add-on MATLAB toolbox).

Detailed documentation for the above given standard functions of the MATLAB system and its popular extensions Control System and System Identification Toolboxes can be found in the MathWorks documentation, e.g. [12, 16].

4   Results

A testing version of the developed application is accessible via the Internet at the following URL: http://matserver.utb.cz/ExpIdent [13]. The application user interface (presentation tier) is designed as a guide and allows entering all the required information for system identification in several interconnected steps. Common control buttons such as *Next/Previous* and *Home* are used to navigate throughout the application; relevant help information is also displayed. The interface consists of these main parts:

- Start-up screen which provides basic information about the web-application together with contact information; a language control button (Czech/English) and also the *Home* button are accessible throughout the whole process of work with this site; testing data in 3 different formats (TXT, MAT and XLS) can be downloaded here;

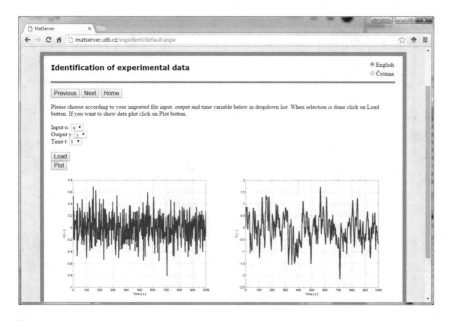

Fig. 5 Selection of variables and graphical display of the data

- In the next step, a user can choose a file with experimental (input/output) data from a process to be identified; currently three different data formats are supported: simple text format (with columns separated by semicolons and names of the columns on the first line), MATLAB binary data-file (with at least 3 vectors of the same length containing input, output and time data) and Excel datasheet with information stored in the first list only and variables names in the first row); after successful data-file import its size is displayed and a user can also delete actual data here;
- Further, corresponding input, output and time variables are selected from a dropdown list and after loading they can be displayed graphically, as illustrated in Fig. 5;
- The fourth step can be used to filter simply the uploaded data; a user can choose from three options—no filtration, filtration using moving average from a given number of samples and filtration using the first-degree filter with a given

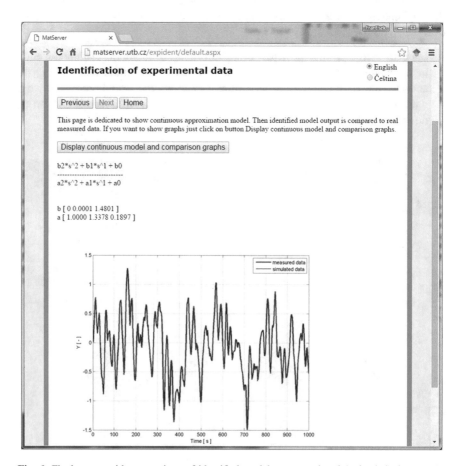

Fig. 6 Final screen with comparison of identified model output and real (uploaded) data

time-constant; filtered and non-filtered data can be compared graphically and a user can choose to work with the filtered or original data;

- In the next screen, it is possible to choose a structure of the identified ARX model, i.e. degrees of numerator and denominator polynomials; then, after clicking the *Compute* button the model is identified and results of the identification are presented, i.e. coefficients of the identified model (in the form of numerator and denominator vectors, as usual in the MATLAB environment); additional information concerning the identification, such as e.g. prediction error and covariance matrix, are also presented; all the results and identification variables can be exported into the MATLAB format easily;

- Finally, in the last step, the identified model can be transformed into its continuous-time counterpart and comparison of measured (uploaded) and simulated output (using the identified model) is also presented, again with the possibility to export the results into the MATLAB format, as seen in Fig. 6. All the presented graphs can be easily downloaded using the right mouse click in the PNG image format and in every part of the application it is possible to return to previous steps to correct the entered information or directly back home for a new session.

5   Conclusion

This paper has presented the process of developing a web-application connected to the MATLAB computing system with the aim to provide simple user-friendly interface for system identification from experimental data. The result is easily accessible using the Internet and standard web browsers which can help students, teachers and researchers in the field of control engineering with (at the moment) limited access to the MATLAB and its toolboxes. At present the application is in a testing mode to debug and tune their functions so experiences, remarks and suggestions of prospective users are welcome. The application is built in a modular way so it will not be difficult to extend more its capabilities, e.g. number of supported data-file formats, additional filters or other identification methods and models. Preliminary results indicate that the application output is comparable to the MATLAB output of similar functions for system identification.

References

1. Attaway, S.: Matlab: A Practical Introduction to Programming and Problem Solving. Butterworth-Heinemann, Boston (2013)
2. MATLAB—The Language of Technical Computing. http://www.mathworks.com/products/matlab
3. Wolfram, S.: The MATHEMATICA Book. Cambridge University Press, Cambridge (1999)

4. Wolfram Mathematica: Definitive System for Modern Technical Computing. http://www. wolfram.com/mathematica
5. MATLAB Online. http://www.mathworks.com/products/matlab-online
6. Wolfram Mathematica Online—Bring Mathematica to Life in the Cloud. http://www.wolfram. com/mathematica/online
7. Math Apps for Students—Online Calculators and Math Apps for Students. http://www. maplesoft.com/products/mobiusproject/studentapps
8. Gazdos, F.: Direct methods of controller design and tuning. The Czech Science Foundation project no. GACR-102/07/P148 (2007–2009)
9. Gazdos, F., Rakus, D.: Web-application for direct controller design and tuning from experimental data. In: Technical Computing Prague 2009—17th Annual Conference Proceedings, pp. 1–8. HUMUSOFT, Prague (2009)
10. Gazdos, F., Facuna, J.: Web Application for LTI Systems Analysis. In Silhavy, R. et al. (eds.) Intelligent Systems in Cybernetics and Automation Theory: Proceedings of the 4th Computer Science On-line Conference 2015 (CSOC2015), vol 2: Intelligent Systems in Cybernetics and Automation Theory, pp. 101–109. Springer, Heidelberg (2015)
11. Micola, P.: Web-application for System Identification form Experimental Data. Master's thesis, Tomas Bata University in Zlin, Zlin (2014)
12. The MathWorks Inc: Control System Toolbox: User's Guide. Natick, USA (2015)
13. MAT Server—Identification of experimental data, http://matserver.utb.cz/ExpIdent
14. Ljung, L.: System Identification: Theory for the User. Prentice Hall, New Jersey (1999)
15. Bobal, V., Bohm, J., Fessl, J., Machacek, J.: Digital Self-tuning Controllers: Algorithms, Implementation and Applications. Springer, London (2005)
16. Ljung, L.: System Identification Toolbox: User's Guide. The MathWorks Inc, Natick, MA, USA (2015)

On the Intrinsic Relation of Linear Dynamical Systems and Higher Order Neural Units

Peter Benes and Ivo Bukovsky

Abstract This paper summarizes the fundamental construction of higher-order-neural-units (HONU) as a class of polynomial function based neural units, which are though non-linear discrete time models, are linear in their parameters. From this a relation will be developed, ultimately leading to a new definition for analysing the global stability of a HONU, not only as a model itself, but further as a means of justifying the global dynamic stability of the whole control loop under HONU feedback control. This paper is organised to develop the fundamentals behind this intrinsic relation of linear dynamic systems and HONUs accompanied by a theoretical example to illustrate the functionality and principles of the concept.

Keywords Adaptive control · Higher-order neural units (HONUs) · Gradient descent (GD) · Levenberg-Marquardt (LM) · Batch propagation through time (BPTT)

1 Introduction

Due to the expanding complexities of our technological society, adaptive control has become a rapidly growing area of study in the field of modern engineering and computational science. Since as early as the 1960s, adaptive control has taken big leaps not only in the sense of adaptive tuning methods for conventional forms of industrial controllers, but also the conception of entirely different architectures of adaptive control methods. One such discipline, which has received particularly rapid growth in theoretical studies as well as positive applications in the

P. Benes (✉) · I. Bukovsky
Department of Instrumentation and Control Engineering,
Czech Technical University in Prague, Prague, Czech Republic
e-mail: petermark.benes@fs.cvut.cz

I. Bukovsky
e-mail: ivo.bukovsky@fs.cvut.cz

© Springer International Publishing Switzerland 2016
R. Silhavy et al. (eds.), *Automation Control Theory Perspectives in Intelligent Systems*, Advances in Intelligent Systems and Computing 466,
DOI 10.1007/978-3-319-33389-2_23

engineering field, is the use of polynomial function based neural networks (PNN) and Higher-Order-Neural-Networks (HONNs) furthermore Higher-Order-Neural-Units (HONUs) as fundamental non-linear feedback controllers. The advantage of standalone HONUs is clear customizable non-linearity of input-output mapping via its polynomial order while these neural architectures being linear in parameters, and furthermore, every neural weight is unique in its function within HONU structure, contrary to layered networks. The conception of HONU as standalone neural units was predominantly introduced in [1–3]. Some of the first works on HONU as standalone neural controllers appeared in publications from [4]. In these works a comparison was drawn between the use of conventional neural networks as such that of MLP architecture and the use of HONUs. A drawback that may be ascertained from using MLP architectures as a means of system modelling or extension to control is their necessity to use rather complex training algorithms in lieu with longer training runs to achieve adequate convergence to a minimal square error. In contrast to this, the approximation strength of such conventional neural networks may be improved by adding more neurons or even additional layers, which can provide better approximation for non-linear systems. However, as can be seen in the work [5], for efficient real time learning algorithms as such that of the gradient descent algorithm (GD) [6] and the batch propagation through time (BPTT) algorithm as an extension of the gradient descent algorithm with the famous Levenberg-Marquardt (L-M) equation [7], HONUs are computationally faster in achieving adequate convergence in square error whilst achieving desirable control performance for both non-linear unknown systems as well as linear systems of SISO structure. Some more recent applications of HONUs for real-time adaptive control may be found in the works [8, 9], where HONUs were applied in extension to the previously employed conventional control loops for successful optimisation on various SISO engineering processes.

With the ever-extending innovations in adaptive control comes the challenge of guaranteeing stability of the proposed adaptive model or adaptive control loop. With regards to NN forms of adaptive control numerous works have been published with concern to learning stability for online adaptive control algorithms. However a further area which has received particular attention in recent times, is the assessment of global asymptotic stability (GAS) of the adaptive control loop as a whole, with further intention led towards minimising the amount of online computations during application of the controller algorithm. Earlier methods for global stability analysis of neural networks have been focussed on linearization analysis and Lyapunov function based techniques for a class of Hopfield neural networks [10, 11]. Some recent examples include the use of homeomorphism and Lyapunov function based techniques by [12] where global robust exponential stability of the equilibrium point for delayed neural networks is developed. In [13] problems concerning stability for static neural networks with time delays are investigated, via a delay partitioning technique a novel stability criteria justified via the Lyapunov-Krasovskii criteria is investigated to warrant the global stability of the investigated static NNs. An earlier method necessary to remark maybe found in the work [14] and his further works there in. Here, the assessment of GAS of the respective equilibrium points for all non-decreasing neural network activation functions is presented for a class of dynamic layered neural networks. Should the neural model be defined in state matrix form, the negative of the positive diagonal matrix T serving as an interconnection matrix with respect to a non-linear diagonal mapping matrix g, if proven to be quasi-diagonally column-sum dominant the authors proof yields existence and uniqueness of such equilibrium point along with justification of global asymptotic stability. Till this date one of the most common and readily employed approaches for assessing asymptotic stability of recurrent neural networks, is that of the Linear Matrix Inequality (LMI), an example of such method maybe found in [15] and further [16] where the Lyapunov-Krasovskii stability theory along with the linear matrix inequality approach are used to determine stability of both constant and time-varying delayed neural networks. Motivated by the above listed works, this paper extends on the consequences of Lyapunov's matrix inequality condition for discrete time systems along with adding a novel representation of a HONU model of general n-th orders, advantaged by its property of linear parameterisation to define a method for assessing the global stability of a non-linear HONU adaptive model and further determining the global stability of a non-linear HONU controlled dynamic process.

In this paper, the proposed derivations are accompanied by a theoretical example incorporating fundamental learning algorithms namely via GD and the BPTT algorithm. The presence of capital letters denote matrices, small bold letters stand for vectors and small italic ones denotes scalars. The meaning of the most frequent notation and symbols is given in the accompanying text, with the remaining being evident at first appearance.

2 Preliminaries Behind Higher Order Neural Units

The classical notation of HONUs is the summation of multiplication terms that is for example for a QNU, i.e. HONU for $r = 2$, represented as follows

$$\tilde{y} = \sum_{i=0}^{n_x} \sum_{j=i}^{n_x} \mathbf{W}_{i,j}.x_i.x_j = \mathbf{w}.\mathbf{colx} \tag{1}$$

and for CNU, i.e. HONU for r = 3, it is extended as follows

$$\tilde{y} = \sum_{i=0}^{n_x} \sum_{j=i}^{n_x} \sum_{k=j}^{n_x} \mathbf{W}_{i,j,k}.x_i.x_j.x_k = \mathbf{w}.\mathbf{colx} \tag{2}$$

where $\mathbf{W}$ denotes a multidimensional array of the HONU neural weights. Conveniently as used e.g. in [2], where the multidimensional neural weights may further be translated into a flattened representation $\mathbf{w}$ and further, an augmented input vector of neural inputs $\mathbf{x}$ yields a long-column vector form $\mathbf{colx}$.

The weight update system of such HONU models are thus generally given as follows

$$\mathbf{w}(k + 1) = \mathbf{w}(k) + \Delta\mathbf{w}; \Delta\mathbf{w} = \Delta\mathbf{w}(\mathbf{w}(k)) \tag{3}$$

For the scope of this paper, a focus towards the GD learning algorithm and BPTT learning algorithm are emphasised. In the sense of both forms, e denotes neural unit error that may be defined as follows.

$$e(k) = y(k) - \tilde{y}(k) \tag{4}$$

regarding an incremental form of training the neural weights the GD weight update rule may be denoted as follows

$$\Delta\mathbf{w} = \mu.e(k).\mathbf{colx}^T \tag{5}$$

where μ corresponds to the learning rate. However, should a batch form of neural weight training be employed the BPTT training algorithm, dictates the respective weight update rule where $\mathbf{J}$ is Jacobian matrix and $\mathbf{I}$ is the identity matrix

$$\Delta\mathbf{w} = (\mathbf{J}^T.\mathbf{J} + \frac{1}{\mu}.\mathbf{I})^{-1}.\mathbf{J}^T.\mathbf{e} \tag{6}$$

Regarding the extension of an adaptive HONU controller to the previously identified HONU model, denoted via the above relations. The output value computed by the HONU feedback controller q is given by a summation of multiplication terms. Thus, the HONU controller output for a QNU (i.e. HONU for $r = 2$), may be denoted as follows

$$q = \sum_{i=0}^{n_{xi}} \sum_{j=i}^{n_{xi}} \mathbf{V}_{i,j}.\xi_i.\xi_j = \mathbf{v}.\mathbf{col}\xi \tag{7}$$

or for a CNU, i.e. HONU for $r = 3$ we may denote the HONU controller output as follows

$$q = \sum_{i=0}^{n_x} \sum_{j=i}^{n_x} \sum_{k=j}^{n_x} \mathbf{V}_{i,j,k}.x_i.x_j.x_k = \mathbf{v}.\mathbf{col}\xi \tag{8}$$

where $\mathbf{V}$ represents the r-dimensional array of neural weights corresponding to the HONU feedback controller and $\mathbf{v}.\mathbf{col}\xi$ is the respective long-vector multiplication form. Here the variable $\mathbf{col}\xi$ now comprises of the variables ξ_i, ξ_j in the sense of a QNU or further ξ_k in the sense of a CNU. These terms thus comprise of the previous outputs of the identified HONU plant model and either the difference between the reference model and output of the neural model $(y_{ref} - y)$, denoted in short as $e_{ref,}$ or shortly, the desired value of the process d. Further to this, the new inputs u that are fed into the plant or existing control loop for optimisation (e.g. featuring a PID controller) may thus be as follows

$$u = d - r_q.q \tag{9}$$

Thus, analogical to the adaptive identification process the adaptive controller may also incorporate a sample-by-sample update or batch form of neural weight training. The overall weight update mechanism may therefore be summarized as follows

$$\mathbf{v}(k + 1) = \mathbf{v}(k) + \Delta\mathbf{v}; \Delta\mathbf{v} = \Delta\mathbf{v}(\mathbf{v}(k)) \tag{10}$$

3 Relation and Extension of Linear Dynamic Theory to Higher Order Neural Units Stability

Following the review of the works [10–16] motivation arises to define an efficient approach for assessing the global stability of an identified HONU adaptive model, further a HONU adaptive controlled loop model. As may be recalled from (1)–(2) a HONU is a conjunction of multidimensional neural weights which may translated into a flattened representation $\mathbf{w}$, in conjunction with augmented input vector of neural inputs $\mathbf{colx.}$ Therefore such model may also be classified as a discrete difference equation with a constant time-step resembled by the variable k. On expansion of (1) for a HONU or r = 2, i.e. a QNU, the following general expanded form for length n of the input vector $\mathbf{x}$ may be given as follows

$$
\begin{aligned}
y(k+1) =\, & w_{0,0} + w_{0,1}.y(k) + w_{0,ny+1}.y(k-ny) + \cdots + w_{0,n}.u(k-nu) \\
& + w_{1,1}y(k).x_1(k) + \cdots + w_{1,n}y(k).x_n(k) + w_{2,2}y(k-1).x_2(k) \\
& + \cdots + w_{2,n}y(k-1).x_n(k) + \cdots + w_{ny+2,ny+2}u(k).x_{ny+2}(k) \\
& + \cdots + w_{ny+2,n}u(k).x_n(k) + \ldots + w_{n,n}u(k-nu).x_n(k)
\end{aligned} \tag{11}
$$

From relation (11), each state variable $y(k)...y(k\text{-}ny)$ will individually feature a combination of neural weights $w_{i,j}$ where $i = 0, 1,...,n$ and $j = i,...,n$. and input vector value x_j which corresponds to a previous output value of the plant or model, further an input value u of the plant. Recalling on classical linear theory, the general form of a linear system State Space model may be denoted as follows

$$
\begin{aligned}
\mathbf{x}(k+1) &= M.\mathbf{x}(k) + N.\mathbf{u}(k) \\
\mathbf{y}(k) &= C.\mathbf{x}(k)
\end{aligned} \tag{12}
$$

Thus, on analysis of relation (11), it yields that since such HONU difference equation features a clear non-linearity of input-output mapping via its polynomial order whilst being linear in parameters, such matrix representation as in (12) may be extended to HONU adaptive models via the following general form

$$\mathbf{x}(k+1) = M_r.\mathbf{x}(k) + N_r.\mathbf{u}(k) + \mathbf{w}_0$$
$$\mathbf{y}(k) = C_r.\mathbf{x}(k) \tag{13}$$

where M_r corresponds to the matrix of dynamics corresponding to coefficients of the difference equation, more explicitly a matrix of HONU weights in the sense of an LNU architecture (HONU, r = 1). The sub-index r thus denotes a given order of the HONU model i.e. $r = 2$ for QNU and $r = 3$ for CNU. Analogically, N_r corresponding to the coefficients of inputs matrix for r-th order of the HONU model. Furthermore, $\mathbf{x}(k)$ denotes a vector of previous system or model outputs y and $\mathbf{u}(k)$ denotes the samples of previous plant inputs u.

Therefore the above general form may be expressed via the following matrix representation

$$
\begin{bmatrix} y(k-ny-1) \\ . \\ . \\ . \\ y(k-1) \\ y(k) \\ y(k+1) \end{bmatrix} =
\begin{bmatrix} 0 & 1 & \cdots & 0 \\ : & & & : \\ : & & & : \\ : & & & : \\ 0 & \cdots & 1 & 0 \\ 0 & 0 & \cdots & 1 \\ a_{x_{ny}} & a_{x_{ny-1}} & \cdots & a_{x_1} \end{bmatrix}
\begin{bmatrix} y(k-ny) \\ y(k-ny-1) \\ . \\ . \\ y(k-2) \\ y(k-1) \\ y(k) \end{bmatrix}
$$

$$
+ \begin{bmatrix} 0 & 0 & 0 & 0 \\ : & & & : \\ : & & & : \\ : & & & : \\ 0 & 0 & 0 & 0 \\ a_{u_1} & a_{u_2} & \cdots & a_{u_{nu}} \end{bmatrix} .[u(k) \quad u(k-1) \quad \cdots \quad u(k-nu)] + w_{0,0}
\tag{14}
$$

$$
y(k) = [0 \quad 0 \quad \cdots \quad 1].
\begin{bmatrix} y(k-ny) \\ y(k-ny-1) \\ . \\ . \\ y(k-2) \\ y(k-1) \\ y(k) \end{bmatrix}
\tag{15}
$$

where now the variables a_{x1}, a_{xny-1}, a_{xny} are in fact a HONU of one order lower in dimension as applied for HONU adaptive identification. These coefficients correspond to the combination of coefficients represented by each element of the input vector $\mathbf{x}$ i.e.; x_i where $i = 1, 2, 3,...,n_y$. Furthermore, the variables a_{u1}, a_{u2}, a_{unu} represent the combination of variables for x_i where $i = n_y + 2,...,n_u$. Given the expression in (14)–(15) and returning to the notion that such HONU models are although collectively non-linear in representation, they hold the property of linear parameterisation. Therefore the output of the State Space Representation for a HONU model as in (14)–(15) may be reformulated in a general linear transformation as follows

$$y(k+1) = w_{0,0} + a_{x1}.y(k) + a_{xny-1}.y(k - ny - 1)...a_{ny}.y(k - ny) + a_{u1}.u(k) + ...a_{unu}.u(k - nu) \tag{16}$$

Given the expression (16) it may be stated that the matrix of dynamics M_r resembling the coefficients of dynamics for a HONU difference equation may also be evaluated under the consequences of Lyapunov's stability criteria, thus the condition for stability of this newly shaped HONU State Space model is also dependant on the following discrete time linear dynamic model condition, which is that the modulus of each respective eigenvalues must lie fully within the unit circle, centred at the origin of the Imaginary and Real axes i.e.

$$|\lambda_i|, (\text{where} \quad i = 1...n_{eig}) < 1 \tag{17}$$

Mathematically, this may be justified via the principles of the Lyapunov matrix inequality stability equation for discrete-time models. For a given dynamical system with an equilibrium point at $x = 0$. Where V is a continuous Lyapunov function such that $V(0) = 0$ and further that $V(x) > 0$ across all other values of x then given this the following holds,

$$\Delta V(x(k)) \doteq V(x(k)) - V(x(k - 1)) < 0, \forall x(k) \in \mathbb{R}$$
$$V(x) \to \infty \tag{18}$$

Thus the origin, corresponding to the equilibrium point of the system is in fact globally asymptotically stable. Furthermore, a consequence of Lyapunov's stability method yields that two equally justified statements go hand in hand.

1. The matrix M_r is asymptotically stable if and only if the relation (17) holds.
2. Given any matrix $Q = Q^T > 0$ there exists a positive definite matrix $P = P^T$ such that satisfies the proceeding relation.

$$M_r^T.P.M_r - P = -Q \tag{19}$$

Thus, if the Lyapunov matrix equation is satisfied for our discrete-time linearized representation of a HONU model, it may then be concluded that M_r is in fact a matrix with all eigenvalues present within the unit circle. On selection of the following Lyapunov candidate function $V(x) = x^T Px$, where $Vx(0) = 0$, the idea is then to prove the statement in Eq. (18). Consider a non-linear system of a HONU model in linearized state space representation as follows

$$x(k+1) = M_r.x(k),$$
$$x(k) = M_r.x(k-1) \tag{20}$$

Now, substitution of the chosen Lyapunov function candidate into the expression V (x(k)) – V(x(k-1)) therefore yields the following

$$V(x(k)) - V(x(k-1)) = x(k)^T P.x(k) - x(k-1)^T P.x(k-1)$$
$$\therefore if \ x(k+1) = M_r.x(k) \ then, \tag{21}$$
$$V(x(k+1)) - V(x(k)) = x(k+1)^T P.x(k+1) - x(k)^T P.x(k)$$

Therefore, given the matrix property that $x(k+1)^T = x(k)^T. M_r^{\ T}$ then,

$$x(k)^T M_r^T.P.M_r x(k) - x(k)^T P.x(k)$$
$$= x(k)^T.(M_r^T.P.M_r - P).x(k) \tag{22}$$

However, according to the Lyapunov matrix equation in (19) it yields that,

$$= - \sum_{k=0}^{N-1} x(k)^T.Q.x(k) \tag{23}$$
$$where \quad N \to \infty$$

Therefore, the sequence denoted in relation (23) is strictly decreasing for all values of x(k) being either both positive or negative, as Q is a positive matrix where $Q = Q^T > 0$. Thus, it may be concluded that the eigenvalues of the matrix M_r corresponding to the matrix of dynamics for the HONU state space model, must have eigenvalues satisfying (17) i.e. being fully within the unit circle and hence, globally asymptotically stable. This is of course with the assumption that the HONU model weights are trained such to correctly model the engineering system and that the newly measured system data corresponds to the true working data of the engineering process. As a further remark, relations (14)–(15) respectively (17) similarly yield a condition for global stability of the whole adaptively controlled loop with the extension of a HONU adaptive feedback controller. Considering the substitution of relation (9), it can be found that the dimension of the matrix M_r expands by n_u coefficients of $a_{xi,}$ respectively too, the number system eigenvalues. Furthermore, the previous variable u, is now replaced with the desired value of the process d, which indeed corresponds to the newly fed inputs to the engineering process.

4 Experimental Analysis

In this section let us consider a conditionally stable, non-linear integrating process. On extending a proportional controller in feedback with a gain value K the system may be expressed via the following relation

$$\frac{K(s+1)^2}{s^3 + K(s+1)^2} \tag{24}$$

In addition a non-linearity being an introduced saturation in the range of –0.8 to 0.8 applied at the plant input will be considered. The interesting property the system (24) holds, is that for high values of the static gain K the system is indeed relatively stable. Should K be of a lower order e.g. K < 0.5 corresponding to a frequency w < 1 it may be found that the system indeed undergoes a degree of instability. Given this, the system may be adaptively identified within the stable working region where K = 25 and the magnitude of the system input pulse will be r = 3. From here applying a QNU model trained via the BPTT algorithm with input values of $n_y = 4$ and $n_u = 3$, we yield an adequate dynamic identification of the above system (24).

On re-simulation of the previously derived HONU model, in the region where the real plant data corresponds to the training data for the neural model. The HONU is able to quite adequately model the dynamic behaviour of the new system data. However, as can be seen in Fig. 1, if the gain K of the system is decreased to K = 0.25, immediately a zone of instability for the system is reached. As a result the previously, well identified HONU model is not able to function in this region as the dynamics are unknown within the previously identified neural weights and thus, the model along with the real system data falls into a mode of instability.

Thus, on testing the performance of the relations (14)–(15) respectively (17) it can be expected that the eigenvalues corresponding to the HONU matrix of dynamics should yield all values within the unit circle for the first 95 s of

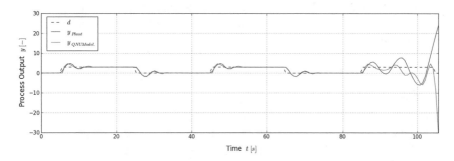

Fig. 1 HONU Simulation on New Experimental Data for Non-linear Integrating Process (24), 0–95[s] the Process is Stable K = 25. 95–120 [s] the Process and HONU Model enters a region of Instability where K = 0.25

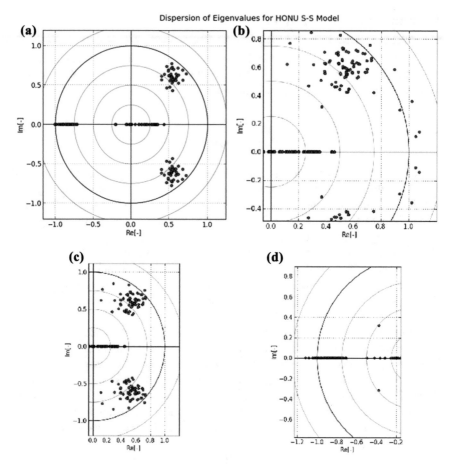

Fig. 2 Dispersion of HONU Eigenvalues. **a** 0–95 [s], **b** RHS: 110–120 [s], **c** 95–110 [s], **d** LHS: 110–120 [s]

simulation for the HONU model with the newly tested system data and eigenvalues beyond in the remaining stages of simulation, corresponding to an instable plant and hence too HONU model (Fig. 2).

5 Conclusion

In this paper the relation between linear dynamic systems and higher order neural units has be shown, due to the advantage of standalone HONUs having a clear customizable non-linearity of input-output mapping via its polynomial order whilst being linear in parameters, it was shown that such model may also be classified as a discrete difference equation with a constant time-step resembled by the variable k.

Following further derivation the relations (14)–(15) summarise the matrix representation of a discrete-time HONU equation which is in fact its corresponding state-space representation. Given this, the ideas of linear stability theory were extended to the non-linear HONU discrete model and justified via the principles of Lyapunov's matrix inequality condition. A clear advantage of this method contrary to those reviewed is that only the engineering process data and a correctly identified adaptive HONU model is necessary for evaluating the stability of the model and further global control loop. This method is with the absence of any unknown stability functions and further may be computed offline with respect to the working range of the engineering process data, minimising the need for online computations during operation. As a further, the experimental example of a conditionally stable non-linear system indeed justifies the concepts of global stability for such approach to HONUs. Should a HONU adaptive feedback controller be applied in feedback and extended on an existing HONU model of an engineering system, the relations (14)–(15) respectively (17) with substitution of relation (9) yield a condition for global stability of the whole adaptively controlled loop, advantageous for application to non-linear process control where precise mathematical descriptions of the process dynamics including their non-linearities as necessary for other conventional stability methods may be difficult to define.

Acknowledgments The authors of this paper would like to acknowledge the following study grant for its support SGS12/177/OHK2/3T/12

References

1. Nikolaev, N.Y., Iba, H.: Adaptive Learning of Polynomial Networks Genetic Programming, Backpropagation and Bayesian Methods. Springer, New York (2006)
2. Bukovsky, I., Hou, Z.-G., Bila, J., Gupta, M.M.: Foundation of notation and classification of nonconventional static and dynamic neural units. In: 6th IEEE International Conference on Cognitive Informatics, pp. 401–407 (2007)
3. Solo, G.M.: Fundamentals of higher order neural networks for modeling and simulation. Artif. High. Order Neural Netw. Model. Simul. (2012)
4. Song, K.-Y., Redlapalli, S., Gupta, M.M.: Cubic neural unit for control applications. In: Fourth International Symposium on Uncertainty Modeling and Analysis, ISUMA 2003, pp. 324–329 (2003)
5. Bukovsky, I., Redlapalli, S., Gupta, M.M.: Quadratic and cubic neural units for identification and fast state feedback control of unknown nonlinear dynamic systems. In: Fourth International Symposium on Uncertainty Modeling and Analysis, ISUMA 2003, pp. 330–334 (2003)
6. Williams, R.J., Zipser, D.: A learning algorithm for continually running fully recurrent neural networks. Neural Comput. **1**(2), 270–280 (1989)
7. Werbos, P.J.: Backpropagation through time: what it does and how to do it. Proc. IEEE **78**(10), 1550–1560 (1990)
8. Benes, P., Bukovsky, I.: Neural network approach to hoist deceleration control. In: 2014 International Joint Conference on Neural Networks (IJCNN), pp. 1864–1869 (2014)

9. Bukovsky, I., Benes, P., Slama M.: Laboratory systems control with adaptively tuned higher order neural units. In: Silhavy, R., Senkerik, R., Oplatkova, Z.K., Prokopova, Z., Silhavy, P. (eds.) Intelligent Systems in Cybernetics and Automation Theory. Springer International Publishing, pp. 275–284 (2015)

10. Chu, T.: An exponential convergence estimate for analog neural networks with delay. Phys. Lett. A **283**(1–2), 113–118 (2001)

11. Zhang, Z.: Global exponential stability and periodic solutions of delay Hopfield neural networks. Int. J. Syst. Sci. **27**(2), 227–231 (1996)

12. Zhao, W., Zhu, Q.: New results of global robust exponential stability of neural networks with delays. Nonlinear Anal. Real World Appl. **11**(2), 1190–1197 (2010)

13. Wu, Z.-G., Lam, J., Su, H., Chu, J.: Stability and dissipativity analysis of static neural networks with time delay. IEEE Trans. Neural Netw. Learn. Syst. **23**(2), 199–210 (2012)

14. Arik, S.: Global asymptotic stability of a class of dynamical neural networks. IEEE Trans. Circuits Syst. Fundam. Theory Appl. **47**(4), 568–571 (2000)

15. Liao, X., Chen, G., Sanchez, E.N.: LMI-based approach for asymptotically stability analysis of delayed neural networks. IEEE Trans. Circuits Syst. Fundam. Theory Appl. **49**(7), 1033–1039 (2002)

16. Liao, X., Chen, G., Sanchez, E.N.: Delay-dependent exponential stability analysis of delayed neural networks: an LMI approach. Neural Netw. **15**(7), 855–866 (2002)

WiFi Multi Access Point Smart Home IoT Architecture

Piotr Lech

Abstract The paper presents the concept of the wireless sensor network dedicated to smart home applications. The innovative characteristic of this network is related to its use for information transfer and routing of selected sensors that simultaneously serve as Access Points and network clients. As a result, we obtain an alternative for the popular solutions based on AdHoc idea. Network traffic creates a cascading path for messages with optional selection of routes dependent on the strength of the received signal. Messages are transmitted between the AP using store and forward methods. Optionally, in order to save the consumed energy, customers transmitters can be switched solely during the transmission time. Sample implementations have been made on the basis of low-cost microprocessor controllers and communication modules commonly used in IoT. To verify correct implementation, some tests have been conducted, examining the quality and latency.

Keywords Smart home · Iot · WiFi WSN

1 Introduction

Technology Home Automation and Smart Home are booming application domains binding together elements of sensor networks, embedded systems, IP networks and cloud computing. In such solutions the sensor networks or elements of the IoT technologies are successfully implemented. The integration of these technological areas has three potential directions of development: measuring and control devices and telecommunication infrastructure for the M2M (machine-to-machine) communication model and information processing as well as convergent control applications in the H2M (human-to-machine) model. In practice, all major players

P. Lech (✉)
Faculty of Electrical Engineering, Department of Signal Processing
and Multimedia Engineering, West Pomeranian University of Technology,
Sikorskiego 37, 70-313 Szczecin, Poland
e-mail: piotr.lech@zut.edu.pl

© Springer International Publishing Switzerland 2016
R. Silhavy et al. (eds.), *Automation Control Theory Perspectives
in Intelligent Systems*, Advances in Intelligent Systems and Computing 466,
DOI 10.1007/978-3-319-33389-2_24

of the ICT sector have in their offer some products related to IoT technology. Unfortunately commercial solutions are usually closed-form and high costs. The appearance of low-cost equipment such as: Intel Galileo, Raspberry PI or ESP2866 system allowed the dissemination of Smart Home ideas based on open architecture. The use of WiFi networks allows the use of existing infrastructure and a wide range of equipment causing significant reduction of the equipment costs.

Considering three-level IoT architecture, research areas can be divided into: data center/cloud computing, gateway and devices network. The successful implementation of the IoT solution depends on the efficient and reliable operation of each element in the individual layers. Figure 1 shows the migration from the Home Automation [1] systems based on industry automation standards by IoT technologies to the WoT (Web of Things) technology [2]. The role of the Home automation is to ensure the domestic control device for the comfort and safety of the inhabitants and their energy management. The main manageable elements of the system are lighting, alarm, heating and similar installations implemented in various standards unfortunately not always compatible with each other. These elements are based on wired sensor network (C-Bus) or WSN—wireless sensor networks (e.g. ZigBee or Z-Wave) [3]. Multiple of technologies and sensors and a large range of closed industrial solutions makes it difficult to integrate services and realize the full range of services expected in the case of implementation of Smart Home ideas. A greater flexibility in the creation of multi-standard Smart Home implementations can be provided by the application of IoT technology. Integration of standards occurs through the use of the IP Backbone and bridges binding different subsystems. It is worth to notice some trends related to moving IoT deployments to the WoT

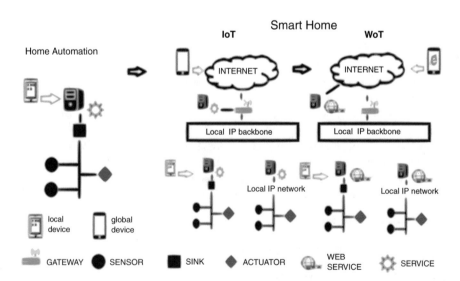

Fig. 1 Smart home evolution

standard by means of migration of bridges to smart gateways and the transfer of control applications and data storing towards Web applications. Network topology, routing and transport protocols are designed to provide high network reliability, expandability and low power consumption. It seems that all these conditions are satisfied by the Wireless Sensor Network implemented in WiFi standard and Smart Home solutions applied in IoT or WoT technology [4–9].

2 Motivation

Assurance of WiFi network coverage for a significant area (in a large space) of the building in terms of transmission quality and costs of investment are currently the only limitations which are responsible for slowing down the dynamic deployment of Smart Home systems. Wireless sensor networks in building installations are carried out usually in a mesh topology where WiFi network is implemented in the AdHoc configuration. More rarely a star or tree topologies with the access point can be found. In all those solutions coverage of a large area of the building with guaranteed access to the network with good performance transmission is a serious problem. The matter is complicated by the fact that the area around transmitters is surrounded by walls having different signal attenuation resulting in uneven coverage by good quality signal.

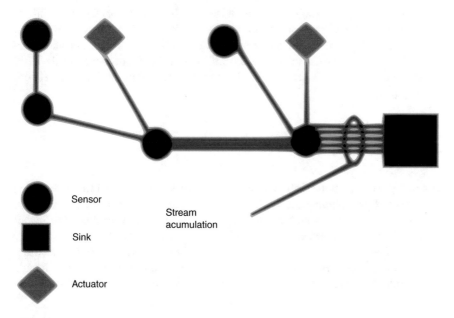

Fig. 2 Multi-sensor data streaming

The mesh network topology requires constant activity of points involved in the network traffic which greatly reduces the operating time of battery powered systems. Unevenly generated network traffic from multiple sensors (Fig. 2) can lead to the overflow and loss of stability of the link.

3 Concept of Smart Home Network Based on Mesh WiFi Access Point

A wide range of low-cost communication modules in WiFi standard allows easy construction of sensors and actuators working in this standard. The basic elements of the WSN are typically equipped with a single communication system for connecting with the nearest statically defined AP. These modules usually work in sleep mode to save energy and are activated only when there is a need to send a message in the network. In the proposed network (named as mesh AP) a special group of equipment (sensors or actuators) can be distinguished with two communication modules. One of them is defined as an Access Point and the second acts as a client for another AP in the network. In this case, in order to optimize energy consumption, client module is also in sleep state and is activated when the collection of messages received from the sensors should be forwarded to the next AP in the cascade of the AP route. The path in cascade route is determined on the base of known adjacency of APs and direction of information flow. The path is chosen dynamically in some fragments. Allowing the broadcast for the AP will be implemented on the base on signal quality. In its simplest form, as the most appropriate AP the one with the stronger signal will be selected. An additional advantage of the proposed solution is the ability to work independently of the WiFi infrastructure providing access to the Internet by increasing the level of network security [10].

4 Implementation

During the construction of an experimental network it is assumed that investment costs should be minimized. Additionally only the publicly available software libraries are used to create the software. In order to minimize costs the sensor network is build from sensors used in cheap and widely available microprocessor modules: Intel Galileo 2, Raspberry Pi 2, Arduino Nano (listed in the order determined by the cost of purchasing a single module) [11]. The WiFi module responsible for communication is ESP8266-01 with integrated TCP/IP stack and mode selection.

From the family of communication modules ESP8266 a particular attention can be given to its development version nodemcu, expanding the capabilities of the tasks typical for the IoT along with shared communication protocols and application examples. Due to the integration of the micro-controller with the communication module unit it has become a favorite one as it has the highest ratio of potential to its price. The structure of the sensor has been significantly simplified. Four types of sensors have been built belonging to the AP cascade (4—Intel Galileo, 4—Rasspberry, 4—Arduino Nano and 4—nodemcu)—similar topology is shown in Fig. 3. By combining cascades, you can easily get a multi-platform mesh structure. The 5—nodemcu modules, 14—standard PCs and 4—Odroid M1 (all of them as sensor simulators) are used to generate traffic in network.

Physically data traffic via the APs is conducted in a cascading way and we have information about possible routes, however, the algorithm determines a route accessible only through the AP with the strongest signal. Messages are transmitted by the AP using the store and forward method. Transportation of messages and acknowledgments in the IP network is conducted using the UDP protocol. The diagram of communication in network is shown in Fig. 3.

The algorithm of the route for a single mesh from the sensor to the recipient of a message can be summarized as follows (Fig. 4):

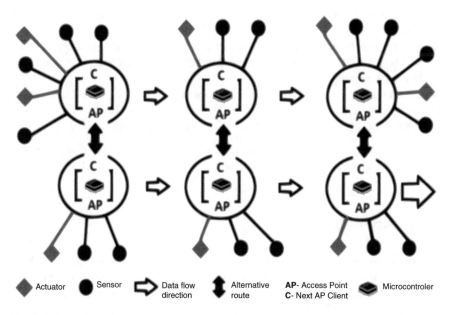

◆ Actuator   ● Sensor   ⇨ Data flow direction   ↕ Alternative route   **AP**- Access Point   **C**- Next AP Client   ◆ Microcontroler

Fig. 3 Proposed topology

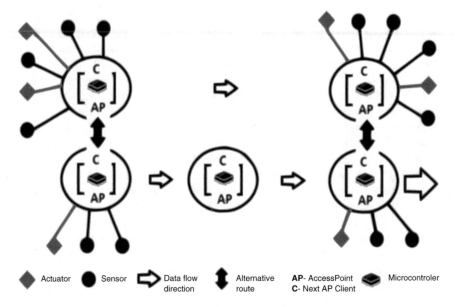

| ◆ Actuator | ● Sensor | ⇨ Data flow direction | ⬍ Alternative route | **AP**- AccessPoint **C**- Next AP Client | ◈ Microcontroler |

Fig. 4 Test topology

- create a collection of messages from the closest sensors and neighboring mesh element;
- when the collection grows to the assumed size the client communication module will be launched and it will be checked which of the known cascade route APs has the strongest signal;
- a collection of messages with repetition is sent to the AP with the strongest signal until a positive confirmation of transmission is received;
- the destination AP returns data to confirm the correct transmission.

The control of actuators is similar, except that we do not create data collection but send the message immediately without any delays.

5 Experiments

The basic test was to verify the quality of the transmission in the network with an extended range relatively to the network with a single AP. The results are shown in Figs. 5 and 6 illustrating two first cases in which the two APs are placed in a maximum possible distance from which user can log on to the network and the second involving additional AP arranged in the middle of this distance. The first graph shows the number of repetitions of transmission of collection of communications in tome between the particular APs. The second represents the delays occurring during transmission for 10 kb of data collection (measured delay times

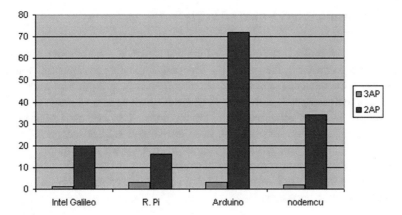

Fig. 5 Registered number of transmission errors

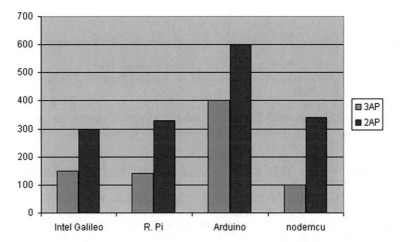

Fig. 6 Delays in network (values in (ms))

include also the time required for preparation of collection and sending the confirmations). The results confirmed the expectations that adding the additional AP will significantly improve the reliability of the network [12].

Additional study was designed to test the correctness of routing messages with the choice of the strongest AP in the network. A study in which the mesh APs subsequently set are the same as in the previous experiment at the maximum usable distance diminished by half has been conducted. In this study, we recorded the signal strength from the AP from the defined route and the number of retransmissions. With so selected conditions of the experiment there were no anomalies in the operation of the routing algorithm.

6 Conclusions and Future Work

The modern technological solution Multi-WiFi Access Point Smart Home IoT Architecture proposed in the paper is a low cost alternative to industrial solutions. The use of dual devices (sensors and actuators), which are both APs and network clients of another network allows creating mesh surrounded by sensors with single transmission modules arranged in a star topology. The proposed algorithm for routing messages in the network has been tested, and studies have confirmed its usefulness. The experiment used a variety of microprocessor equipment. Their uncontested work during testing demonstrates their potential and possibilities of integration. In the future it is planned to run WoT implementations on the developed and tested infrastructure.

References

1. Chong, C.-Y., Kumar, S.P.: Sensor networks: evolution, opportunities, and challenges. Proc. IEEE **91**(8), 1247–1256 (2003)
2. Guinard, D., Vlad, T.: Building the web of things. Manning (2015)
3. Dargie, W., Poellabauer, C.: Fundamentals of Wireless Sensor Networks: Theory and Prac-tice. Wiley (2010)
4. Heesuk, S., Byoungoh, K., Taehun, K., Dongman, L., Soon-Joo, H.: Sherlock-SD: a light-weight universal service discovery for web of things (WoT) services. In: Proceedings of 12th Annual IEEE Consumer Communications and Networking Conference (CCNC), pp. 276–282 (2015)
5. de Sousa, Al-Ali, A.R., Aburukba, R.: Role of internet of things in the smart grid technology. J. Comput. Commun. **3**(05), 229 (2015). M. Internet of Things with Intel Galileo. Packt Publishing Ltd
6. Choubey, P.K., Pateria, S., Saxena, A., Vaisakh Punnekkattu Chirayil, S.B., Jha, K.K., Sharana Basaiah, P.M.: Power efficient, bandwidth optimized and fault tolerant sensor management for IOT in smart home. In: Proceedings of IEEE International Advance Computing Conference (IACC), pp. 366–370 (2015)
7. Amadeo, M., Campolo, C., Iera, A., Molinaro, A.: Information centric networking in IoT scenarios: the case of a smart home. In: Proceedings of IEEE International Conference on Communications (ICC), pp. 648–653 (2015)
8. Brennan, C.P., McCullagh, P.J., Galway, L., Lightbody, G.: Promoting autonomy in a smart home environment with a smarter interface. In: Engineering in Medicine and Biology Society (EMBC)—Proceedings of 37th Annual International Conference of the IEEE, pp. 5032–5035 (2015)
9. Abdmeziem, M.R., Tandjaoui, D., Romdhani, I.: Architecting the internet of things: state of the art. In: Robots and Sensor Clouds, pp. 55–75. Springer (2016)
10. Santoso, F.K., Vun, N.C.: Securing IoT for smart home system. In: Proceedings of IEEE International Symposium on Consumer Electronics (ISCE), pp. 1–2 (2015)
11. Maksimovic, M., Vujovic, V., Davidovic, N., Milosevic, V., Perisic, B.: Raspberry Pi as internet of things hardware performances and constraints. In: Proceedings of IcETRAN: First International Conference on Electrical, Electronic and Computing Engineering, pp. 1–6 (2014)
12. Hussain, M.I., Dutta, S.K., Ahmed, N., Hussain, I.: A WiFi-based reliable network architecture for rural regions. ADBU J. Eng. Technol. **3**, (2016)

Self-Organizing Migrating Algorithm Used for Model Predictive Control of Semi-batch Chemical Reactor

Lubomír Macků and David Sámek

Abstract The current availability of powerful computing technologies enables using of complex computational methods. One of such complex method is also the self-organizing migrating algorithm (SOMA). This algorithm can be used for solving of various optimization problems. It may be used even for such complex task, as the non-linear process control is. In this paper, the capability of using SOMA algorithm for the model predictive control (MPC) of semi-batch chemical reactor is studied. The MPC controller including self-organizing migrating algorithm (SOMA) is used for the optimization of the control sequence. The reactor itself is used in chromium recycling process in leather industry.

Keywords Model predictive control · SOMA · Chemical reactor · Exothermic reaction · Mathematical modeling

1 Introduction

An enzymatic dechromation is a waste free technology which recycles waste originated during chrome tanning process and also waste generated at the end of the final product lifetime (used leather goods).

Part of the recycling process includes also an oxidation-reduction reaction which is strongly exothermic and can be controlled by the chromium filter sludge into the hot reaction blend of chromium sulphate acid dosing [1]. Optimization of this process control is studied in this paper.

L. Macků (✉) · D. Sámek
Faculty of Applied Informatics, Department of Security Engineering,
Tomas Bata University in Zlin, Nám. T.G. Masaryka 5555,
760 01 Zlín, Czech Republic
e-mail: macku@fai.utb.cz

© Springer International Publishing Switzerland 2016
R. Silhavy et al. (eds.), *Automation Control Theory Perspectives in Intelligent Systems*, Advances in Intelligent Systems and Computing 466,
DOI 10.1007/978-3-319-33389-2_25

255

2 Semi-batch Chemical Reactor

Process itself runs in a semi-batch chemical reactor. The chemical reactor is a vessel with a double wall filed with a cooling medium. It has a filling opening, a discharge outlet, cooling medium openings and a stirrer.

The reactor is filled with initial filling given by the solution of chemicals without the chromium sludge (filter cake). The sludge is fed into the reactor to control the developing heat since the temperature has to stay under a certain critical level ($T(t) < 373.15$ K), otherwise the reactor could be destroyed. On the other hand it is desirable to utilize the maximum capacity of the reactor to process the maximum amount of waste in the shortest possible time (higher temperature is desirable). Therefore an optimal control strategy has to find a trade-off between these opposite requirements.

3 Mathematical Model

The above mentioned reactor can be described by the mathematical model shown below. Under usual simplifications, based on the mass and heat balance, the following 4 nonlinear ordinary differential equations can be derived [2]:

$$\frac{d\,m(t)}{d\,t} = F_I \tag{1}$$

$$\frac{d\,a(t)}{d\,t} = \frac{F_I[1 - a(t)]}{m(t)} - A \cdot e^{-\frac{E}{R\cdot T(t)}} \cdot a(t) \tag{2}$$

$$\frac{d\,T(t)}{d\,t} = \frac{F_I \cdot c_I \cdot T_I}{m(t) \cdot c} + \frac{A \cdot e^{-\frac{E}{R\cdot T(t)}} \cdot \Delta H_r \cdot a(t)}{c} - \frac{K \cdot S \cdot T(t)}{m(t) \cdot c} + \frac{K \cdot S \cdot T_C(t)}{m(t) \cdot c} - \frac{T(t)F_I}{m(t)} \tag{3}$$

$$\frac{d\,T_C(t)}{d\,t} = \frac{F_C \cdot T_{CI}}{m_C} + \frac{K \cdot S \cdot T(t)}{m_C \cdot c_C} - \frac{K \cdot S \cdot T_C(t)}{m_C \cdot c_C} - \frac{F_C \cdot T_C(t)}{m_C} \tag{4}$$

Individual symbols have the following meaning: m is the total weight of reaction components in the reactor, a is the mass concentration of the reaction component in the reactor, $c = 4500$ J·kg·K^{-1} is the specific heat capacity of the reactor content and T its temperature. F_I, $T_I = 293.15$ K and $c_I = 4400$ J·kg·K^{-1} is the reaction component input mass flow rate, temperature and specific heat capacity. $F_C = 1$ kg·s^{-1}, $T_{CI} = 288.15$ K, T_C, $c_C = 4118$ J·kg·K^{-1} and $m_C = 220$ kg is the cooling water mass flow rate, input temperature, output temperature, specific heat capacity and weight of the cooling water in the cooling system of the reactor, respectively. Other constants: $A = 219.588 \ s^{-1}$, $E = 29967.5087 \ J \cdot mol^{-1}$, $R = 8.314 \ J \cdot mol^{-1} \cdot K^{-1}$, $\Delta H_r = 1392350 \ J \cdot kg^{-1}$, $K = 200 \ kg \cdot s^{-3} \cdot K^{-1}$, $S = 7.36 \ m^2$.

The fed-batch reactor use jacket cooling, but the effective heat-transfer area ($S = 7.36 \ m^2$) in the mathematical model was treated as constant, not time varying. The initial amount of material placed in the reactor takes about two-thirds of the in-reactor volume and the reactor is treated as ideally stirred, so we can do this simplification.

Variables F_I, F_C, T_I, T_{CI}, can serve as manipulated signals. However, from practical point of view, only F_I and F_C are usable. The T_I or T_{CI} temperature change is inconvenient due to the economic reasons (great energy demands).

3.1 Technological Limits and Variables Saturation

Maximum filling of the reactor is limited by its volume to $m = 2450$ kg approximately. The process of the chromium sludge feeding F_I has to be stopped by this value. Practically, the feeding F_I can vary in the range $F_I \in \langle 0; 3 \rangle$ kg·s^{-1}. As stated in the system description, the temperature $T(t)$ must not exceed the limit 373.15 K; this temperature value holds also for the coolant (water) but it is not so critical in this case as shown by the further experiments.

4 SOMA Algorithm

The Self-Organizing Migrating Algorithm is used for the above mentioned system optimization. SOMA algorithm can be used for optimizing any problem which can be described by an objective function. This algorithm optimizes a problem by iteratively trying to improve a candidate solution, i.e. a possible solution to the given problem. The SOMA has been successfully utilized in many applications [3–5], while interesting comparison to with simulated annealing and differential evolution is provided by Nolle et al. [6].

5 Control of the Reactor

The state of art of chemical reactors control presents Luyben in [7, 8], control and monitoring of batch reactors describes Caccavale et al. [9]. Generally, it can be stated that chemical reactors controllers uses various control methods, such as PI controllers, adaptive control methods, robust approaches, predictive control and the like [10–19]. The model predictive control [20–22] belongs to the one of the most popular and successful approaches for semi-batch reactors control. However, this methodology brings some difficulties in finding optimal control sequence especially when complex nonlinear model is utilized. Interesting way how to cope with the optimization problem offers the usage of evolutionary algorithms [23, 24].

In this paper two different approaches to the model predictive control of the given plant are introduced. At first, the model predictive controller that uses self-organizing migrating algorithm for the optimization of the control sequence is presented. This methodology ensues from model predictive control method [25] while it uses same value of the control signal for whole control horizon. This modification was applied in order to reduce computational demands of the controller. The classic MPC controller, which uses Matlab Optimization Toolbox for obtaining the optimal control sequence, is used as the comparative method. This MPC controller computes the control sequence at every sampling period but only first value is applied.

5.1 Notation

From the systems theory point of view the reactor manipulated variables here are input flow rates of the chromium sludge F_I and of the coolant F_C. Input temperatures of the filter cake T_I and of the coolant T_{CI} can be alternatively seen as disturbances. Although the system is generally MIMO, it is assumed just as a single input—single output (SISO) system in this chapter. Thus, the only input variable is the chromium sludge flow rate F_I and the coolant flow rate F_C is treated as a constant. This input variable, often called control action, is in the control theory denoted as u. The output signal to be controlled is again the temperature inside the reactor T. The output signal is usually in the control theory denoted as y and consequently the desired target value of the output signal as y_r.

5.2 Model Predictive Control

The main idea of MPC algorithms is to use a dynamical model of process to predict the effect of future control actions on the output of the process. Hence, the controller calculates the control input that will optimize the performance criterion J over a specified future time horizon [26]:

$$J(k) = \lambda \cdot \sum_{i=N_1}^{N_2} (y_r(k+i) - \hat{y}(k+i))^2 + \rho \cdot \sum_{i=1}^{N_u} (u_t(k+i-1) - u_t(k+i-2))^2$$

(5)

where k is discrete time step, N_1, N_2 and N_u define horizons over which the tracking error and the control increments are evaluated. The u_t variable is the tentative control signal, y_r is the desired response and $\hat{y}$ is the network model response. The parameters λ and ρ determine the contribution that the sums of the squares of the future control errors and control increments have on the performance index.

Typically, the receding horizon principle is implemented, which means that after the computation of optimal control sequence only the first control action is implemented. Then, the horizon is shifted forward one sampling instant and the optimization is again restarted with new information from measurements.

5.3   Model Predictive Control Using SOMA

All simulations with algorithm SOMA were performed in the Mathematica 8.0 software. Here the algorithm SOMA was used for the cost function (5) minimization and was set as follows: Migrations = 25; AcceptedError = 0.1; NP = 20; Mass = 3; Step = 0.3; PRT = 0.1; Specimen = {0.0, 3.0, 0.0}; Algorithm strategy was chosen All To One. First two parameters serve for the algorithm ending. Parameter "Migrations" determines the number of migration loops, "AcceptedError" is the difference between the best and the worst individuals (algorithm accuracy). If the loops exceed the number set in "Migrations" or "AcceptedError" is larger than the difference between the best and the worst individuals, the algorithm stops. Other parameters influence the quality of the algorithm running. "NP" is the number of individuals in the population (its higher value implicates higher demands on computer hardware and can be set by user), "Mass" is the individual distance from the start point, "Step" is the step which uses the individual during the algorithm, "PRT" is a perturbation which is similar to hybridizing constant known from genetic algorithms or differential evolutions. "Specimen" is the definition of an exemplary individual for whole population. For details see [27].

There were performed seven different simulations using SOMA algorithm described in this section. First tree simulations (SOMA1–SOMA3) were done to study the control horizon N_u influence, next three (SOMA4–SOMA6) the prediction horizon N_2 influence and the last one (SOMA7) is the simulation with an optimal setting. All settings can be seen in Table 1.

Table 1 SOMA controller settings

	λ	ρ	N_2	N_u
SOMA1	1	1	300	30
SOMA2	1	1	300	60
SOMA3	1	1	300	90
SOMA4	1	1	200	60
SOMA5	1	1	280	60
SOMA6	1	1	360	60
SOMA7	1	1	320	60

Fig. 1 Results of SOMA7
simulations

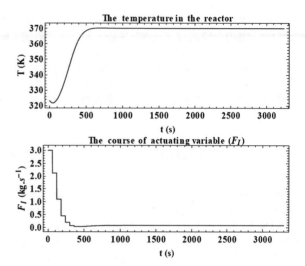

Graphical output of SOMA7 (the optimal settings) simulation is depicted in Fig. 1. Only two most important dependencies are shown—the in-reactor temperature and the chromium sludge dosing development. As was already said, the temperature has to stay under critical point 373.15 K. The chromium sludge dosing shouldn't embody any rapid changes.

The control horizon (N_u) actually means the time interval, for which the actuating variable (F_I) has constant value. It is generally better to set it as short as possible because of more rapid influence on the system, but on the other hand it increases the computing time during the calculations. So it is necessary to find the control horizon value, which balance between these two requirements.

The prediction horizon (N_2) determines how forward controller knows the system behavior. If the horizon is too short, the controller doesn't react in time and the system may become uncontrollable. Long horizon means again the more demanding computation, i.e. the need of more powerful computer hardware.

5.4   Conventional Model Predictive Approach

The comparative method is based on the classical MPC strategy described in the Sect. 5.2. The optimization utilizes standard Matlab Optimization Toolbox function `fmincon` and receding control strategy was implemented. The `fmincon` function used trust-region-reflective algorithm [28].

For the reason that the compared controllers SOMA applied the same control action for whole length of the control horizon $N_u = 60$ (with the sample time = 1 s), the sample time of the comparative controllers was set to 60 s. The control

Table 2 Settings of the controllers

	λ	ρ	γ	γ_c	N_1	N_2	N_u
MPC1	1	100	0	0	1	10	10
MPC2	1	100	2000	100	1	10	10
MPC3	1	100	2000	200	1	10	10
MPC4	1	100	1500	100	1	10	10

horizon N_u was set to 10 as well as the prediction horizon N_2. The rest of the controller design remained same—the predictor was based on the white-box model described by Eqs. (1)–(4), cost function of MPC was the same.

The control loop was simulated using Matlab/Simulink, the controller was built-in as the Matlab M-File and predictor used Matlab S-Function technology.

However, such controller (Table 2 MPC1 settings) that follows Eq. (5), did not provide acceptable results. It was not possible to avoid either the overshoot of the in-reactor temperature or the permanent control error using the control configuration. It results from simulations that the biggest problem comes in the very beginning of the control, when the control error is the highest and the controller performs enormous control actions. This is caused by the fact that the reaction is strongly exothermic and even small concentration of the filter cake causes steep rise of the temperature. Thus, the enhancement of the criterion (5) that penalizes values of the control signal in the beginning of semi-batch process was necessary. Also, at the same time the penalization has to decrease taperingly (6) and (7):

$$J(k) = \lambda \cdot \sum_{i=N_1}^{N_2} (y_r(k+i) - \hat{y}(k+i))^2 + \rho \cdot \sum_{i=1}^{N_u} (u_t(k+i-1) - u_t(k+i-2))^2$$

$$+ \gamma(k) \cdot \sum_{i=1}^{N_u} u_t(k+i) \tag{6}$$

$$\gamma(k) = \gamma(k) - \gamma_c \tag{7}$$

where γ_c is the parameter which defines the speed of the decrement in γ. This enhancement brought possibility to influence more intensely the speed of dosing of the filter cake with two parameters. In other words γ parameter defines the level of penalization of the control signal, while the ratio $\gamma/\gamma c$ specifies the length of the penalization interval. However, too high γ parameter or γ/γ_c ratio caused unavailing delays or oscillations (the settings *MPC2* in Table 2). On the other hand, small γ/γ_c ratio led to overshoots of the temperature (the settings *MPC3* in Table 2). The best result which was obtained using this approach was obtained for *MPC4* settings and is presented in Fig. 2.

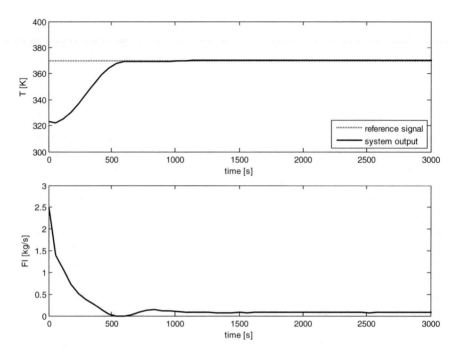

Fig. 2 Control of the semi-batch reactor using *MPC4* controller

6 Comparison

The best representatives of the two different approaches were selected to be compared in this section. In order to describe the control error the criterion function S_y was defined:

$$S_y = \sum_{i=1}^{t_f} (y_r(i) - y(i))^2 \tag{8}$$

The criterion S_u describes the speed of the control signal changes. Monitoring of it is very important, because lifetime of the mud pump that injects the filter cake solution to the reactor would be significantly shortened due to steep changes of delivery.

$$S_u = \sum_{i=1}^{t_f} (u(i+1) - u(i))^2 \tag{9}$$

The t_f defines number of steps for that are the criterions S_y and S_u computed. In this paper the t_f was set to 50 steps, because only the first 3000 s provide

Table 3 Comparison of the controllers

Controller	S_y (K^2)	S_u (kg$^2 \cdot$s^{-2})	y_{max} (K)	t_b (s)
SOMA7	9.2571×10^3	2.3200	370.1740	3242
MPC4	1.0333×10^4	1.5500	370.2358	3300

information about control. The dosing (control) commonly finishes shortly after 3000 s and after that only cooling is performed.

For the reason that the plant is very exothermic and it is very sensitive to the exceeding of the desired value of the temperature (y_r = 370 K), it was necessary to observe the maximum overshoot of the output value y_{max}. Furthermore, it is essential to observe the time of the reaction (dosing) t_b.

As can be seen from Table 3, the in-reactor temperature overshoots were in both cases quite similar. Nevertheless, the overshoot obtained by using SOMA algorithm is a little bit lower. As far as the criterion S_y, we need as low value as possible—the lower value the better. Even if the results were again close, the SOMA provided the lower value, which means that quality of the control was better too. The time of dosing achieved by SOMA was shorter approximately for 1 min (58 s).

The only criterion which was better in MPC control was the S_u. Here the value 2.3200 is quite higher than 1.5500. For the actuating device it means higher loading and shorter lifetime.

7 Conclusion

The paper presented comparison of two approaches to semi-batch process model predictive control. Firstly, the Self Organizing Migrating Algorithm implemented in Mathematica software was used for cost function optimization. Secondly, as the comparative method model predictive controller implemented in Matlab was selected. This controller used built-in function `fmincon` from Matlab Optimization Toolbox instead of SOMA. It can be concluded that both algorithms provided applicable results. Nevertheless, SOMA provided slightly faster semi-batch process. What is more, the comparative method based on Matlab Optimization Toolbox required modification of the cost function in order to obtain appropriate control.

The comparison itself was made from four points of view. Firstly, the value of the in-reactor temperature overshoot and the related quality of the in-reactor temperature course were observed. Secondly, the time of processing which is important for effectiveness of a real plant and also the course of the actuating signal that is important from the practical point of view were monitored.

The paper showed that Self Organizing Migrating Algorithm successfully improved the control performance of the selected plant. It can be concluded that evolutionary algorithms should be considered as good way how to optimize control sequence in model predictive control, especially in case of complex nonlinear systems.

References

1. Kolomaznik, K., Adamek, M., Uhlirova, M.: Potential danger of chromium tanned wastes. In: HTE'07: Proceedings of the 5th IASME/WSEAS International Conference on Heat Transfer, Thermal Engineering and Environment, pp. 136–140, WSEAS, Athens, Greece (2007)
2. Gazdos, F., Macku, L.: Analysis of a semi-batch reactor for control purposes. In: Proceedings of 22nd European Conference on Modelling and Simulation ECMS 2008, pp. 512–518. ECMS, Nicosia, Cyprus (2008)
3. Coelho, L.S., Marianib, V.C.: An efficient cultural self-organizing migrating strategy for economic dispatch optimization with valve-point effect. Energy Convers. Manage. **51**(12), 2580–2587 (2010)
4. Deep, K.: A self-organizing migrating genetic algorithm for constrained optimization. Appl. Math. Comput. **198**(1), 237–250 (2008)
5. Singh, D., Agrawal, S.: Log-logistic SOMA with quadratic approximation crossover. Paper presented at the International Conference on Computing, Communication and Automation, ICCCA **146–151**, 2015 (2015)
6. Nolle, L., Zelinka, I., Hopgood, A.A., Goodyear, A.: Comparison of an self-organizing migration algorithm with simulated annealing and differential evolution for automated waveform tuning. **36**(10), 645–653 (2005)
7. Luyben, W.L.: Chemical Reactor Design and Control. Wiley, Hoboken, NJ (2007)
8. Luyben, W.L.: Process Modeling, Simulation and Control for Chemical Engineers. McGraw-Hill, New York (1996)
9. Caccavale, F., Iamarino, M., Pierri, F., Tufano, V.: Control and Monitoring of Chemical Batch Reactors. Springer, London (2011)
10. Aguilar-Garnica, E., Garcia-Sandoval, J.P., Gonzalez-Alvarez, V.: PI controller design for a class of distributed parameter systems. Chem. Eng. Sci. **66**(15), 4009–4019 (2001)
11. Aguilar-Lopez, R., Martinez-Guerra, R., Maya-Yescas, R.: Temperature regulation via PI high-order sliding-mode controller design: application to a class of chemical reactor. Int. J. Chem. Reactor Eng. **7**(1) 2009
12. Vojtesek, J., Dostal, P.: Simulation of Adaptive Control Applied on Tubular Chemical Reactor. WSEAS Trans. Heat Mass Transf. **6**(1), 1–10 (2011)
13. Vojtesek, J., Dostal, P.: Two types of external linear models used for adaptive control of continuous stirred tank reactor. In: Proceedings of the 25th European Conference on Modelling and Simulation, pp. 501–507, ECMS, Krakow, Poland (2011)
14. Leosirikul, A., Chilin, D., Liu, J., Davis, J.F., Christofides, P.D.: Monitoring and retuning of low-level PID control loops. Chem. Eng. Sci. **69**(1), 287–295 (2012)
15. Du, W., Wu, X., Zhu, Q.: Direct design of a U-model-based generalized predictive controller for a class of nonlinear (polynomial) dynamic plants. Proc. Inst. Mech. Eng. Part I: J. Syst. Control Eng. **226**(1), 27–42 (2012)
16. Zhang, R., Xue, A., Wang, S.: Dynamic Modeling and Nonlinear Predictive Control Based on Partitioned Model and Nonlinear Optimization. Ind. Eng. Chem. Res. **50**(13), 8110–8121 (2011)
17. Matusu, R., Zavacka, J., Prokop, R., Bakosova, M.: The Kronecker summation method for robust stabilization applied to a chemical reactor. J. Control Sci. Eng. **2011**, article ID 273469 (2011)
18. Hosen, M.A., Hussain, M.A., Mjalli, F.S.: Control of polystyrene batch reactors using neural network based model predictive control (NNMPC): An experimental investigation. Control Eng. Pract. **19**(5), 454–467 (2011)
19. Rani, K.Y., Patwardhan, S.C.: Data-driven model based control of a multi-product semi-batch polymerization reactor. Chem. Eng. Res. Des. **85**(10) (2007)
20. Xaumiera, F., Le Lann, M.V., Cabassud, M., Casamatta, G.: Experimental application of nonlinear model predictive control: temperature control of an industrial semi-batch pilot-plant reactor. J. Process Control (2002). doi:10.1016/S0959-1524(01)00057-9

21. Hvala, N., Aller, F., Miteva, T., Kukanja, D.: Modelling, simulation and control of an industrial, semi-batch, emulsion-polymerization reactor. Comput. Chem. Eng. **35**(10), 2066–2080 (2011)
22. Oravec, J., Bakošová, M.: Robust model-based predictive control of exothermic chemical reactor. Chem. Pap. **69**(10), 1389–1394 (2015)
23. Dao, T.T.: Investigation on evolutionary computation techniques of a nonlinear system. Model. Simul. Eng. **2011**, Article ID 496732 (2011)
24. Zelinka, I., Davendra, D., Šenkeřík, R., Pluháček, M.: Investigation on evolutionary predictive control of chemical reactor. J. Appl. Logic **13**(2), 156–166 (2015)
25. Bouhenchir, H., Cabassud, M., Le Lann, M.V.: Predictive functional control for the temperature control of a chemical batch reactor. Comp Chem Eng **30**(6–7), 1141–1154 (2006)
26. Camacho, E.F., Bordons, C.: Model Predictive Control in the Process Industry. Springer, London (2004)
27. Zelinka, I.: SOMA—Self Organizing Migrating Algorithm. In: Onwubolu, G., Babu, B.V. (eds.) New Optimization Techniques in Engineering, pp. 167–217. Springer, London, UK (2004)
28. Coleman, T.F., Zhang, Y.: Fmincon [online]. Mathworks, Natick. http://www.mathworks.com/help/toolbox/optim/ug/fmincon.html. Accessed 25 Sept 2011

Model of Surveillance System Based on Sound Tracking

Martin Papez and Karel Vlcek

Abstract Research in the audio and video surveillance is gaining more popularity due to its wide-spread applications as well as social impact. There have been considerable efforts which are focused on developing various algorithms and models for surveillance systems. Recently, the attention in audio and video surveillance has turned towards to design autonomous detection system. In this paper a surveillance system based on sound source localization by a microphone array is presented. The design of microphone array has an important issue in the accuracy of source localization. For the surveillance application, beamforming algorithms are implemented to increase the tracking and detection performance. The real-time capabilities of an automated surveillance system are analyzed, providing a performance analysis of the localization system under different acoustic conditions.

Keywords Beamforming · Microphone array · Signal processing · Surveillance

1 Introduction

In recent years, there has been an increasing interest in the automated surveillance systems. The particular interest in this area has dramatically increased with the advances of information technology. In particular, embedded devices has substantially improved in the last few years, finally reaching an affordable price. Audio and video surveillance is a multidisciplinary area, involving several fields such as artificial intelligence, data mining, pattern recognition and signal processing. In effect, the combination of all these fields is necessary to tackle challenging tasks, such as real-time source surveillance.

M. Papez (✉) · K. Vlcek (✉)
Faculty of Applied Informatics, Department of Computer and Communication Systems, Tomas Bata University in Zlin, Nad Stranemi 4511, 760 06 Zlin, Czech Republic
e-mail: papez@fai.utb.cz

K. Vlcek
e-mail: vlcek@fai.utb.cz

© Springer International Publishing Switzerland 2016
R. Silhavy et al. (eds.), *Automation Control Theory Perspectives in Intelligent Systems*, Advances in Intelligent Systems and Computing 466,
DOI 10.1007/978-3-319-33389-2_26

267

Automated video surveillance is a rapidly evolving area in which hardware components are controlled by system intelligence rather than manual intervention. These systems allow more efficient and effective surveillance by employing intelligent audio and video processing algorithms. As with many signal processing applications, there is a need for high performance systems to support real-time automated video surveillance. In recent years, an emerging trend is the use of multi-processor architectures to support computationally intensive applications such as audio and video processing applications in real-time [1]. In contrast, automated surveillance generally requires an embedded system for low cost, low power consideration.

In the previously published studies on video surveillance, the relative importance of microphone array processing has been a subject of considerable discussion [1–4]. Research on the subject has been mostly limited by an insufficient performance in computing demands. Recently investigators have examined the effects of microphone array processing on surveillance system performance [1, 5]. Although some research has been carried out on microphone array processing, only few studies have attempted to investigate in automated audio and video surveillance systems. Currently, the problem of audio and video signal processing by multiple sensors has established relatively low costs solutions especially due to utilizing the Micro-Electro-Mechanical Systems (MEMS). MEMS achieved even with sufficient power and are used for challenges such as locating and filtering incident broadband signal [5, 6].

In this paper the video surveillance system based on microphone array processing is presented. The microphone array design and implementation is discussed. In addition, this paper describes DAS and MVDR beamforming as essential locating and tracking algorithm. The results of adjusted surveillance system based on source localization are obtained on the basis of MATLAB framework.

2   Microphone Array

Microphone array technology has been widely used for the localization of sound sources [7–11]. The localization of sound sources in acoustic environments with noise and reverberation is a challenging task for applications arising in a wide range of areas. The emergence of array signal processing techniques is offering improved system performance for multiple input systems. A comprehensive overview about the field can be found in [12–14]. The multi-channel system allows to solve problems, such as source localization and tracking, which is difficult with single-channel systems.

Lets assume an ULA (Fig. 1) microphone array consisting of number of elements M with microphone spacing d. By summarizing all time-shifted signals captured by all microphones the array output $y[n]$ is given by formula (1),

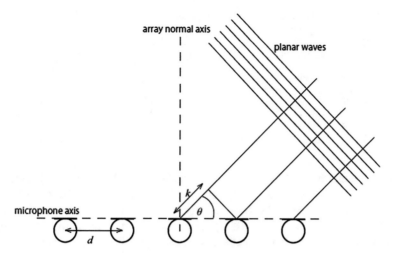

Fig. 1 An ULA, illustration of DoA estimation based on distance between the microphones d

$$y[n] = \sum_{m=0}^{M-1} x[n - m\tau] \tag{1}$$

Assume that, $x[n] = d[n]$, then $y[n] = h[n]$ is the impulse response of the system. Frequency response $h[n]$ can be determine by taking its discrete Fourier transform using Eq. (2).

$$H(\omega, \Theta) = \sum_{m=0}^{M-1} e^{-j2\pi\omega(m\tau)} = \sum_{m=0}^{M-1} e^{-j2\pi\omega\left(m\left(\frac{d\cos(\Theta)}{c}\right)\right)} \tag{2}$$

The spatial response can be visualized by plotting the magnitude of $H(\omega, \Theta)$ as a function of Θ, while d, M, and ω are held fixed. Such a representation, plotted in polar coordinates is called a beampattern.

3   Beamforming

Beamforming is a signal processing technique used in sensor arrays for directional signal transmission or reception. This spatial selectivity is achieved by using adaptive or fixed receive beampattern. The beampattern is formed by adjusting complex weights of the array elements so that the beam is addressed in the direction of interest [15].

Adaptive Beamforming [16] is a technique in which an array of sensors is exploited to achieve maximum reception in a specified direction by estimating the

Table 1 Examples of generalized cross-correlation functions

Cross-correlation method	Weighting function
Maximum likelihood weighting	$\hat{W}_{ML}(t,\omega) = \dfrac{|x_1(t,\omega)||x_2(t,\omega)|}{|N_1(t,\omega)|^2|X_2(t,\omega)|^2 + |N_2(t,\omega)|^2|X_1(t,\omega)|^2}$
Phase transform weighting	$W_{PHAT}(t,\omega) = \left|X_1(t,\omega)X_2^*(t,\omega)\right|^{-1}$
Bandpass weighting	$W_{BP}(t,\omega) = \begin{cases} 1, & 2\pi\,300\,\text{Hz} \le \omega \le 2\pi\,6000\,\text{Hz} \\ 0, & \text{otherwise} \end{cases}$
Cross power spectrum phase	$S(t,\omega) = \dfrac{X_1(t,\omega)X_2^*(t,\omega)}{|X_1(t,\omega)||X_2^*(t,\omega)|}$

arrival of signal from a desired direction while signals of the same frequency from other directions are rejected. This is achieved by varying the weights of each of the sensors used in the array. The type of filtering or weighting functions used with the Generalized Cross-Correlation (GCC) function is crucial to the performance. Some of the well-known weighting functions proposed in the literature are outlined below (Table 1).

Beamforming techniques performance determination is based on the array gain G, which represents the improvement of the SNR of the array compared to an individual sensor (3),

$$G = \frac{SNR_{array}}{SNR_{sensor}} \tag{3}$$

By optimizing the array gain, more sophisticated methods, known as super-directive beamformers, can be used to improve the beamformer's directional selectivity, further cancelling undesired sources. In the almost 40 years of development of acoustic array technology, numerous beamforming modifications, as well as array geometries, have been suggested to improve the overall performance of array systems [11, 17, 18].

3.1 Delay-and-Sum Beamforming (DAS)

The simplest approach of the sound source localization methods is based on DAS beamforming. DAS beamforming provides an elegant way to extract the signal from a desired source through spatial filtering.

The principle behind DAS beamforming is shown in Fig. 2 in the presence of a propagating wave, the signals captured by the microphones are delayed by a proper amount before being added together, to strengthen the resulting signal with respect to noise or waves propagating in other directions (4),

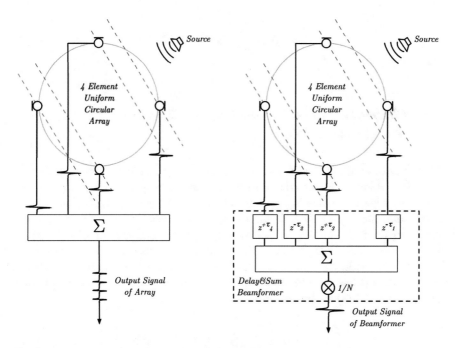

Fig. 2 **a** Array processing without a beamformer. **b** Delay-and-Sum beamforming

$$y(t, \hat{\kappa}) = \sum_{m=0}^{M-1} x_m \omega_m \big(t - \tau_m(\hat{\kappa})\big), \tag{4}$$

where M is the number of microphones, x_m is the signal amplitude measured with the m microphone, ω_m is its associated amplitude weighting and $\tau_m(\hat{\kappa})$ is the delay applied. Weight coefficients for a single microphone are calculated in frequency domain according to formula (5) [5, 8] (Table 2),

Table 2 DAS beamforming—algorithm description

DAS beamformer algorithm
1: **for** $j=1:N$ **do**
2:   **for** $k=1:N$ **do**
3:     $\mathbf{w}(j,:,k) = \dfrac{1}{N_m} e^{-i\frac{2\pi f(k)}{c}\frac{d}{2}\cos(\varphi(j)-\phi_n)}$ $\triangleright$ **w** is a $\big(N_\varphi \times N_m \times N_f\big)$-matrix
4:   **end for**
5: **end for**

$$\mathbf{w}(\omega) = \frac{1}{N} \sum_{n=1}^{N} e^{-j\frac{\omega d}{c2}\cos(\varphi - \phi_n)} \tag{5}$$

3.2 Minimum Variance Distortionless Response Beamformer (MVDR)

MVDR beamforming is a well-known and extensively used beamforming technique that offers a sufficient spectral characteristic of the output and is therefore well suited to acoustic beamforming and speech enhancement [7–10]. This method results from modifying DAS Beamforming in the frequency domain which offers much higher dynamic range than other techniques. The aim of MVDR beamforming is to minimize the power of the output signal of the array while maintaining unity gain in the look direction and also maximizing the white noise gain. MVDR beamforming is based on filter-delay-sum beamforming and its frequency domain output signal $Y(e^{j\omega})$ is defined as:

$$Y\left(e^{j\omega}\right) = \sum_{m=0}^{M-1} w_m\left(e^{j\omega}\right) X_m\left(e^{j\omega}\right) = \mathbf{w}^{\mathbf{H}}\mathbf{X}, \tag{6}$$

where $w_m(e^{j\omega})$ denotes the filter coefficients of the beamformer for sensor m at frequency ω, $X_m(e^{j\omega})$, are the microphone input signals and $[\ ]^H$ denotes the matrix transpose conjugate. In addition to is suggested to minimize the total output power $Y_m(e^{j\omega})$ under the assumption of a diffuse noise field in order to optimize spatial filtering with respect to reverberant environments. This leads to the super-directive beamformer with weight vector (7),

$$\mathbf{w}(\omega) = \frac{T^{-1}(\omega)\mathbf{v}(\omega)}{\mathbf{v}^H(\omega)T^{-1}(\omega)\mathbf{v}(\omega)}, \tag{7}$$

where $T_{i,j}(\omega)$ denotes the coherence of an isotropic noise field with absolute magnitude matrix of the microphone inter-distance. MVDR beamformer, therefore optimizes its steering vector, beamforming frequency spectrum and spatial characteristic for any input signal that is not considered uniformly distributed white noise [1, 5, 11, 19] (Table 3).

Table 3 MVDR beamforming—algorithm description

Algorithm	MVDR beamformer
1: $\gamma = \mathbf{x}$	
2: $\lambda = a_s$	
3: **for** $j = 1 : N_\varphi$ **do**	
4: **for** $k = 1 : N_f$ **do**	
5: $d(:) = e^{i\frac{2\pi f(k)}{c}\frac{d}{2}\cos(\varphi(j) - \phi_n)}$	
6: $\boldsymbol{R}_{xx} = \lambda \cdot \boldsymbol{R}_{xx} + (1 - \lambda) \cdot n(\omega) n^H(\omega) + \gamma \boldsymbol{I}$	
7: $\mathbf{w}(j,:,k) = \dfrac{\boldsymbol{R}_{xx}^{-1} \cdot d}{d^H \boldsymbol{R}_{xx}^{-1} \cdot d}$ $\triangleright$ $\mathbf{w}$ *is a* $\left(N_\varphi \times N_m \times N_f\right) - matrix$	
8: **end for**	
9: **end for**	

4 MATLAB Model

An evaluation of proposed audio surveillance system is on basis of MATLAB framework. The aim of this investigation is to create a prototype of surveillance system based on proposed design of microphone array structure and implementation of detection algorithm. For further performance analysis of audio surveillance, MATLAB simulations of localization and detection algorithms are carried out. The results are discussed below.

4.1 Design of Microphone Array

In this section, design of wideband beamformer and simulation results are presented. MEMS microphone array was assembled in MATLAB simulations to verify receiver characteristics, which is used for spatial filtering of the processed signal are obtained from the simulations. Increasing of the number of MEMS microphones in array lead to more accurate signal processing and localization properties. However, it also increasing computing demands (Fig. 3).

The required relative bandwidth of processed wideband signal was the main criteria for design of MEMS array. The relative bandwidth was selected to 0.8–4 kHz Another demand was placed on the size of the device, which should not exceed the maximum size of $S = 0.06\,\text{m}^2$ and the maximum cylinder diameter $D = 0.088\,\text{m}$. Accordingly to these conditions the minimum distance between individual MEMS sensors was adjusted to $d \geq 0.03\,\text{m}$. Due to this conditions, the simulations of array has been limited to a maximum number of MEMS sensors that

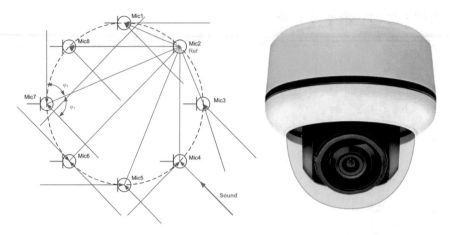

Fig. 3 The draft microphone array distribution for audio and video surveillance device

were used. In the simulations of the uniform array were established the maximum of MEMS sensor to $N = 8$.

The proposed MEMS microphone array distribution is illustrated in the Fig. 4, where the heterogeneous array structure is used. The array geometry was chosen in order to allow beamsteering on the horizontal and vertical planes. Selected array is considered from 4 MEMS sensors with the microphone spacing d {42.85 mm, 74.25 mm}.

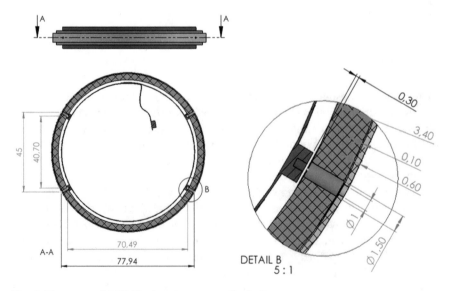

Fig. 4 The proposed MEMS microphone array distribution

4.2 Beamforming

MATLAB simulations have been carried out for proposed heterogeneous microphone array. Beamformers (DAS, MVDR) were considered for incoming signal sources from various angles. In the simulations, the numerous sampling frequencies {8 kHz, 16 kHz, 48 kHz, 96 kHz, 192 kHz} were adjusted. Also the length of window function (FFT modulator) were set up from 256 to 4096 samples. Accordingly to microphone array structure, three Butterworth bandpass filters [950 Hz, 1150 Hz], [1950 Hz, 2150 Hz], [2550 Hz, 2750 Hz] were composed. The estimation of the weight vector of a null steering beamformer was calculated by using weight Eqs. (5) and (7). SNR were assumed to uniformly distributed for all incoming directions. Based on these conditions and acquaintances wideband signal, array simulations were performed.

The simulation results showed that proposed design of microphone array provides high accuracy while processing the wideband signal. The relative bandwidth of microphone arrays is much larger than that of wideband arrays operating at the range of frequencies 500 Hz–5 kHz. As a consequence, microphone arrays are a special challenge for wideband beamformer designs. Two different beamforming

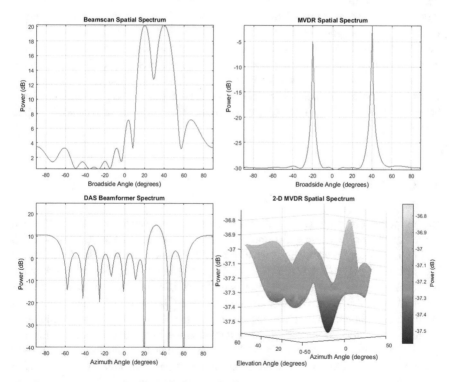

Fig. 5 MATLAB simulation for signals departing the microphone array. *Left* DAS beamformer. *Right* MVDR beamformer

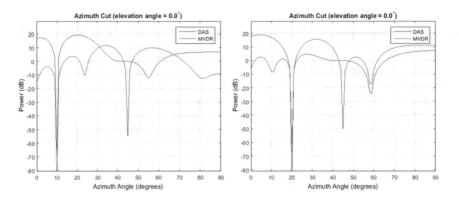

Fig. 6 Comparison of DAS and MVDR beamformer

techniques have been examined in MATLAB simulations. Sampling frequency were adjusted to 8096 samples and the length of FFT modulator were 1024. The normalized outputs obtained for DAS and MVDR beamformers are shown in Fig. 5, where the array is processing signals from various sources (Fig. 5a, DAS beamformer from 20° and 40°; Fig. 5b, MVDR beamformer from −20° and 40° with angle elevation from 0° to 45°). The comparison of proposed algorithms— DAS and MVDR beamformer can be better seen in Fig. 6, where both techniques have been examined. Two interference signals were arriving −5° and 10° elevation along the directions [10°; 50°] and [20°; −60°] degrees, respectively. Adaptive beamformers provides superior interference rejection compared to that offered by conventional beamformers. When the knowledge about the target direction is not accurate, the MVDR beamformer is preferable because it prevents signal self-nulling.

5 Conclusion

In this paper the proposal of microphone array processing for audio and video surveillance system is analyzed and described. The numerous simulation of DAS and MVDR beamformer were carried out in MATLAB framework. Furthermore, the results show that microphone array processing is capable of performing admirable achievement for automated video surveillance when the correct recognition algorithm is used. The results indicate, overall, that accordingly with recent studies microphone array processing can significantly improve the performance of source surveillance. This research paper demonstrates that sound source localization can be efficiently performed in real-time surveillance applications. In further research, our objective is to create a prototype of an embedded audio and video surveillance system with proposed microphone array structure.

Acknowledgments This work was supported in frame of Internal Grant Agency of Tomas Bata University in Zlin, Faculty of Applied Informatics IGA/CebiaTech/2015/005 and IGA/FAI/2016/005.

References

1. Ngo, H.T., Ives, R.W., Rakvic, R.N., Broussard, R.P.: Real-time video surveillance on an embedded, programmable platform. Microprocess. Microsyst. **37**(6–7), 562–571 (2013). doi:10.1016/j.micpro.2013.06.003
2. Habib, T.: Doctoral thesis auditory inspired methods for multiple speaker localization and tracking using a circular microphone array (2011)
3. Denman, S., Kleinschmidt, T., Ryan, D., Barnes, P., Sridharan, S., Fookes, C.: Automatic surveillance in transportation hubs: no longer just about catching the bad guy. Expert Syst. Appl. **42**(24), 9449–9467 (2015). doi:10.1016/j.eswa.2015.08.001
4. Kim, H., Lee, S., Kim, Y., Lee, S., Lee, D., Ju, J., Myung, H.: Weighted joint-based human behavior recognition algorithm using only depth information for low-cost intelligent video-surveillance system. Expert Syst. Appl. **45**, 131–141 (2016). doi:10.1016/j.eswa.2015.09.035
5. Zwyssig, E.: Speech processing using digital MEMS microphones. Doctor of Philosophy Centre for Speech Technology Research School of Informatics, University of Edinburgh (2013)
6. Papez, M., Vlcek K.: Enhanced MVDR beamforming for MEMS microphone array, MCSI, 2015. In: 2015 International Conference on Mathematics and Computers in Sciences and in Industry (MCSI). ISBN: 978-147994324-1
7. Bitzer, J., Simmer, K.U.: Superdirective microphone arrays. In: Microphone Arrays, pp. 19–38. Springer (2001)
8. Elko, G.W., Meyer, J.: Microphone arrays. In: Springer Handbook of Speech Processing, pp. 1021–1041. Springer (2008)
9. Balasem, S., Tiong, S., Koh, S.: Beamforming algorithms technique by using MVDR and LCMV. World Appl. Program. **2**(5), 315–324 (2011). http://waprogramming.com/papers/50af7cd413e266.17413401.pdf
10. Doblinger, G.: An adaptive microphone array for optimum beamforming and noise reduction. In European Signal Processing Conference (2006)
11. Doclo, S., Moonen, M.: Superdirective beamforming robust against microphone mismatch. IEEE Trans. Audio Speech Lang. Process. **15**(2), 617–631 (2007). doi:10.1109/TASL.2006.881676
12. Krim, H., Viberg, M.: Two decades of array signal processing research: the parametric approach. IEEE Signal Process. Mag. **13**, 67–94 (1996)
13. Brandstein, M.S., Ward, D.B. (eds.): Microphone Arrays: Signal Processing Techniques and Applications. Springer-Verlag, Heidelberg (2001)
14. Benesty, J., Chen, J., Huang, Y.: Springer Topics in signal processing: microphone array signal processing. In: Benesty, J., Kellermann, W. (eds.). Springer, Heidelberg (2008)
15. Compton, R.: Adaptive Antennas Concept and Performance. Prentice Hall (2011)
16. Bellofiore, S., Foutz, J., Balanis, C.A., Spanias, A.S.: Smart-antenna system for mobile communication networks Part 2: beamforming and network throughput IEEE Antennas Propag. Mag. **44**(4) (2002)
17. Chen, Z., Li, Z., Wang, S., Yin, F.: A microphone position calibration method based on combination of acoustic energy decay model and TDOA for distributed microphone array. Appl. Acoust. **95**, 13–19 (2015). doi:10.1016/j.apacoust.2015.02.013

18. Friedland, G., Janin, A., Imseng, D., Anguera, X., Gottlieb, L., Huijbregts, M., Knox, M., Vinyals, O.: The ICSI RT-09 speaker diarization system. IEEE Trans. Audio Speech Lang. Process. **20**(2), 371–381 (2012)
19. Zhu, G., Xie, H., Zhou, Y., Xie, H., Zhou, Y.: Author ' s accepted manuscript to appear in : signal processing (2015)

Adaptive Decentralized Controller for Regulating an Elastic Coupled Multi-motor System

Essam A.G. El-Araby, Mohammad A. El-Bardini
and Nabila M. El-Rabaie

Abstract Elastic coupled multi-machine system (ECMMS) is a one of the MIMO nonlinear complicated systems in industry. In such system, mechanical vibrations are resulted due to mechanical elastic coupling through shafts of the multiple motors of the system. One solution is to apply local controller for each motor of the system. The local controllers have to be adaptive to maintain the alignment irrespective of multiple or shared load variations and other variable disturbances. So, this paper presents a method to equalize against the resulted mechanical oscillation, and to maintain the alignment among motors in the ECMMS by designing a decentralized control system associated with an adaptation method of the parameters of local controllers to be automatically adapted with variable operating conditions. An experimental set-up of ECMMS is used to apply the proposed approach practically. The simulation data are compared to the real time data provided from the experimental set-up.

Keywords Elastic coupled multi-motor system · Adaptive PID · Decentralized control

1 Introduction

Speed control of motor drives in ECMMS is a very important task. Coupled motors are commonly used in multiple fields in industry. Each motor in ECMMS has to be speed controlled at the same speed of the other motors in order to maintain constant

E.A.G. El-Araby (✉) · M.A. El-Bardini · N.M. El-Rabaie
Faculty of Electronic Engineering, Menoufia University,
Shibin Al Kawm, Al Minufiyah, Egypt
e-mail: essam.elarabi@el-eng.menofia.edu.eg

M.A. El-Bardini
e-mail: dralbardini@ieee.org

N.M. El-Rabaie
e-mail: nabila2100@gmail.com

© Springer International Publishing Switzerland 2016
R. Silhavy et al. (eds.), *Automation Control Theory Perspectives
in Intelligent Systems*, Advances in Intelligent Systems and Computing 466,
DOI 10.1007/978-3-319-33389-2_27

279

tension on the material used for coupling in the ECMMS. This case is required in industries like tissue paper machines, textile, wire, and others. In other cases it is required to control each motor at a different speed in order to have graduated tension on the processed material. A common example is the multi-motor system driving the rolling mills in the steel industry to product out different thickness sheet of steel along the production line. Other robotic application which uses a variable tension rope for robot-human interaction to compromise between wide bandwidth and low stiffness transmission on one hand and achieving a precise interaction force response and vibration free performance on the other hand [1]. This task was achieved via control of two identical direct drive motors which are coupled to each other through the controlled tension rope.

Costas A. Michael et al. have investigated in their work the influence of stiffness and damping of real elastic shafts on the behavior of multi-drive system driving a tissue paper machine [2]. In the area of industrial electro-mechanical coupled systems, the torsional resonance problems in ECMMS were noted first in rolling mill applications [3–7].

Different configurations and control methodologies of PID controllers are still giving satisfied results in the current research work. Relating the hydraulic systems, a decentralized PID controller gave reliable results in regulating hydraulic elements at irrigation channel [8–12].

The work which is presented in this paper focuses on regulating and reducing the effects of mechanical elastic coupling in ECMMS by maintaining the alignment among interactively driven motors of ECMMS. The proposed technique depends on applying decentralized control system which comprises of multiple local controllers which work on the mechanically coupled motors of ECMMS. The core controllers employed in the proposed decentralized control system are the well-known PID controllers. PID control algorithm is simple and effective when the operating conditions are fixed or of limited variations, but it may fail when the operating point of the system changes widely. However, if the parameters of the PID are adapted continuously using a proper adaptation technique, the output result of the control will be extensively improved.

The paper proposes a technique to search the optimal parameters of the included PID controllers off-line. Then, to apply the resulted optimal parameters as a starting parameters to control the ECMMS in the real time. The application of the control parameters in the real time is associated with a continual adaptation of the parameters of local controllers by another integrated adaptation technique.

2 System Description

Figure 1 shows a schematic diagram illustrating the main structure and the way of elastic coupling within the ECMMS. The system is comprised of three controlled DC motors. Two motors (specified as motor 2 and motor 3 or M2 and M3) are mechanically coupled to the third motor (specified as motor 1 or M1) through

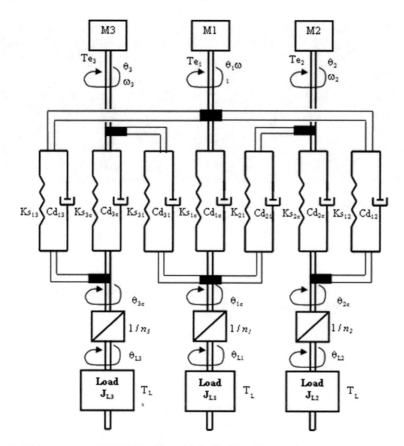

Fig. 1 Main structure of ECMMS with multiple distributed loads

elastic belts and coupling reels attached to motor shafts. Moreover, every motor shaft is also mechanically coupled to a variable load specified to such motor through a gear box of conversion ratio ni ('i' denotes the motor number 1 or 2 or 3). The name plate of each motor has the following electrical specifications: 230 V dc supply voltage, 8.5 A armature current, 2.5 HP motor power and 1500 rpm the motor speed.

An experimental set-up is illustrated as snapshot camera pics in Fig. 2. The set-up represents the ECMMS in which motor 1 (in the middle position among motor 2 and motor 3) is elastically coupled to each of motor 2 and motor 3 through elastic belts and reels attached to the shafts of motors in the system. One common variable load is attached mechanically to motor 1. The load is comprised of one DC generator and switchable electrical loads connected to it. Each motors of the practical set-up has the following electrical specifications: 38 V dc armature voltage, 1.9 A armature current, 53 W output power and 3700 rpm mechanical output speed.

Fig. 2 Camera snapshot of the ECMMS prototype

3 System Model

The behavior of the ECMMS is described by the following equations:
Relating motor i of the ECMMS system, $(i = 1, 2, 3)$:

$$u_{ai} = i_{ai}R_{ai} + L_{ai}\frac{di_{ai}}{dt} + K_i\phi_i\omega_i \tag{1}$$

$$T_{ai} = K_i\phi_i i_{ai} \tag{2}$$

$$J_i\ddot{\theta}_i = T_{ei} - MMS_{1i} - MMS_{2i} - MMS_{3i} \tag{3}$$

$$MMS_{1i} = [K_{si}(\theta_i - \theta_{ei}) + C_{di}(\dot{\theta}_i - \dot{\theta}_{ei})] \tag{4}$$

$$MMS_{2i} = [K_{sij}(\theta_i - \theta_{ej}) + C_{dij}(\dot{\theta}_i - \dot{\theta}_{ej})] \tag{5}$$

$$MMS_{3i} = [K_{sik}(\theta_i - \theta_{ek}) + C_{dik}(\dot{\theta}_i - \dot{\theta}_{ek})] \tag{6}$$

$$J_{Li}\ddot{\theta}_{Li} = n_i[MMS_{1i} + MLS_{2i} + MLS_{3i}] - T_{Li} \tag{7}$$

$$MLS_{2i} = [K_{sji}(\theta_j - \theta_{ei}) + C_{dji}(\dot{\theta}_j - \dot{\theta}_{ej})] \tag{8}$$

$$MLS_{3i} = [K_{ski}(\theta_k - \theta_{ei}) + C_{dki}(\dot{\theta}_i - \dot{\theta}_{ek})] \tag{9}$$

where: $(i = 1, 2, 3)$ denotes the motor label of the ECMMS.
$(j = 2)$ for $\{(i = 1)\}$. However, $(j = 1)$ for $\{(i = 2)$ and $(i = 3)\}$.
$MLS_{3i} = MMS_{3i} = 0$ for $\{(i = 2)$ and $(i = 3)\}$. $(k = 3)$ for $(i = 1)$.
$\theta_{ei} = n_i\theta_{Li}$

4 Problem Formulation

Actually, the ECMMS system forms mechanical resonator. The inertias of the rotors of the mechanically coupled motors within the ECMMS system as well as the inertias of the loads driven by the system motors, in conjunction with the shaft stiffness (shaft diameter, thickness and length) constitute the main reasons of the mechanical resonance that may occur in the system either during its startup or during its normal operation. This resonance, when occurred, causes undesirable results in the system such as wearing and/or mechanical damage of some parts of the system.

Figure 3 shows the response curves of the ECMMS. The figure shows the curves of armature current 'Ia' and output mechanical speed 'ω' of motor 1 from the starting instant of step input application to the system. Curves denoted 'a1' and 'b1' cover the time range from the start to sample No 1000 of the time axis for Ia and ω. The curves denoted 'a2' and 'b2' cover the time range from the start up to sample No 25000 on the time axis. Incremental oscillations associated to the motor current and speed responses. Similar responses are resulted from simulation programs related to motor 2 and motor 3.

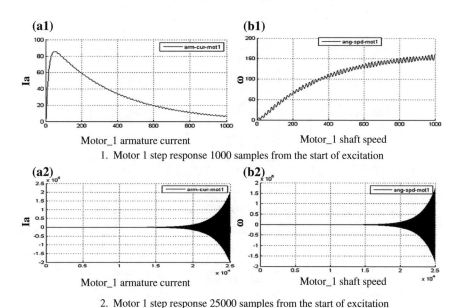

1. Motor 1 step response 1000 samples from the start of excitation

2. Motor 1 step response 25000 samples from the start of excitation

Fig. 3 Open loop step response curves of motor 1

5 Control Algorithm

In this paper, a decentralized control system (DCS) is proposed to control the ECMMS. The DCS control scheme is depicted in Fig. 4. Some advantages of using DCS with the ECMMS, are as follows:

1. Each local controller deals with subsystem of the whole ECMMS leading to reducing the order and the dynamics which are dealt with by each local controller;
2. The simple structure of the PID occupies a smaller size of memory for on-line implementation of the control system;
3. The implemented control algorithm permits application of shorter sampling periods. Thus, contributing to faster response leading to good tracking of the varied operating conditions.

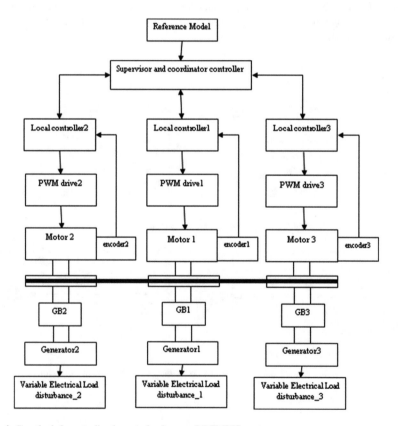

Fig. 4 Practical decentralized control scheme of ECMMS system

4. The performance of the PID to track continual variation conditions is improved by applying adaptation technique of the controller parameters.
5. The proposed adaptation technique is a combination between off-line searching mechanism and on-line continual adaptation mechanism which are interacted to result in good control performance.

5.1 Off-Line Searching Technique

The proposed technique works on the system model by simulation programs (Matlab 2010). By this technique, the whole operating range is partitioned into multiple sub-ranges. The number of the included sub-ranges depends on:

1. The permitted size of memory of the platform on which the controller will be implemented.
2. The behavior of the system through its whole operating range. The evaluation of this behavior depends on the experience relating to system operation and reaction to the applied disturbances.

The control signal of the local PID controller of motor (i) of the ECMMS is given by the incremental expression of the PID control signal:

$$U_i^k = U_i^{k-1} + \Delta U_i^k \tag{10}$$

$$\Delta U_i^k = K_{Pi}^k \left(e_i^k - e_i^{k-1} \right) + K_{Ii}^k e_i^k + K_{Di}^k \left(e_i^k - 2e_i^{k-1} + e_i^{k-2} \right) \tag{11}$$

5.2 On-Line Searching Technique

The platform on which the on-line control algorithm is implemented is selected to be the ATmega 2560 microcontroller which is supported by Arduino Mega 2560 kit [13]. Each local controller of the DCS is implemented on a separate kit of Arduino Mega 2560. The Arduino kits are inter-communicated through the serial COM port 3 of each one. The sampling period Ts is selected to be 15 ms.

The optimal parameters which are estimated through the preceded stage are passed to the platform on which the control algorithm is implemented. However, the on-line technique starts from the estimated control parameters for each sub-range of ECMMS operating range. Within each sub-range, the proportional gain parameter of each local controller is continually adapted. The whole range expected to the error value in each sub-range is divided into positive and negative sub-regions around the mid-point of the related sub-region. The size with which the controller parameter is adapted depends on the sub-region in which the error lies, at every sample during the control operation. The equations describing the adaptation technique in each sub-region:

$$K_{Pi}^k = K_{Pi}^{k-1} + ccl_n \quad \text{for} \quad ll_{n-1} \leq \omega_i^k < ll_n \tag{12}$$

$$K_{Pi}^k = K_{Pi}^{k-1} - cch_n \quad \text{for} \quad ll_{5-m} \leq \omega_i^k < lh_{4-m} \tag{13}$$

where: $ll_n = \omega_{ri}^k - kl_n$, $(n = 1, 2, 3, 4)$, $ll_0 \equiv (\omega_{ri}^k = \omega_{ri}^{min})$
$lh_m = \omega_{ri}^k + kh_m$, $(m = 1, 2, 3, 4)$, $lh_0 \equiv (\omega_{ri}^k = \omega_r^{max})$.

6   Result Assessment

Through this section, both the simulation results recorded during the off-line searching mechanism and the results provided from the application of on-line control mechanism are discussed as follows:

6.1   Simulation Results

Two cases are recorded through running of simulation programs:

1. Measuring the speed of tracking of the system responses to the desired input variation.
2. Measuring the capability of the controller with the optimal parameters to compensate against variable loading disturbances efficiently.

The first measure is illustrated by Fig. 5. This figure presents the response of motor 1 of the ECMMS. The figure presents the stepwise variation of the desired of

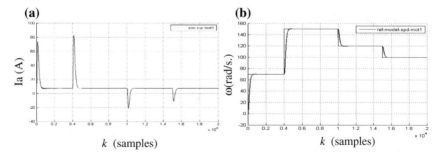

Fig. 5 Closed loop step response curves of motor 1 of the ECMMS. **a** Motor_1 armature current in ampere. **b** Motor_1 shaft speed in rad/s

(a)   **(b)**

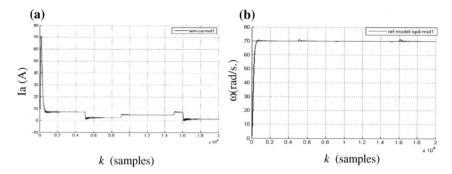

k (samples)   *k* (samples)

Fig. 6 Closed loop variable load response curves of motor 1 of the ECMMS. **a** Motor_1 armature current in ampere. **b** Motor_1 shaft speed in rad/s

speed motor 1 (at 65–150–120–100 rad/s) through total period of 300 s (corresponding to $2 * 10^4$ samples, where Ts = 15 ms) on the horizontal time axis.

The disturbance load torque is kept fixed at its full value corresponding $I_{a1} = 8.5$ A, while the desired speed is switched between the values 70 rad/s ($\approx$700 rpm), 150 rad/s ($\approx$1500), 120 rad/s ($\approx$1200), 100 rad/s ($\approx$1000 rpm).

Figure 6 depicts the response of motor 1 of the ECMMS after applying the optimal controller parameters which are resulted from the off-line technique. Fixed value of the desired speed is applied to the system at 70 rad/s. The load torque is switched at the samples 5000, 9000, 15000, 16000 as shown in the figure the effect of disturbance on the speed curve is limited between 0.7 % minimum and 2.9 % for maximum.

6.2 Practical Results

The same experiments are performed practically through the application of the described adaptive technique. The control algorithms are programmed and tested on-line with the aid of Arduino 2560 kits as hardware local controllers. The sampling period Ts is selected to be 15 ms.

Four cases are tested experimentally, for the purpose of comparison and measuring the performance of the proposed adaptive control technique. The first two cases (case 1 and case 2) are illustrated in Fig. 7 and in Fig. 8 respectively. Figure 7 illustrates the response of motor 1 and motor 2 when the control parameters are fixed and the control is applied to only motor 1 of the ECMMS, however, motor 2 and motor 3 are uncontrolled. The response in this case is very slow. The developed torque of motor 1 is not sufficient to drive the entire ECMMS. As a result, the tracking of the desired speed of motor 1 and motor 2 is bad. In Fig. 8, motor 1 only is still controlled; however the control parameters are now adaptive according to the proposed technique. As a result the speed of the response is greatly improved. However, the problem of being not sufficient torque to drive the entire ECMMS is

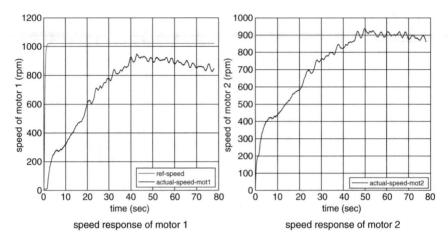

Fig. 7 Response of motor 1 and motor 2 of the ECMMS. Case 1: motor 1 is only controlled, while motor 2 and motor 3 are driven by motor 1. The DCS has fixed parameters

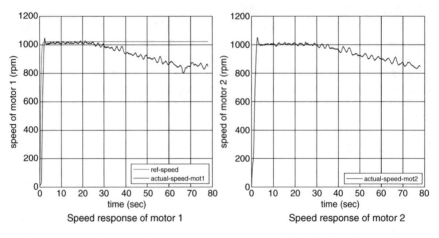

Fig. 8 Response and the control signals of motors 1, 2 of the ECMMS. Case 2: motor 1 is only controlled, while motor 2 and motor 3 are driven by motor 1. The DCS has adaptive parameters

still existed. So, the tracking operation fails to continue, this is noted after the instant 30 on the time axis for both motor 1 and motor 2 as shown in Fig. 8.

The other two cases (case 3 and case 4) are illustrated through Fig. 9 and Fig. 10 respectively. Figure 9 shows the speed of tracking of the response of motor 1 to the desired output (depicted by 9-1 of Fig. 9), as well as the recovery time after the instant of applying the load to the entire ECMMS (depicted by 9-2 of Fig. 9). The control parameters in this case are fixed at its optimal values provided from the offline searching technique. The same test is performed in case 4 to result out

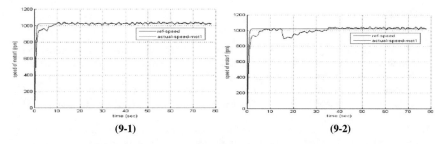

(9-1)   (9-2)

Fig. 9 Response of motor 1 of the ECMMS for a step excitation of 1020 rpm. Case 3: *9-1*—No load is applied; *9-2*—20 % load applied at instant 15, control parameters are fixed

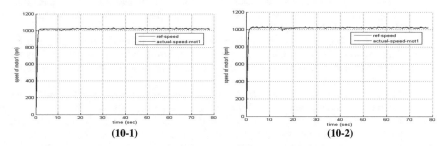

(10-1)   (10-2)

Fig. 10 Response of motor 1 of the ECMMS for a step excitation of 1020 rpm. Case 5: *10-1*—No load is applied; *10-2*—20 % load is applied at instant 15, control parameters are adaptive

the response of Fig. 10. Case 4 is characterized by applying the adaptive technique during the control of ECMMS. The comparison of the responses of motor 1 of the ECMMS provided by Figs. 9 and 10 reveal the improvements in control due to the application of the adaptive technique to the controller parameters.

7 Conclusion

The proposed off-line auto-searching algorithm is experienced on the ECMMS system. The ECMMS is characterized by higher vibrations which affect its proper operation and cause instability effects in some cases. This paper proposes a decentralized PID control system for regulating the motors of the ECMMS. The performance of the control is proved by the application of the practical imple-mentation of the adaptive DCS control. The simplicity of each local controller does not degrade the control performance. The on-line adaptation technique enhances the tracking and compensation against variable operating conditions and disturbances. The improvement is measured through measuring the rate of tracking and recovery after loading, by comparing the responses both of fixed and adaptive parameters.

References

1. Mitsantisuk, Chowarit, Ohishi, Kiyoshi, Katsura, Seiichiro: Control of interaction force of twin direct-drive motor system using variable wire rope tension with multi-sensor integration. IEEE Trans. Ind. Electron. **5**, 498–5109 (2012)
2. Michael, C.A., Safacas, A.N.: Dynamic and vibration analysis of a multi-motor DC drive system with elastic shafts driving a tissue paper machine. IEEE Trans. Ind. Electron. **54**, 2033–2046 (2007)
3. Ohmae, T., Matsuda, T., Kanno, M., Saito, K., Sukegawa, T.: A microprocessor-based motor speed regulator using fast-response state observer for reduction of torsional vibration. IEEE Trans. Ind. Electron. **5**, 863–871 (1987)
4. Ozaki, S.: A microprocessor-based dc motor drive with a state observer for impact drop suppression. In: Conference Record-IEEE-IAS Annual Meeting (1983)
5. Brandenburg, G., Wolfermann, W.: State observers for multi motor drives in processing machines with continuous moving webs. In: Proceedings of Power Electronics and Application, vol. 16. Brüssel (1985)
6. Dhaouadi, R., Kubo, K., Tobise, M.: Robust speed control of rolling mill drive systems using the loop transfer recovery design methodology. In: 1991 International Conference on Industrial Electronics. Control and Instrumentation, 1991 Proceedings. IECON'91. IEEE (1991)
7. Dhaouadi, R., Kubo, K., Tobise, M.: Two-degree-of-freedom robust speed controller for high-performance rolling mill drives. IEEE Trans. Ind. Appl. **29**, 919–926 (1993)
8. Lacasta, A., Morales-Hernández, M., Brufau, P., García-Navarro, P.: Simulation of PID control applied to irrigation channels procedia. Engineering **70**, 978–987 (2014)
9. Leena, G., Ray, G.: A set of decentralized PID controllers for an n–link robot manipulator. Sadhana **37**, 405–423 (2012)
10. Santibañez, V., Camarillo, K., Moreno-Valenzuela, J., Campa, R.: A practical PID regulator with bounded torques for robot manipulators. Int. J. Control Autom. Syst. **8**, 544–5552010))
11. Das, Saptarshi, Acharya, Anish, Pan, Indranil: Simulation studies on the design of optimum PID controllers to suppress chaotic oscillations in a family of Lorenz-like multi-wing attractors. Math. Comput. Simul. **100**, 72–87 (2014)
12. Ali, M.: An extended PID type iterative learning control. Int. J. Control Autom. Syst. **11**, 470 (2013)
13. Arduino and Genuino products > Arduino MEGA 2560 and Genuino MEGA 2560. https://www.arduino.cc/en/Main/ArduinoBoardMega2560. Last update (2016)

Part II
Intelligent Information Technology, System Monitoring and Proactive Management of Complex Objects

Computer-Based Ground Motion Attenuation Modeling Using Levenberg-Marquardt Method

E. Irwansyah, Rian Budi Lukmanto, Rokhana D. Bekti and Priscilia Budiman

Abstract In this paper, we present the results of research on the optimization modeling of ground motion attenuation of the two establish models by Youngs et al. [25] and the model of Lin and Lee [13] using the Levenberg-Marquardt method. This modeling is particularly important in the case of ground motion given that it takes a good model for predicting the strength of earthquakes in order to reduce the risk of the impact of the natural disaster. There are two main contributions of this research is the optimization of ground motion attenuation models with Levenberg-Marquardt method on two models that have been extensively used and the development of computer applications to help accelerate modeling, especially on large data with an area of extensive research. Levenberg-Marquardt method proved to give a good contribution to the modeling of ground motion attenuation that is indicated by the very small deviations between the predicted values with the actual value.

Keywords Ground motion attenuation · Levenberg-Marquardt · Earthquake

E. Irwansyah (✉) · R.B. Lukmanto
School of Computer Science, Bina Nusantara University, Jl. KH. Syahdan No. 9,
Kemanggisan, Jakarta 11480, Indonesia
e-mail: eirwansyah@binus.edu
URL: http://binus.ac.id/

R.B. Lukmanto
e-mail: rianbl@binusian.org

R.D. Bekti
Department of Statistics, Institut Sains and Teknologi AKPRIND, Jl. Kalisahak
No. 28 Kompleks Balapan Tromol Pos 45, Yogyakarta 55222, Indonesia
e-mail: rokhana@akprind.ac.id
URL: http://www.akprind.ac.id/

P. Budiman
Department of Statistics, Bina Nusantara University, Jl. KH. Syahdan No. 9,
Kemanggisan, Jakarta 11480, Indonesia
e-mail: priscilia@binus.ac.id
URL: http://binus.ac.id/

© Springer International Publishing Switzerland 2016   293
R. Silhavy et al. (eds.), *Automation Control Theory Perspectives*
in Intelligent Systems, Advances in Intelligent Systems and Computing 466,
DOI 10.1007/978-3-319-33389-2_28

1   Introduction

Indonesia is a country in the world that is prone to natural disasters due to its position earthquake located at the confluence of three tectonic plates, the Australian plate, the Asian plate and the Pacific plate, which is moving on the different direction. In 2004, an earthquake and tsunami in Nanggroe Aceh Darussalam as impact of the plate tectonic movement, which takes a lot of lives, loss of property, and severe damage to the environment.

In order to overcome or reduce the risk of natural disasters, we need a model that can predict how big consequences or risks caused by the earthquake that occurred at the point of the epicenter and surrounding areas. The estimate of ground motion attenuation can be done through modeling the value of peak ground acceleration (PGA). PGA is the acceleration that occurs on the earth surface from rest until exposed to shocks which in this case is an earthquake. PGA unlike the Richter scale or magnitude scale, PGA measure how strong the earth's surface moves in the earthquake which occurred in a region [24].

Research that has been conducted in order to find the nonlinear model of the PGA has been done by the scientists but the results obtained every model is different and not necessarily compatible and can't be used in certain area. The most commonly used equation to find the value of PGA is the equation [25] developed from the data of earthquakes in Alaska, Chile, Cascadia, Japan, Mexico, Peru, and the Solomon Islands with a magnitude ranging from 5.0–8.2 on a Richter scale. Several other researchers have done a case study to develop a more general model of PGA as [2] which uses the same area with [24] to get a PGA equation but the strength of the earthquake was raised to 5.0–8.3 on the Richter scale. Some others model are developed for calculating the value of PGA among others such as [10, 12, 14, 20]. Paper [14, 25] using nonlinear least square (NLS) for modeling PGA. Research published in [15, 8], conducted modeling which aims to predict earthquakes at various location. Paper [15] improved the speed and stability of back propagation neural network (BPNN) using LMM and paper [7] conducted LMM and ANN for predicting seismic-induced damage using PGA data.

Along with the development of Levenberg-Marquardt method for estimation and modeling in various fields, on the other hand, the computer as a computing tool is also experiencing rapid development both hardware and software. Computer technology is already widely used in various fields because it provides ease of use and speed up the time to complete a job mainly related to the modeling process. PGA computer-based modeling will result in the model and the predictive value of the PGA of an area more easily and quickly. In the conditions of a shift or change in the value of a variable, then the value of the PGA of an area would be faster counted. In this study, conducted modeling PGA values using nonlinear models with Levenberg-Marquardt optimization method that uses computer-based seismic data in the region of Aceh and surrounding areas, Aceh Province, Indonesia.

2   Ground Motion Attenuation Model

Attenuation relationship is one of the key components of earthquake or seismic hazard assessment an area [17]. Currently various attenuation function has no good function for shallow seismic sources, seismic source with deep background and attenuation function to the source of the earthquake due to the earthquake in sub-duction zone as published by [2, 5, 10, 14, 20, 25].

Paper [25] has proposed an attenuation function regression using the data catalog inter-plate earthquakes with magnitude variations of 5–8.2 recorded in the area of sub-duction in the area of Alaska, Chile, Cascadian, Japan, Mexico, Peru and the Solomon Islands. This attenuation function modified by [20] by comparing the results of observations and predicted using data catalog of earthquakes in 1991 and 2001 in the wider area include New Ireland, New Britain, Kamchatka, Santa Cruz Is, Peru, Is Kurile, Japan and Sumatra. Modifications attenuation function performed mainly in cases with earthquake source within more than 200 km with earthquake magnitude 6.8–8.3 Mw.

Paper [10] to develop an attenuation function to the location in the Cascadian sub-duction of the same functions as proposed by [25] used a stochastic model of finite-fault ground motion from [3] with a variation on the high magnitude of 8.0–9.0 Mw. The advantage of using this model is that unlike the empirical attenuation relationships, which require field samples and geometry based on strong-motion series of data available, the effect of such finite-fault rupture propagation, direction and resources to site geometry, stochastic models finite- fault can be systematically calculated.

Maximum similarity regression method with moment magnitude 5.0–8.3 Mw in various sub-duction areas in the world such as Alaska, Japan, Mexico and Central America are used by [2] to develop the ground motion attenuation function. Regional variability analysis result of the amplitude of the ground motion using a global database available to support the fact that there are significant differences between regional as shown by the amplitude difference of more than two factors among the Cascadian area and the area of Japan. This model uses only the shortest distance from the source of the earthquake at a distance of 10–500 km as used by [10, 25]. In the same year, [5] using a combination of empirical models that use the estimated value of both the stochastic ground motion and theoretical, to develop typical regression model to be implemented in the zone of Eastern North America (ENA) using a relational model that has previously been developed using the data seismic data for the Western part of North America (WNA). Campbell attenuation function 2003, was developed especially for earthquakes with a magnitude of variation 5.0–7.5 with the straight line distance nearest to the location of the source of the earthquake was at 1–1000 km.

Paper [14] with the reference of the previous attenuation function developed by [6, 25], develop models of other regression attenuation function using the recording seismic movements in the bedrock in the area between the plates at sub-duction zone in the Northeast Taiwan and other regions of the value of the low magnitude of about 4.1–8.7 on the Richter Scale. The use of a low magnitude value <5.0 Mw this is something different from the attenuation function has been developed by researchers before. Reference [14] noted that the usage of their attenuation function to compute the peak ground acceleration (PGA), relatively higher than the value that is generated using attenuation equations that had previously been used in Taiwan and tend to lower than attenuation equations that are used globally especially in earthquake zones due to tectonic sub-duction.

3   Levenberg-Marquardt Method

Nonlinear regression equation has many methods to estimate the parameters. The estimation method includes the Ordinary Least Square (OLS), Nonlinear Least Square (NLS), generalized Nonlinear Least Square and Maximum Likelihood. NLS is a form of least squares analysis used in modeling of nonlinear regression by minimizing the Residual Sum of Squares-RSS [23]. A method to minimize the value of RSS is the parameter optimization. Some of these methods include the Gauss-Newton method, Hartley's, Powell's Hybrid, Quasi Newton, Brute-Force and Levenberg-Marquardt Method. Levenberg-Marquardt method is a method of combination between Gauss-Newton method and gradient decrease method (gradient descent).

Levenberg-Marquardt method [13, 16] commonly known as damped least squares method (DLS) which produces a numerical solution to minimize a non-linear function of the parameters in the function. Levenberg-Marquardt method is the result of interpolation between the Gauss-Newton method and method Gradient-Descent. The main application of the Levenberg-Marquardt method is the least squares problem that aims to optimize the parameter β from the model f(x_i, β), so that the RSS in (1) be minimal value.

$$S(\beta) = \sum_{i=1}^{n} \epsilon_i^2 = \sum_{i=1}^{n} (y_i - f(x_i, \beta))^2 \tag{1}$$

where ϵ is residuals, y is dependent variable, x_i is independent variable, β is parameter model, i is observation ($i = 1, 2, 3, \ldots n$).

Levenberg-Marquardt method is using the iterative procedure. To begin the process of minimization with the first step is to estimate the value of the parameter vector, β. At each stage of iteration, the vector parameter, β, will be replaced with a new estimated value, i.e. $\beta + \delta$ to find the value of δ, the function $f(x_i, \beta + \delta)$ is approached in the way making it linear as in (2).

$$f(x_i, \beta + \delta) \approx f(x_i, \beta) + J_i\delta \tag{2}$$

where δ is increment on β. J_i is the gradient (row vector) of f to the parameter β. It calculate by (3),

$$J_i = \frac{\delta_f(x_i, \beta)}{\delta \beta} \tag{3}$$

Approximation of $f(x_i, \beta + \delta)$ will produce new function (4),

$$S(\beta + \delta) \approx \sum_{i=1}^{n} [y_i - f(x_i, \beta) - J_i \delta]^2 \tag{4}$$

or in vector notation becomes (5),

$$S(\beta + \delta) \approx ||y - f(\beta) - J\delta||^2 \tag{5}$$

Levenberg-Marquardt method is modifying the Gauss-Newton step into (6)

$$(J'J + \lambda I)\delta = J'[y - f(\beta)] \tag{6}$$

where J is the Jacobian matrix that has rows J_i and where f and y is a vector with components $f(x_i, \beta)$ and y_i is dependent variable and much as i. δ values are values that give direction down (descent direction) of the vector parameter β. λ value is the damping parameter that must not be negative and it will be adjusted in each iteration. Damping parameter λ, will be adjusted in each iteration. If S is drastically declined or fast, we can use a small λ value, which would make this method becomes similar to the Gauss-Newton method, which iterations will give small residual results. λ value can be enlarged that will have an impact on the direction of decreasing gradient with a gradient S of the β equal to $-2(J^T[y - f(\beta)])^T$. Therefore, for large λ value, the stages will be carried out in the direction of the gradient approximation. Iterations stop if the number of stages, δ, or reduction Sum of Squares of the last parameter vector, $\beta + \delta$, is below a predetermined limit. Based on [19], the last parameter, β, is a solution of the Levenberg-Marquardt method can be written in as (7)

$$\beta^{(j+1)} = \beta^{(j)} + (J'J + \lambda_j I)^{-1} J'(y - f(\beta)) \tag{7}$$

According with [9], iteration in the Levenberg-Marquardt method also be determined by two limits, namely by:

1. *First convergence test*. Iteration will be stop if it meets (8),

$$|fvec| < (1 + ftol)|fvec_0| \tag{8}$$

where *fvec* are residuals and *ftol* is a non-negative numbers. Iteration will stop if both the relative reduction (actual and forecast) the sum of squares over *ftol* value.

2. *Second convergence test*. Iteration will be stop if it meets (9),

$$|D(par - par_0)| < ptol|Dpar_0| \tag{9}$$

where *par* is the best parameter obtained and *ptol* is a non-negative numbers. Iteration will stop if the value of the relative error between two consecutive iterations over *ptol* value.

4 Methodology

4.1 Ground Motion Attenuation Model Optimization

Research conducted using secondary earthquake data consist of distance of the location of the epicenters (Km), the depth of the earthquake (Km), magnitude (Mw), and the value of the PGA (gals) from year 2005 through 2007 are derived from the meteorology, climatology, and geophysical (BMKG) central government office. Population and sample used was the West Coast of Sumatra, Indonesia in the area in a radius of 500 km from the center of Banda Aceh municipality, Aceh province, Indonesia.

The study consisted of two main stages, optimization modeling and computer application development stage. Optimization modeling consists of four stages (Fig. 1), namely: (1) To test the linearity of the data with the test Ramsey's RESET [22], (2) determining the PGA nonlinear models for Youngs et al. [25] and the Lin and Lee [14] model:

1. Youngs et al. [25] model in (10):

$$In(PGA) = C_1 + C_2M + C_3In[R + e^{c_4 - (\frac{c_2}{c_3})M}] + C_9H \tag{10}$$

2. Lin dan Lee [14] model in (11):

$$In(PGA) = C_1 + C_2M + C_3In(R + C_4e^{C_5M}) + C_6H \tag{11}$$

(3) doing parameters model estimation for Youngs et al. [25] and Lin and Lee [14] using the Levenberg-Marquardt method. Modeling conducted with the stage (a) determining the starting value for each parameter from each model, (b) determine the boundaries of iterations (c) perform iterations to obtain parameters that minimize the residual sum of squares (RSS) and (d) testing the residual assumptions (identical, independent, and normal distribution) for the model is formed. In detail, to four stages of modeling is as shown in the picture (4) compare the modeling result.

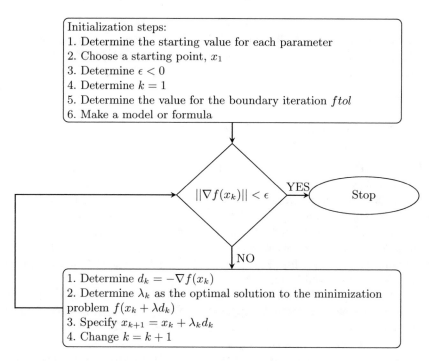

Fig. 1 Modeling stages using Levenberg-Marquard method

4.2 Application Development

Development of computer applications aimed at assisting the modeling of ground motion attenuation was done in two stages: (1) the design of a computer program with the waterfall model is a series of stages defining program requirements, designing systems and software, implementation and unit testing, integration and test of the system operational and (2) the design of the interface (gradient descent) [21].

5 Result and Discussion

5.1 Nonlinearity Test

Nonlinearity test in this study was conducted to determine whether the data used to follow the pattern of nonlinear models or not. Nonlinearity test conducted by plotting the data between seismicity variable such as distance from earthquake center, depth of the earthquake and earthquake magnitude with the PGA and the nonlinearity pattern. Nonlinearity test can also be done by the method of Ramsey's RESET. The nonlinearity test generate result of $F_{value} = 10.5256$ with $df_1 = 3$ and

$df_2 = 16$ with $p - value = 0.0004572$. Testing concluded that the data follows the nonlinear pattern because the value of $F = 10.5256$ greater than the value of $F_{(0.05,3,16)} = 3.24$. $P - value = 0.000452$ also indicates that the data follows the nonlinear pattern because the $P - value$ less than the value of $\alpha = 0.05$.

5.2 Modeling and Model Comparison

Modeling optimization in this study conducted for the model developed by [25] and the model of [14]. Both of these models will be optimized in order to generate value residual sum of square (RSS) is small with Levenberg-Marquardt method. In contrast to research conducted [1] which implements fuzzy logic attenuation model and compared with the results of the model developed by Boore et al. [4]. In Table 1 we can see the details of each parameter estimation in the second model in the optimization. It use Eq. (7) to get the parameter equation based on Young model (10) and Lin and Lee model (11).

After entering the data through the function Import Data, the user can start the process of modeling in computer applications through Analyze function, both to test Ramsey's RESET, Levenberg-Marquardt Regression (Fig. 2a, b).

After modeling the nonlinear regression have conducted, then performed the assumption that residuals must meet with $\epsilon \sim IIND$, namely residual must meet identical test using Glejser test, independent test with Durbin-Watson and lag1-plot, and the normal distribution test using Kolmogorov-Smirnov. Residual assumption test conducted showed that both models tested had fulfilled all residual assumptions, as can be seen in the test results using computer applications summarized in Table 2.

Knowing the best model is generated by comparing the predicted results of the two models was optimized. Comparisons were performed in this research is to compare the descriptive statistics value of data such as average, variance, and residual standard error value. In Table 3, it can be seen that the two models are built have been able to predict the PGA with better value, especially for the nearest average value and residual standard error. The resulting average value of the actual PGA compares Youngs et al. [25] model.

Table 1 Parameter estimation of Youngs et al. [25] and Lin and Lee [14] model with Levenberg-Marquardt method	Youngs et al. [25] model		Lin and Lee [14] model	
	Parameter	Estimation	Parameter	Estimation
	C1	−1.101	C1	−0.7644
	C2	−0.0008244	C2	−0.4649
	C3	0.005139	C3	0.177
	C4	11.83	C4	0.1986
	C5	0.0000283	C5	2.638
			C6	0.00001767

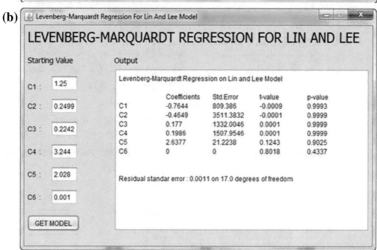

Fig. 2 Window display of Levenberg-Marquardt regression for Youngs et al. [25] (**a**) and Lin and Lee [14] model (**b**)

Table 2 Residual assumption of Young et al. [25] and Lin and Lee [14] model

Residual assumption	Levenberg-Marquardt model	
	Young et al. [25] model	Lin and Lee [14] model ($)
Identical	Yes	Yes
Independent	Yes	Yes
Normal distribution	Yes	Yes

Table 3 Descriptive statistics comparison of Young et al. [25] and Lin and Lee [14]

Descriptive statistics	Actual PGA	Levenberg-Marquardt model	
		Young et al. [25] model	Lin and Lee [14] model
Averages	0.35383	0.35379	0.35473
Variance	0.0000002	0.00000001	0.00000016
Residual standard error		0.001251	0.001102

Plotting PGA predictions against the PGA actual value for each of the model as in Fig. 3, shows that the Youngs et al. [25] model which in the estimation of the Levenberg-Marquardt method produces PGA predictive value which is closer to the actual PGA value as average values in Table 3 are show similarities and differences

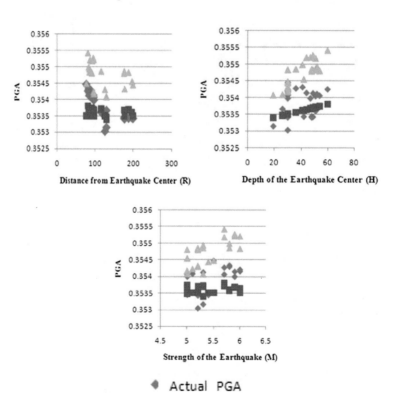

Fig. 3 Comparison between PGA actual and PGA prediction

were very small with PGA predictions value ranging around ±0.0001. Lin and Lee [14] model in which estimation by the same method, showed a significant difference that is supported by the fact that the PGA predictive values that range ±0.0009. Both of models were estimated by Levenberg-Marquardt method, Youngs et al. [25] model, is a model that can predict the PGA value more better and this is shown by the closeness point in scatterplot between the PGA predictions and PGA actual. These findings would suggest that the model developed by Youngs et al. [25] is more suitable for use as a model for the prediction of ground motion attenuation as in inferred by [11] especially for data which is based on the earthquake data in subduction zones on the West Coast of Sumatra, Indonesia.

Development of local attenuation function that uses the data in the same location, has been carried out by [17] with the attenuation function developed by [18] in the form of synthetic regression movement of bedrock in an earthquake zone as a result of subduction of the plates using a kinematic model of finite fault as adopted from [10]. The size of the data from the central radius distance varies between 200 and 1500 km. Model validation is carried out using the data Sumatra, Indonesia megathrust that includes a very large earthquake with strength of up to 9.0 Mw.

6 Conclusion

Modeling of ground motion attenuation generated for Youngs et al. [25] model and the Lin and Lee [14] model which are estimated by Levenberg-Marquardt method is as follows

1. Youngs et al. [25] model:

$$ln(PGA) = -1.101 - 0.00082M + 0.00514ln[R + \epsilon^{11.83 - (\frac{-0.00082}{0.00514})M}]$$

2. Lin and Lee [14] model:

$$ln(PGA) = -0.7644 - 0.4649M + 0.177ln(R - 0.1986\epsilon^{2.638M}) - 0.00001H$$

Levenberg-Marquardt method proved to give a good contribution to the modeling of ground motion attenuation as indicated by the very small deviations both in average value, variance and residual standard error between the predicted values with the actual value. In addition, all assumptions are met and the results of residual plots showed similarities predicted results with actual value.

Development of computer applications can be accelerate the process of modeling the ground motion attenuation by the Levenberg-Marquardt method, especially for the amount of data and a lot more extensive research sites.

References

1. Ahumada, A., Altunkaynak, A., Ayoub, A.: Fuzzy logic-based attenuation relationships of strong motion earthquake records. Expert Syst. Appl. **42**(3), 1287–1297 (2015)
2. Atkinson, G.M., Boore, D.M.: Empirical ground-motion relations for subduction-zone earthquakes and their application to cascadia and other regions. Bull. Seismol. Soc. Am. **93**(4), 1703–1729 (2003)
3. Atkinson, G.M., Silva, W.: Stochastic modeling of california ground motions. Bull. Seismol. Soc. Am. **90**(2), 255–274 (2000)
4. Boore, D.M., Joyner, W.B., Fumal, T.E.: Equations for estimating horizontal response spectra and peak acceleration from western north american earthquakes: a summary of recent work. Seismol. Res. Lett. **68**(1), 128–153 (1997)
5. Campbell, K.W.: Prediction of strong ground motion using the hybrid empirical method and its use in the development of ground-motion (attenuation) relations in eastern north america. Bull. Seismol. Soc. Am. **93**(3), 1012–1033 (2003)
6. Crouse, C.: Ground-motion attenuation equations for earthquakes on the cascadia subduction zone. Earthq. Spectra **7**(2), 201–236 (1991)
11. de Lautour, O.R., Omenzetter, P.: Prediction of seismic-induced structural damage using artificial neural networks. Eng. Struct. **31**(2), 600–606 (2009)
19. do Nascimento, P.F., França, G.S., Moreira, L.P., Von Huelsen, M.G.: Application of gauss-marquardt-levenberg method in the inversion of receiver function in central brazil. Revista Brasileira de Geofsica **30**(3) (2012)
7. Elzhov, T., Mullen, K., Spiess, A., Bolker, B.: minpack. lm: R interface to the Levenberg-Marquardt nonlinear least-squares algorithm found in minpack, plus support for bounds. R Packag version **1**, 1–8 (2013)
8. Gregor, N.J., Silva, W.J., Wong, I.G., Youngs, R.R.: Ground-motion attenuation relationships for cascadia subduction zone megathrust earthquakes based on a stochastic finite-fault model. Bull. Seismol. Soc. Am. **92**(5), 1923–1932 (2002)
9. Irwansyah, E., Winarko, E., Rasjid, Z., Bekti, R.: Earthquake hazard zonation using peak ground acceleration (pga) approach. J. Phys: Conf. Ser. **423**(1), 012067 (2013)
10. Kanno, T., Narita, A., Morikawa, N., Fujiwara, H., Fukushima, Y.: A new attenuation relation for strong ground motion in japan based on recorded data. Bull. Seismol. Soc. Am. **96**(3), 879–897 (2006)
12. Levenberg, K.: A method for the solution of certain non–linear problems in least squares. Q. Appl. Math. (1944)
13. Lin, P.S., Lee, C.T.: Ground-motion attenuation relationships for subduction-zone earthquakes in northeastern taiwan. Bull. Seismol. Soc. Am. **98**(1), 220–240 (2008)
14. Ma, L., Xu, F., Wang, X., Tang, L.: Earthquake prediction based on Levenberg-Marquardt algorithm constrained back-propagation neural network using demeter data. In: Knowledge Science, Engineering and Management, pp. 591–596. Springer (2010)
15. Marquardt, D.W.: An algorithm for least-squares estimation of nonlinear parameters. J. Soc. Ind. Appl. Math. **11**(2), 431–441 (1963)
16. Megawati, K., Pan, T.C.: Ground-motion attenuation relationship for the sumatran megathrust earthquakes. Earthq. Eng. Struct. Dyn. **39**(8), 827–845 (2010)
17. Megawati, K., Pan, T.C., Koketsu, K.: Response spectral attenuation relationships for sumatran-subduction earthquakes and the seismic hazard implications to singapore and kuala lumpur. Soil Dyn. Earthq. Eng. **25**(1), 11–25 (2005)
18. Monahan, J.F.: Numerical Methods of Statistics. Cambridge University Press (2011)
20. Petersen, M.D., Dewey, J., Hartzell, S., Mueller, C., Harmsen, S., Frankel, A., Rukstales, K.: Probabilistic seismic hazard analysis for sumatra, indonesia and across the southern malaysian peninsula. Tectonophysics **390**(1), 141–158 (2004)
21. Pressman, R.S.: Software Engineering: A Practitioner's Approach. McGraw-Hill, NY (2010)

22. Ramsey, J.B.: Tests for specification errors in classical linear least-squares regression analysis. Journal of the Royal Statistical Society. Series B (Methodological) pp. 350–371 (1969)
23. Ritz, C., Streibig, J.C.: Nonlinear regression with R. Springer (2008)
24. Santoso, E., Widiyantoro, S., Sukanta, I.N.: Studi hazard seismik dan hubungannya dengan intensitas seismik di pulau sumatera dan sekitarnya. Jurnal Meteorologi dan Geofisika 12(2) (2011)
25. Youngs, R., Chiou, S.J., Silva, W., Humphrey, J.: Strong ground motion attenuation relationships for subduction zone earthquakes. Seismol. Res. Lett. 68(1), 58–73 (1997)

Method of the Debugging of the Knowledge Bases of Intellectual Decision Making Systems

Olga Dolinina

Abstract The paper describes the method of the debugging of the intellectual decision making systems. It combines the detecting of the structural errors and the testing of the knowledge base. Static analysis allows to detect so called structural errors such as incomplete knowledge, inconsistency, extra rules. Static debugging allows to build the static correct knowledge base. But even the static correct knowledge base can have errors connected with the inconsistency of the subject area which can be detected with the dynamic debugging (testing). The paper shows that the most difficult for the detection is the "forgetting about the exception" type of the errors. There is described the method of the generation of the full test set which allows to detect such types of the errors in the knowledge base. The method is based on the building of the tests for the logic schemes. The method was successfully approved for the testing of the rule-based expert systems and for the artificial network based on the 3-level perceptron.

Keywords Debugging of the intellectual decision making systems · Static analysis · Testing · Generation of the full test set · Rule-based systems

1 Introduction

Intellectual systems (IS) of decision making are used for solving tasks in the poorly formalized fields. In the process of IS development the most difficult stage is the knowledge processing—knowledge extraction, formalization, debugging. Methods of knowledge extraction as well as formalization are well developed, at the same time methods of knowledge bases (KB) debugging are still not formalized. Ways of the software debugging can not be used for the knowledge bases and most of the developers use expert approach for the debugging of the knowledge bases. Knowledge base is developed by the knowledge engineer and is based on the expert

O. Dolinina (✉)
Yury Gagarin State Technical University of Saratov, Saratov, Russia
e-mail: odolinina09@gmail.com

© Springer International Publishing Switzerland 2016
R. Silhavy et al. (eds.), *Automation Control Theory Perspectives in Intelligent Systems*, Advances in Intelligent Systems and Computing 466,
DOI 10.1007/978-3-319-33389-2_29

knowledge; expert can make various errors while designing the knowledge bases. Markott [1] showed that the percentage of the expert's errors in the knowledge base is 15 %. Using of the IS for making solutions in the fields connected with the special demands of the reliability as medicine, diagnostics of the faults in the technical fields, demands the quality of the making solutions and thus the development of the methods and the software of the debugging.

2 Method of the Rule-Based Systems Debugging

Let us use the term "debugging of the IS" introduced in the paper [2] in the narrow sense as a process of detecting, localization of the bugs in the knowledge bases as well as updating the knowledge bases not connected with the changing of the knowledge type.

One can characterize the current stage of the development of the debugging methods as:

In order to achieve full knowledge base debugging it is necessary to use two groups of methods:

- static debugging as the formal checking of the KB, it does not demand the running of the intellectual system;
- testing as a form of dynamic debugging. Testing is a process of the IS running with the set of the test data and compare the results with the reference data.

Rule-based knowledge (RKB) representation can be considered as the most popular model for the knowledge base development. Rule-based knowledge base is set as:

$$P = (F, R, G, C, I), \tag{1}$$

where F is a finite set of the facts in the concrete field about the problem. Each fact can be set up or unknown, the set of the set up facts defines the situation in the field of studies.

R—a set of rules where

$$r_m : IF\ f_i\ and\ f_j \ldots and\ f_n\ THEN\ f_k,, \tag{2}$$

where r_m—is a name of the rule, $r_m \in R$; $f_i, f_j \ldots, f_n$—facts of the condition of the rule; f_k—consequence of the rule,

$f_i, f_j, \ldots, f_n, f_k \in F$; G—is a set of the goals or the IS terminal facts; I is the interpreter of the rules, realizes the process of the goal solution.

C is the set of the permitted combination of the facts.

For effective knowledge bases debugging it is suggested to transform the rule-based knowledge base into a graph model, which is used for detecting of the various types of errors. First of all, it is necessary to guarantee the absence of the so called structural errors that can be detected by the static analysis of the knowledge base graph structure, e.g. incompleteness of KB could be characterized by unattainable missed vertexes, terminating rules (which could be complete but their consequences are not the aim of consultancy and are missed among the other rules conditions).

The redundancy of KB is characterized by duplicating rules and also by the rules which do not lead to any aim.

However, the absence of structural errors is a necessary but insufficient demand of effective debugging. Statistic debugging cannot find the most difficult to detect errors inside the knowledge because of the inner inconsistency of the subject area. This kind of error can be detected with the help of testing which is a kind of dynamic debugging. The method of expert systems' knowledge base debugging is shown in Fig. 1.

It should be noticed that in the process of RKB formation due to psychological features the expert almost always finds the facts which help with installation for the fact f_i given. However the facts, that disturb the installation of the fact f_i viewed, are usually missed by the expert.

As the result there appear errors that have a feature of "forgetting" about rules applicability boarders [3]. The error of forgetting-about-the-exception type is the error that is the most difficult to find and it covers all the other classes of errors (forgetting about several exceptions).

The formal model of the forgetting-about-the-exception errors type is the following:

The rule is always performed, except the case, when the set of facts is installed in RKB

$$\{f_1, f_2 \ldots, f_k\} = S \in C. \tag{3}$$

The logics of RKB, determined by its AND/OR graph G can be set with the corresponding logical network after statistical analysis. For example, the graph shown in Fig. 2 corresponds the logical network shown at Fig. 3.

A wide-spread model of malfunctions in diagnosing digital devices is a so-called constant malfunction [4]. In the logical network, which includes a constant malfunction, one of the outputs is always in "0" state or is always in "1" state and it does not depend on the state of its inputs. The error of the type (3) can be described as a malfunction "constant 0" of logical network, which is revealed only at some set of signal` meaning in LN.

In LN test terms which find the malfunction of type 3 there is a set of determined and undetermined facts which provides the activation of the rule r_i and the activation of the way in LN from the line r_i to one of the output lines, which corresponds with the aim RKB [4]. Herewith there should be established facts $f_1, \ldots, f_k$. If the activation r_i and the activation of the way with the established facts $f_1, \ldots, f_k$ is impossible, then the test doesn't exist.

Any consistent set of facts S can be established in P by defining at least one set of input facts $S \in C$. Herewith the set C should also be defined according to the

Fig. 1 Expert systems
debugging method

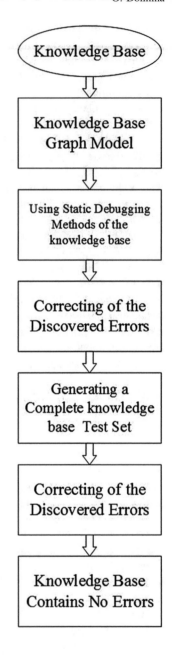

input facts of RKB. This representation of C can be found by doing reverse logical conclusion for all the forbidden combinations which contain inner facts of P. It should be noticed that for the expert forbidden combinations, unlike approved combinations, they are easier to build up.

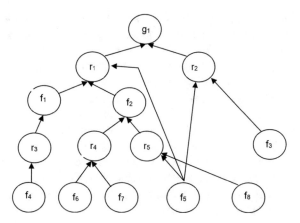

Fig. 2 Example of the graph of rule-based knowledge system

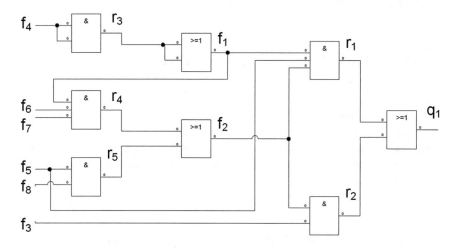

Fig. 3 Logical network that corresponds the graph shown at Fig. 2

The mentioned above is the justification for the building up of RKB tests as the sets of its input facts.

Under simultaneous installation of the set of facts $f_1, ..., f_k$, which are unknown, the rule stops to work. That's why for conducting r_i it is necessary to sort out all the variants of the activization of the rule r_i and all the possible ways of the transportation of corresponding r_i signal of the logical chain to one of its exits. All the facts which state is insignificant for the activization of r_i and for the ways from r_i to exits of the chain should be installed. It should be noticed that the rule r_i, in general case, can be wrong at some combination of the facts. The given approach to building up the set of facts for r_i provides the detection of situations of that kind.

There is a chance of multiple errors in RKB, when several rules might be wrong. A complete RKB test can be obtained with the help of uniting the test sets of all the rules from the R set. Let us denote thus obtained test set T for P.

It is easy to verify that T will ensure the detection of any type multiple errors (3) of the P. Indeed, because the logical network, corresponding to G, contains no inversions, multiples errors cannot mask each other.

To make the RKB tests, you can use the paths sensitization methods developed in technical diagnostics. You need to make all possible kinds of tests for each leaf node. The cubes, conducting the reporting rules activation and paths sensitization overlap with many valid cubes of input C facts. All the X symbols are replaced by "1" in the resulting cubes. It makes most sense to use PODEM algorithm adopted in technical diagnostics [4]. This algorithm ensures return through LN inputs after each signal transportation via next element. In the case of test generation for RKB this approach is the most effective one. The input facts cube resulting from the return intersects with the C set cube. If the intersection is empty, the further activation of the reporting path does not occur.

3 The Method of the Testing of the Three-Layer Perceptron

The proposed method makes it possible to build a complete set of tests for the rule knowledge base (1), as well as for artificial neural network (ANN) based on 3-layer perceptron. Forgetting-about-the-exception error type can be formalized as follows:

For such input data that are exceptions in the subject area, on the basis of the established set of facts

$$f_1, f_2, \ldots, f_n \tag{4}$$

and some observations on X_l ANN concludes that X_l belongs to o_k class;

$$X_l \in X^{o_k} \tag{5}$$

what actually turns out to be wrong due to the fact that X_l is characterized by a certain combination of facts

$$F_x = f_b, f_c, \ldots, f_d \tag{6}$$

and determining this combination clearly indicates that X_l does not belong to o_k class:

$$f_b \wedge f_c \wedge, \ldots, \wedge f_d \Rightarrow X_i \notin X^{o_k} \tag{7}$$

Let us assume that ANN is already trained to make decisions in a particular subject area. Note that to generate a test set and to identify forgetting-about-the-exception error type of ANN, the direct use of the method proposed is not suitable due to the implicit nature of expertise, which the neural network extracts from training set and which is formalized in the form of multiple weighting factors between neurons.

To use the proposed method it is necessary to extract the decision rules from ANN. To extract the rules it is proposed to use the GLARE algorithm. GLARE output information structures are close to the description of the rules in natural language.

By means of GLARE algorithm expert information from ANN is converted into decision rules like this:

$$IF\ criterion_1\ OR\ value_1 = criterion_1 = value_2\ OR....$$
$$AND\ criterion_2 = value_1\ OR\ criterion_2 = value_2\ OR.... \qquad (8)$$
$$THEN\ class_1$$

The paper [5] shows it can be ensured that decision rules, extracted by GLARE would be no less accurate than the accuracy of the source of neural network information structure. GLARE has no restrictions for the volume of training set, as well as for the number of input and output neurons. The execution time of the algorithm is linearly dependent on the above parameters. Also note that GLARE is characterized by the relative simplicity of programming and high-speed performance. Algorithm settings make it possible to vary the complexity and accuracy of the extracted rules widely.

To implement the rule-based knowledge bases test generation method, the special software has been developed that is described in the following paper [6]. This method has made it possible to debug expert systems in medicine [6, 7] compressors troubleshooting [8].

4 Conclusion

There is described the method of debugging the intellectual decision making systems, which allows to detect various types of errors including forgetting-about-the-exception type in the subject area. The method is based on the converting the knowledge base into the graph model for identifying structural errors and then into the logical network for applying the developed method of generating a complete test set. The method is used for debugging rule-based knowledge bases and for artificial neural networks based on the 3-level perceptron from which critical rules are extracted and a decision-making graph is made.

References

1. Marcot, B.: Testing your knowledge base. In: AI Expert, August, pp. 43–47 (1987)
2. Dolinina, O.N.: Razrabotka metoda testirovanija produkcionnyh baz znanij jekspertnyh sistem s uchetom oshibok tipa "zabyvanija ob iskljuchenii": dis…kand. tehn. nauk. Saratov (1999). 171 s
3. Dolinina, O.N.: Informacionnye tehnologii v upravlenii sovremennoj organizaciej [IT in Organisation Management]. Saratov: SSTU, (2006). 160 s. (in Russian)
4. Goel, P., Rosales, B.: PODEM—X: an automatic test generation system for VLSI logic structures. In: Proceedings of 18th IEEE Design Automation Conference, pp. 260–268. IEEE Press Piscataway, NJ, USA. (1981)
5. Gupta, A., Park, S., Lam, S.: Generalized analytic rule extraction for feedforward neural networks. IEEE Trans. Knowl. Data Eng. 11(6), 965–991 (1999)
6. Dolinina, O.N.: Otladka iskusstvennoj nejroseti, osnovannoj na trjohslojnom perseptrone, na primere jekspertnoj sistemy dlja oftalmologii [Debugging artificial neural network based on the 3-level perceptron: a study of the expert system in ophthalmology]/Kuzmin, A.K., Dolinina, O. N.//Vestnik Astrahanskogo gosudarstvennogo tehnicheskogo universiteta. Ser.: Upravlenie, vychislitel'naja tehnika i informatika. 80–90 (2011). (in Russian)
7. Dolinina, O.N.: Otladka nejrosetevoj ekspertnoj sistemy dlja oftalmologii [Debugging neural network expert system for ophthalmology]/Dolinina, O.N., Kuzmin, A.K.//Vestnik Saratovskogo gosudarstvennogo tehnicheskogo universiteta. 4(62). V. 4. S. 248–253(2011). (in Russian)
8. Dolinina, O.N., Antropov, P.G., Kuzmin, A.K., Shvarts, A.Ju.: Ispolzovanie intellektualnyh sistem dlja diagnostiki neispravnostej gazoperekachivajushhih agregatov [Using intellectual systems for compressors troubleshooting]//Sovremennye problemy nauki i obrazovanija 6 (2013). http://www.science-education.ru/113-11252 (data obrashhenija: 23.12.2013). (in Russian)

Motion Strategy by Intelligent Vehicles-Agents Fleet in Unfriendly Environment

Viacheslav Abrosimov and Vladislav Ivanov

Abstract This article considers the territory monitoring problems by air vehicles. Vehicles are considered as intelligent agents. A specific feature consists in the inherent antagonism of the motion environment, which is a common situation in practice. Three major conditions of monitoring are formulated, namely, (1) the necessity of repeated solution of the monitoring tasks with varying routes in each cycle, (2) the necessity of online communication among vehicles under their complete independence in decision-making and (3) the possibility of task failure by some vehicles due to constraints imposed by an unfriendly environment. We introduce a group control strategy for a fleet of vehicles performing monitoring. All vehicles-agents receive a given route from a leading agent or calculate and correct the route in the autonomous mode. The efficiency of the suggested approach is demonstrated by monitoring of an emergency situation, viz., a fire in a forest zone approaching a critical object (a nuclear power plant).

Keywords Intelligent agent · Air vehicle · Group control · Strategy · Monitoring · Unfriendly environment

1   Introduction

In practice monitoring is necessary for investigation of remote settlements, search of people in distress, assessment of extended objects (pipelines, heavy roads) in the conditions of emergency situations (fire, flood, natural disasters, etc.). Monitoring possesses military applications, chiefly in the field of intelligence also.

Territory monitoring can be performed by different technical means. In practice, these are flying vehicles (aircrafts, helicopters, unmanned aerial vehicles (UAV) and

V. Abrosimov (✉) · V. Ivanov
Moscow Aviation Institute (National Research University), Moscow, Russia
e-mail: avk787@gmail.com

V. Ivanov
e-mail: vladislav_ivanov_2015@list.ru

© Springer International Publishing Switzerland 2016
R. Silhavy et al. (eds.), *Automation Control Theory Perspectives in Intelligent Systems*, Advances in Intelligent Systems and Computing 466, DOI 10.1007/978-3-319-33389-2_30

so on) equipped with appropriate facilities. Each vehicle is described by individual characteristics which form a tuple $\{v_i, t_i^{max}, \sigma_i\}$, here v_i denotes the mean speed of the vehicle on its route from one object to another; t_i^{max} specifies the maximum time of vehicle motion defined by its fuel reserves; and finally, σ_i reflect the vehicle's conditional capabilities of object monitoring depending on the surveillance devices mounted on it.

Monitoring often requires involvement of several vehicles [1]. They can differ by the technical capabilities of object monitoring. In an ordinary situation the vehicles receive a task and carry out it individually. But it is inefficient. As far as a fleet fulfils a group mission, vehicles do not enjoy the freedom of actions. They must jointly discuss and solve a common task via interaction and negotiations, with possible redistribution of monitored objects in the course of their motion (in the online mode or with a given frequency). From this point of view they can be considered as the agents possessing the corresponding properties—ability to interaction, pro-activity, desires, intentions, commitments, rationality and so on.

In principle, the solution approaches to this class of problems are developed within the framework of the Travelling Salesman Problem. For a fleet of vehicles, the corresponding class of problems is called the Vehicle Routing Problem [2]. However, the main feature of the territory monitoring problems concerns the need for recurrent flying around object. A common situation in practice is when a series of vehicles still continue their tasks, but an additional fleet of vehicles must be allocated for surveying separate points and domains on a territory. Therefore, we formulate the first major condition of monitoring as the necessity of repeated solution of the monitoring problems with varying routes in each cycle.

Most interest during monitoring belongs to situations, where separate vehicles become unable to perform their tasks in the group mission due to technical failures or the impact of an unfriendly environment. This is inadmissible. In such cases, the rest (operable) vehicles must receive information about the current situation, jointly discuss it and elaborate new strategies of accomplishing the group mission. Hence, we formulate the second major condition of monitoring as the necessity of online communication among vehicles under their complete independence in decision-making.

An external environment of monitoring can be friendly, neutral or unfriendly with respect to a vehicle. Such external environment (especially unfriendly) may impose appreciable constraints on vehicle motion and fulfillment of its tasks. For example the source of active antagonism consists in air space or anti-missile defense systems terminating vehicles. Active antagonism is always created for special purposes; it has an organization and control with counteraction functions.

Different potential obstacles for observation and motion along given routes may occur in the course of monitoring. An external environment of monitoring can be friendly, neutral or unfriendly with respect to a vehicle. Friendly and neutral environments apply no constraints on implementation of monitoring functions. On the other hand, an unfriendly environment either reduces the efficiency of monitoring, or even counteracts its realization. Thus, an external environment may impose appreciable constraints on vehicle motion and fulfillment of its tasks.

There exist passive and active antagonisms of an environment. A passive unfriendly environment is the one characterized by objective factors complicating monitoring. For instance, the matter concerns smoke, fog, rain, high radioactivity, and so on. An active unfriendly environment takes purposeful actions on reducing the efficiency of monitoring functions, right up to the impossibility of their conduct. Such environments often characterize monitoring processes in military-technical problems. These can be, e.g., electronic countermeasures to monitoring devices mounted on vehicles. Another source of active antagonism consists in air or anti-missile defense systems terminating vehicles. Active antagonism is always created for special purposes; it has an organization and control with counteraction functions.

To assess the degree of restricted motion in an environment, let us introduce a complex quantitative characteristic, i.e., the restriction index P_s of the environment. Naturally, such index represents a fuzzy variable in the range $0 < P_s < 1$, between the minimum and maximum degrees of restrictions. During monitoring, restrictions may show themselves on motion routes. These situations will be reflected by the parameter $\omega_{sr} \in [0, 1]$, the motion risk on a route between objects n_s and n_r. Similar situations (a collapse risk) on an object n will be described by the parameter $\rho_n \in [0, 1]$. The value $\rho_n = 1$ means prohibition of the monitoring of object n.

So, the above-mentioned two major conditions of monitoring are supplemented by the third one-the possibility of task failure by some vehicles due to constraints imposed by an unfriendly environment.

2　Methods

Related works. The monitoring in emergency situations is particularly important the literature emphasizes the complexity of such problems caused by uncertainty of the situation, the necessity to obtain information in real time, the obligatoriness to reallocate resources and other. The Article [3] deals with the choice of the helicopter's fleet and ways of their movement for delivery the required materials to the accident site. In [4, 5] we found genetic algorithms methods optimizing route selection and length used in traffic congestion, and physical damage to roads. In the article [6] the class of problems of logistics in various emergency conditions is developing (the earthquakes and typhoons are as examples).

Vehicle as intelligent agent. Clearly, solution of the monitoring problems in an unfriendly environment requires forming a fleet of vehicles. Each vehicle must act as well as autonomous mode with self-control. Moreover, it is necessary to consider the intentions and actions of other vehicles-agents. The performed analysis demonstrates that vehicle strategy design in the monitoring problems may employ the framework of multi-agent intelligent systems [7, 8]. Our primary idea is to endow all vehicles with the properties of an intelligent agent. In such a situation the vehicle become autonomous, process information independently and can interact with other agents for obtaining a common solution of a group problem. In practice, a reasonable approach implies the following. First, construct a general solution on monitoring using a common algorithm implemented by the leading agent and assign motion routes to each vehicle. Second, allow each vehicle to make decisions depending on the current situation in the autonomous mode. A vehicle exchanges information with other vehicles, reports its current state, action strategy and the fulfillment degree of a group mission. A vehicle considers the goals and intentions of other vehicles. Imagine that a vehicle or several vehicles fail to perform the tasks of monitoring due to difficult conditions or the impact of an unfriendly environment. In this case, other vehicles can take new tasks if available resources are sufficient.

For organizing the interaction of vehicles as independent agents performing a group mission one should allocate a web resource (WR) with a given content structure; all vehicles address the resource in the online mode for acquiring and placing relevant information. In our problem, the WR stores group information of three types: (a) actual information about on the real situation on the routes of object monitoring from the surveillance devices of all vehicles; (b) information about the fact of impossible solution of the monitoring tasks by vehicles and (c) information about the residual resources of all vehicles for solution of additional tasks.

The description of monitoring object. The WR stores the description of the state of surveyed object n at moment t as a tuple

$$n(t) : \{x_n, y_n, z_n, \varphi_n, \mu_n, \mu_{rs}, \omega_{rs}, \rho_n, \alpha_i^n, t_i^n\}. \tag{1}$$

where

- x_n, y_n, z_n—space vehicle coordinates,
- $\varphi_n \in [0, 1] \forall n \in N$ the priority levels of objects,
- $\mu_n \in [0, 1] \forall n \in N$ the intensity rates of the current situation on objects,
- $\mu_{rs} \in [0, 1], \forall s \in N, \forall r \in N$ the elements of intensity rate matrix M_{rs} of the current situations on the route between objects r and s,
- $\omega_{rs} \in [0, 1] \forall s \in N, \forall r \in N$—the elements of risk matrix W_{rs} on the route between objects rand s,
- $\rho_n \in [0, 1] \forall n \in N$—the collapse risks in the n-object,
- $\alpha_i^n = 0, 1$—the mark of the i-object survey,
- t_i^n—a predicted time of the i-object survey.

The description of vehicle. The WR characterizes the vehicle state at moment t in the form of a group containing the route on a given operation M_{iH}^χ, the lists of surveyed and unsurveyed objects (M_{iH+}^χ and M_{iH-}^χ, respectively), as well as the residual resource Res_i and the attribute A_i of the current vehicle operability:

$$V_i : \{i, M_{iH}^\chi, M_{iH+}^\chi, M_{iH-}^\chi, Res_i, A_i\}. \tag{2}$$

So WR store (a) actual information about on the real situation on the routes of object monitoring, (b) information about vehicle intentions and the fact of impossible solution of the monitoring tasks by vehicles and (c) information about the resources of all vehicles for solution of additional tasks.

The negotiation model. Scientific literature describes different methods of task allocation and control for a fleet of vehicles. The article [9] separated out four settings of a negotiation environment, namely, the number of issues, the number of participants, environment state and negotiation relationship. Among different types of agents cooperation (coordinated cooperation, simple cooperation, collective/individual competition for resources), we will consider simple cooperation. In the territory monitoring problems vehicles organize negotiations on a single issue (route correction if necessary). The number of participants is the number of vehicles in a fleet after deduction of vehicles that announced their task failure (multilateral participants). We will consider dynamic negotiations of agents: decision rules in negotiations are based on situation forecasting, negotiations run in the online mode and the current situation is continually refined in the course of motion. In the sequel, a negotiation relationship is treated as a solo: negotiations on route correction do not depend on other negotiations. In this case, V_i announces its intentions by allocating relevant information about the given route on the WR, i.e., by assigning $\alpha_i^n = 1$ for object n and specifying the planned time t_i^n of its survey.

The approach to help agents to each other.

Definition 1 "Inclusion level". A vehicle's inclusion level is a resource amount required for monitoring of an additional object by means of this vehicle. Consider the route M_{iH}^χ: $(n_{i1}, \ldots n_{ik})$ and add object n_h to M_{iH}^χ. The inclusion level for $1 \le j \le k - 1$ is defined by

$$\Delta F_i(n_h) = \min_j \{ d(n_h, n_{ij}) + d(n_h, n_{i,j+1}) - d(n_{ij}, n_{i,j+1}) \}/v_i \}, \qquad (3)$$

where κ denotes the number of objects in the route chosen for V_i. In the physical sense, the inclusion level characterizes the most efficient inclusion conditions of monitoring object n_h into the route. It reflects the minimum time required for to leave the route M_i in order to monitor object n_h, with subsequent return to the same route for motion from objects i to j.

Definition 2 The residual resource of a vehicle. The residual resource of a vehicles its active motion period available after monitoring of all objects assigned to this vehicle. The residual resource of a vehicle has the formula

$$Res_i = t_i - D^\Sigma/v_i \qquad (4)$$

where t_i indicates the maximum available motion time of Vehicle$_i$ determined by the fuel balance conditions (fuel reserves and consumption) and D^Σ means the total distance along the planned vehicle route. Therefore, for an arbitrary object n_h it is possible to evaluate the minimum resource (in the present example, time) required for V_i to achieve this object:

$$c_{ih} = \text{Res}_i - \Delta F_i(n_h). \qquad (5)$$

Of course, if $|c_{ih}| \leq 0$, then V_i has insufficient resources for achieving the object n_h. Negotiations engage only V_i possessing available resources and an inclusion level $\Delta F_i(n_h) > 0$.

Definition 3 "Elimination level". An elimination level of a vehicle is a resource amount released via adding an unsurveyed object n_h to a current route and removing (eliminating) the most inefficient object (in the sense of resource consumption) from the list of unsurveyed objects M_{i-}^{χ} of the route M_i. The elimination level for $1 \leq j \leq k - 1$ is calculated by

$$\Delta G(n_{ij}, c_{ih}) = \left\{ d(n_{i,j-1}, n_{ij}) + d(n_{i,j}, n_{i,j+1},) - d(n_{i,j-1}, n_{i,j+1}) \right\}/v_i + c_{ih} \geq 0. \qquad (6)$$

Elimination level is necessary in situations when route correction for V_i brings to the following outcome. It appears reasonable to incorporate an additional object into the route, eliminating an object whose monitoring is impossible due to the insufficient resource of the vehicle. In the physical interpretation, the elimination level determines monitoring objects such that the resource amounts released by their elimination from the route exceed the inclusion level of a new object n_h.

Now, imagine the following situation. In the course of motion along given routes, at moment t_{fix} there occur technical or other problems impeding monitoring of an object n_h. In this case, V_p sets to zero the attribute of current activity $(A_p := 0)$ and the indicator $\alpha_p^h = 0$. Therefore, some surveyed objects $n_{p1}, n_{p2}, n_{ph}, \ldots, n_{pH}$ are no more assigned to this vehicle and have to-be-reallocated. The negotiations of the rest vehicles can be organized in different ways. Analysis demonstrates that the most adequate approach lies in the auction model. Here resource is the time period Res_i.

3 Results

As the result we offer the group strategy formation for vehicle-agents. Under territory monitoring in an unfriendly environment, the group strategy of actions for a fleet of vehicles as the general solution on object allocation among vehicles by the leading agent of a fleet before monitoring and a set of individual actions of separate vehicles during their motion in order to perform monitoring tasks. A typical action strategy of vehicles during their motion in order to perform monitoring has the following form.

1. Simulating the motion of $V_i (\forall i \in I)$ along a route defined by the leading agent from an object n_r to a next object n_s with a speed v_i depending on the design features of V_i.

2. Acquiring information on the intensity rates μ_{rs} of the current situation on the route and information on the intensity rate μ_n of the current situation on object n; this information is received during motion via surveillance devices mounted by a vehicle.

3. Correcting the elements of the intensity matrix M_{rs} of the current situation on the route and the elements of the intensity rate M_n of the current situation on object n using the formulas

$$= \mu_{rs}^t + \sigma\mu_{rs}^\tau, \mu_n^{t+\tau} = \mu_n^t \pm \sigma\mu_n^\tau. \tag{7}$$

4. Placing the obtained information on the WR.

5. Assessing the current situation by analyzing the indicators $\alpha_j^h (\forall j \in I, \forall h \in N)$.

6. If $\alpha_j^h = 0$ at least for single j or single h, announcing an auction for inclusion of an object n_h into the route.

7. For each $h \in H$ and each V_i engaged in monitoring, calculating the inclusion level $\Delta F_i(n_h)$. Adding the point n_h to the route of V_i which simultaneously meets the following conditions:

$$\Delta F_i^*(n_h) = \min_I \{\Delta F_i(n_h)\} \text{ and } Res_i - \Delta F_i(n_h) \geq 0. \tag{8}$$

8. If $\alpha_i^n = 1 \forall n \in N$, form the attribute of route correction finalization (all $n \in N$ surveyed objects are assigned to appropriate vehicles $i \in I$).

9. If $\alpha_i^n = 0$ at least for single $n \in N$, then repeating the route correction process according to the algorithm below:

   (a) for each V_i, calculate the inclusion levels $\Delta F_i(n_h)$ and the resource required for achieving the object n_h:

   $$c_{ih} = res_i - \Delta F_i(n_h). \tag{9}$$

   (b) find V_f such that $|c_{ih}|_f = \min_I |c_{ih}|$. If this minimum exists at least for single V_f, then assign object n_h to V_f and $\alpha_f^h := 1$.

   (c) using the inclusion level $\Delta G(n_{ij}, c_{ih})$, design a new route: add the object n_h and eliminate all objects requiring higher resources of monitoring than V_i actually has.

The example. Let us consider monitoring of an emergency situation, i.e., a fire in a large-scale forest zone, which approaches a critical object–a nuclear power plant (see NPP in Fig. 1) near Desnogorsk. Monitoring is performed by drones equipped with special fire-fog detectors. According to the task, it is necessary to survey the current situation in Prismara and Novoselki, as well as a series of seats of fire, see white circles.

We have conducted simulation using Any Logic System [10]. Figure 1 demonstrates a computer display with the allocated routes of the drones. Here thin dashed lines indicate the routes of the three drones, which have been designed in

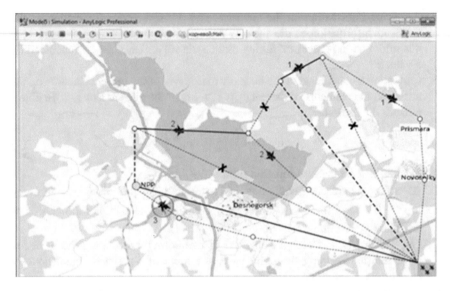

Fig. 1 Corrections in the monitoring routes of an emergency situation based on the results of negotiations between vehicles 1 and 2 under an abrupt failure of vehicle 3 responsible for monitoring of a critical object (a nuclear power plant)

advance by the leading agent and assigned to appropriate vehicles. Motion starts along these routes. In the course of motion, a failure occurs in vehicle 3 whose route passes through the nuclear power plant. Vehicle 3 uses the WR for announcing that the nuclear power plant is no more assigned to any of the vehicles. The rest vehicles (1 and 2) carry out negotiations by the above-stated algorithm, reallocate the surveyed objects between them and follow the new routes (see solid lines). Thick dashed lines mark the segments of the routes originally designed for one vehicle, but actually followed by another vehicle due to object reallocation. And finally, thin dashed lines with X sign correspond to the segments of the routes eliminated from consideration as the result of negotiations.

4 Discussions

The suggested approach to territory monitoring has been developed in the conditions of different constraints. A major constraint concerns the antagonism of an environment. In this paper, the antagonism is that the environment disables a vehicle or several vehicles in a fleet so that their functions cannot be fulfilled. Moreover, each vehicle from the fleet knows nothing about the unfriendly environment. At the same time, analysis of an unfriendly environment during motion seems an interesting problem.

In this context, an efficient approach consists in employing the concept of an environment developed in the [11]. It is introduced as a virtual environment that offers a medium for agents and transport agents to exchange information and to coordinate. Besides, the virtual environment serves the communication of messages and the physical control of vehicles and their behavior. The antagonism of an environment can be assessed only by preliminary reconnaissance. In this respect, it seems efficient to utilize the idea of intelligence ants (feasibility ants) in the terminology of [12]. They explore possible routes on a graph, where nodes are surveyed objects and branches are paths to them, as well as assess their antagonism. A feasibility ant transmits all collected information to the WR, correcting the corresponding values of ρ_n and ω_{sr}. This can be considered during route design and choice of negotiation strategies.

The role of such feasibility ants grows at the second and subsequent iterations of territory monitoring: they follow the antagonism of an environment and, furthermore, the intensity rates of all situations on the routes of vehicles. Therefore, a reasonable solution is to include vehicles with different characteristics and capabilities into a fleet. In this case, they have different behavioral strategies. Owing to feasibility ant usage, one can also reduce the number of vehicles required for monitoring and send them along the routes preliminarily assessed by the intensity rates of the current situation, as well as eliminate from the routes all branches with high risks.

The present paper has been dedicated to the territory monitoring problems by a group of intelligent vehicles. In practice, these problems arise in such fields as maintenance of spatially distributed objects in natural gas industry and oil industry, flying around remote settlements and expeditions, cargo drop for people in distress, maintenance of military objects and so on.

A specific feature of monitoring consists in its implementation in an unfriendly environment. Passive antagonism reduces the efficiency of monitoring. Active antagonism is always created for special purposes, has an organization and control with motion counteraction functions.

We have formulated the three major conditions of monitoring, namely,

- the necessity of repeated solution of the monitoring problems with varying routes in each cycle;
- the necessity of online communication among vehicles under their complete independence in decision-making;
- the possibility of task failure by some vehicles due to constraints imposed by an unfriendly environment.

It has been demonstrated that, in such conditions, one should develop a group control strategy for a fleet of intelligent control object performing monitoring. For this, vehicles are considered as intelligent agents, with the capability of autonomous activity, independent acquisition and transmission of information, interaction with other agents for joint accomplishment of a group mission.

Next, the author has proposed the negotiation algorithms of vehicles. In the conditions of an unfriendly environment, a group behavior strategy for a group of control objects includes the feasibility of route correction. This happens in a situation when one or several vehicles are incapable of performing their tasks due to the impact of an unfriendly environment. In this case, the web-resources serves for organizing negotiations of vehicles. Subsequently, vehicles submit their residual resources (after route completion) to an auction to make a joint decision on monitoring of additional objects.

The results of this work can be adopted in practical logistic problems and transport problems (under existing obstacles to vehicle motion), as well as in military applications.

References

1. Abrosimov, V.K.: Group Motion of Intelligent Aircrafts in an Unfriendly Environment. Nauka Publishing House, Moscow (2013). (in Russian)
2. Golden, B.L., Raghavan, S., Wasil, E.A. (eds.) The Vehicle Routing Problem: Latest Advances and New Challenges. Springer (Operations Research/Computer Science Interfaces Series.) (2008)
3. Barbarosogu, G., Ozdamar, L., Cevik, A.: An interactive approach or hierarchical analysis of helicopter logistics in relief operation. Eur. J. Oper. Res. **140**, 118–133 (2002)
4. Barrie, B.M., Ayechew, M.A.: A genetic algorithm for the vehicle routing problem. Comput. Oper. Res. **30**, 787–800 (2003)
5. Du, M., Yi, H.: Research on multi-objective emergency logistics vehicle routing problem under constraint conditions J. Ind. Eng. Manage. JIEM **6**(1), 258–266 (2013)
6. Hsueh1, C.-F., Chen, H.-K., Chou, H.-W.: Dynamic vehicle routing for relief logistics in natural disasters in vehicle routing problem. In: Caric, H., Gold, H. (eds.), pp 71–84. I-Tech Education and Publishing KG, Vienna, Austria (2008)
7. Weiss, G. (ed.): Multiagent Systems: A Modern Approach to Distributed Artificial Intelligence. MIT Press (2000)
8. Systems, Multiagent: Algorithmic, Game-Theoretic and Logical Foundations. Cambridge University Press, Hardcover (2008)
9. Fenghui, R.: Autonomous agent negotiations strategies in complex environment. PhD, School of Computer Science and Software Engineering, Faculty of Engineering, University of Wollongong (2010)
10. Any Logic System. https://plus.google.com/102823219950373512862/posts
11. Holvoet, T., Valckenaers P.: Exploiting the environment for coordinating agent intentions. In: Proceedings of Third International Workshop on Environments for Multi-Agent Systems (E4MAS06). Hakodate, Japan, Springer. (2006)
12. Weyns, D., Holvoet, T.: Architectural design of a situated multi-agent system for controlling automatic guided vehicles. Int. J. Agent-Oriented Softw. Eng. **1**(2), 90–128 (2008)

Significant Simulation Parameters for RESTART/LRE Method in Teletraffic Systems of Network of Queues

Elena Ivanova, Teodor Iliev, Grigor Mihaylov and Radomir Rashkov

Abstract The blocking events, overflows and losses in telecommunications, especially in systems with very low probability, are one of the major estimation problems. Speed-up simulation is the most convenient method for estimation of rare events. The speed-up simulation in this investigation is done with one of the splitting methods—RESTART with implementation of LRE. The main purpose of the research is to determine the significant parameters for speed-up simulation of network of queues: tandem and parallel queues.

Keywords RESTART · LRE · Rare event · Tandem queues · Parallel queues

1 Introduction

Estimating the probability of a rare event has applications in reliability, telecommunications, insurance, and several other areas. For complicated models, this could be calculated done in principle by the Monte Carlo simulation, but when the event of interest is rare [1], straightforward simulation would in most cases require an excessive number of runs for the rare event to happen frequently enough so that the estimator is meaningful. Rare event simulation is still an open.

RESTART (Repetitive Simulation Trials After Reaching Thresholds) is an accelerated simulation method, which belongs to the importance splitting methods,

E. Ivanova (✉) · T. Iliev · G. Mihaylov
Department of Telecommunication, University of Ruse, Ruse, Bulgaria
e-mail: epivanova@uni-ruse.bg

T. Iliev
e-mail: tiliev@uni-ruse.bg

G. Mihaylov
e-mail: gmihaylov@uni-ruse.bg

R. Rashkov
Inter Engineering 10 Ltd, Razgrad, Bulgaria
e-mail: rashkov@gmail.com

© Springer International Publishing Switzerland 2016
R. Silhavy et al. (eds.), *Automation Control Theory Perspectives in Intelligent Systems*, Advances in Intelligent Systems and Computing 466,
DOI 10.1007/978-3-319-33389-2_31

used to speed-up the rare event simulation. Combination of RESTART and LRE (Limited Relative Error) is described by Görg [2]. The LRE measures the complementary distribution function of the queue occupancy and performs the Run Time Control (RTC) of the simulation. The LRE performs the Run Time Control with two conditions: the Large Sample Conditions and the Relative Error Condition. The first condition assures, that the queuing system has reached the steady state. The second condition, Relative Error Condition, represents a measure to estimate the relative error at the current state of the simulation.

In order to be able to compare the project measures with known results, the part of the project is study different types of models of networks of queues. Evaluation of blocking probability for different teletraffic systems is simulated [3, 4]. RESTART/LRE is implemented the in Ptolemy, a powerful, object-oriented simulator in Microsoft Visual Studio 3.0, which includes the Visual C++. The multifactor regression analysis is based on initial simulation results, which are manipulated with program Statistica v10. The purpose of the analysis is to explore the relationship among the variables of algorithm RETART/LRE and parameters of the teletraffic system, to achieve more blocking events based on received significant parameter.

2 Overflow Probabilities for Networks of Queues

The overflow probability of tandem and parallel queues are investigated with speed-up method for rare event simulation. Majority models, used for queuing systems for tandem queues performance evaluation, are (or can be converted to) discrete-time Markov chains (DTMCs) [3, 5]. Note that some other performance measures, like delays, cannot be obtained from the discrete-time Markov chain description. A DTMC for a Markovian queuing model has a highly regular structure. It is used for analytical verification of the models and implemented algorithm [2, 3, 6].

2.1 Tandem Queues

Consider a complex model of teletraffic systems with two tandem queues, with non-priority classes of arrivals, buffer capacity is N, LIFO or FIFO service. Arrival and service distributions could be chosen Normal, Exponential, Pareto, Geometrical and Poisson (Fig. 1). The Exponential and Poisson distributions are used for verifying the model and implemented algorithm RESTART/LRE for tandem queues [3, 7].

First of all, the states in tandem queue typically could be arranged typically on a grid with as many dimensions as the number of queues. Each coordinate represents

Fig. 1 Two-queue tandem
network

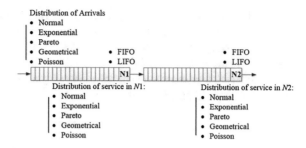

the number of customers in each queue. Second, every transition in the DTMC
corresponds to an elementary event in the queuing model: an arrival or a service
completion at one of the queues. For convenience, in the article will henceforth
refer to such events as "transition events". It should be noted, that these transition
events are defined independently of the state; i.e., there is only one transition event
for a service completion at a given queue, and this single transition event corresponds to a transition out of every state in the DTMC, in which this particular queue
is non-empty. Third, it is obvious that not all transition events are "enabled" in
every state: e.g., in a state where a particular queue is empty, the service completion
event of that particular queue is not possible, i.e., not enabled.

In the project are used reference queue models, to verify the simulation method.
The properties of the method can be described by analytically derived mathematically derived formulas [8].

As an example, consider the overflow probability of the total population in a
(Jackson) network consisting of two queues in tandem, like the one depicted in
Fig. 1. Customers arrive at the first queue according to a Poisson-process with rate
λ. Both servers have exponentially-distributed service times. Overflow probabilities
in simple Jackson networks with rates μ_1 and μ_2. The state of the system at any time
is given by the two integer values N_1 and N_2, which are the number of customers in
the first and second queues (Fig. 2), respectively.

Fig. 2 State space of two
queues in tandem network

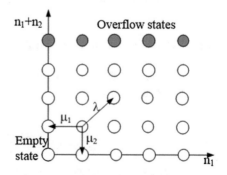

The difficulty of applying accelerated simulation techniques arises, when the first queue is the bottleneck and the rare set definition is related to the value of n_2. If splitting or RESTART is applied, a more detailed choice of the importance function has to be made in this case [3].

Considering the two-queue tandem network presented as reference model for the following three definitions of the rare set A:

$$
\begin{aligned}
n_1 + n_2 &\geq L, \\
n_2 &\geq L, \\
\mathrm{Min}(n_1, n_2) &\geq L
\end{aligned}
\tag{1}
$$

The numbers of retrials were chosen to have values of n_i as close as possible to those given by [2, 5].

2.1.1 Rare State Defined as $n_1 + n_2 \geq L$

For this definition of the rare set, the function of highest importance is $\Phi = n_1 + n_2$. The possible states at event B_i are $(0, n_1 + n_2)$ or $(0, (n_1 + n_2 - 1))$. The importance of these states is different. The higher the value of n_1 (for $n_1 + n_2$) is, the higher the importance of the state, given that customer at n_1 has to be served by both servers before leaving the system, while a customer at n_2 has to be served only at the second one. Here the bottleneck is the first queue, so the states with high value of n_1 and low value of n_2 have the highest probability.

2.1.2 Rare State Defined as $n_1 \geq L$

For this definition of the rare set $n_1 \geq L$, the most natural importance function is not $\Phi = n_2$. To obtain states with approaching importance, it is that weight, albeit smaller than the weight given to n_2, must be given to n_1, so the most natural importance function is not $\Phi = an_1 + n_2$, with $0 \leq a \leq 1$.

2.1.3 Rare State Defined as $\mathrm{Min}(n_1, n_2) \geq L$

For $n_1 \leq L$ and $n_2 \leq L$, the definition of Φ is $\Phi = a\Phi_1 + \Phi_2$. The definition of the importance function Φ_i is (2).

$$
\Phi_i = \begin{cases} n_i & \text{if} \quad n_i \leq L \\ L + b(n_1 - L), & \text{if} \quad n_i > L \end{cases}
\tag{2}
$$

The values of coefficient b are smaller than 1.

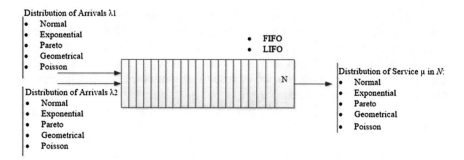

Fig. 3 Two parallel queues

2.2 *Parallel Queues*

Consider a complex model of teletraffic systems with two parallel queues, non-priority classes of arrivals, buffer capacity is N, LIFO or FIFO service. Arrival and service distributions could be chosen Normal, Exponential, Pareto, Geometrical and Poisson (Fig. 3) [3, 5].

The Exponential and Poisson distributions are used for verifying the model and implemented algorithm RESTART/LRE for parallel queues.

The overflow and loss probability or rare event probability is interested in in this case is the steady state probability that the queue is full.

3 RESTART and LRE

The traffic simulation in this analysis is done with one of the splitting methods RESTART with implementation of LRE (Limited Relative Error) [1, 2].

The RESTART method is a variant of splitting, where every chain is splitted by a fixed factor, when it hits a level upward. One of the copies is kept as the original for that level. When any of those copies hits that same level downward, if it is the original one, it continues its path, otherwise it is interrupted.

This rule applies recursively, and the method is implemented in a depth-first fashion, as follows: whenever there is a split, all the non-original copies are simulated completely, one after the other, then simulation continues for the original chain [8].

Applying the RESTART means to split the possible range of values of λ into regions with different importance. The given set of thresholds is L_i, $i = 0, 1, 2,...,$ m. The system reaches the state L_0, while evaluating the first interval $[0, L_0]$ the system states are saved. Once the values for the first interval are established at the time t_0, the simulation restarts from one of the previously saved states at threshold L_0. Each time the threshold L_0 would be crossed when a restart is performed: one of the saved states is randomly chosen and the system is reloaded with this state. Most significant choice for the threshold is the number of customers in the queue. Clearly, this choice satisfies all requirements: it is never incremented by more than 1, it is k in the overflow states, and it becomes zero at the end of the busy period (i.e., in the absorbing state). The initial level is 1: the level immediately after the first arrival to the empty queue.

RESTART method uses the estimation of Cell Loss Ratio. The Cell Loss ratio $P_L = G_L$ can be derived as (3).

$$G_l = P\{\lambda \geq L\} = P\{\lambda \geq L | \lambda \geq L_{m-}\} P\{\lambda \geq L_{m-1} | \lambda \geq L_{m-2}\}$$
$$\ldots P\{\lambda \geq L_1 | \lambda \geq L_0\} P\{\lambda \geq L_0\} \tag{3}$$

The relative error RE_{maxi} is a function of the overall error RE_{max} and the number of thresholds m (4).

$$RE_{maxi} = (1 + RE_{max})^{\frac{1}{m+1}} - 1 \tag{4}$$

The correct choice of thresholds L_i leads to a maximum speed-up according to (5).

$$\frac{G_i}{G_{i-1}} = e^{-2} \tag{5}$$

Since the CCDF is monotonically decreasing, and the thresholds are correctly chosen, one gains a large speed-up factor using RESTART method [5].

The Limited Relative Error [2, 7] approach is based on the construction of independent samples from a given time series by building batches that tend to have lesser autocorrelation than the original series. So essentially it aims on the reduction of autocorrelation. It results in confidence intervals for the mean of the simulation results.

In this investigation RESTART is combined with the third version of the LRE, optimized for discrete, correlated processes [2]. For the implementation of LRE III algorithm it has to be fulfilled the Large Sample Conditions and the Relative Error Condition and afterwards the results have to be summarized.

There are three versions of the LRE. We use the LRE III, since it is optimized for discrete, correlated processes. The LRE III algorithm has the function in this context: the estimation of the CCDF. The LRE is implemented with counters to measure the following quantities [2]:

- n—number of trials;
- h_i—state frequency: number of times the queue was in state i when a cell arrived;
- a_i—transition frequency for decreasing: number of transitions from j to i between two arrivals, with $j > I$;

From these counters, quantities l_i and d_i can be derived as (6) and (7), respectively.

$$d_i = \sum_{j=1}^{k} h_j, \quad \text{with } d_0 = n \tag{6}$$

$$l_i = n - d_i \tag{7}$$

If the Large Sample Conditions (8) and Relative Error Condition (9) are fulfilled the results can be summarized in the end of the algorithm.

The Large Sample Conditions are:

$$n \geq 10^3$$
$$(l_i, d_i) \geq 10^2 \tag{8}$$
$$(a_i, d_i - c_i, l_i - a_i) \geq 1$$

Relative Error Condition:

$$RE_{\mathrm{maxi}} = \frac{1 - d_i/n}{d_i} . \mathrm{Cor}_i \tag{9}$$

The LRE starts with index $i = 1$, tests the conditions (8) and (9). If the conditions are fulfilled the LRE advances to the next index i. The simulation is terminated for the last index $i = L_m$.

4 Simulation Results

The main purpose of our research is to define significant parameters for every one of the systems and parameters of RESTART. The analyzed investigated parameters, which impact on the simulation results of respective RESTART algorithm, are relative error (RE), number of thresholds L and number of trials n. Conducted simulations, compared with analytical probabilistic parameters, prove that RE has insignificantly impacted the simulation probability. Therefore, in some analyzes this parameter will be skipped. Multifactor regression analysis and correlation matrix [4] are used to define the impact of parameters on investigated value/parameter—blocking and simulation blocking rate [9–11]. The aim of conducted simulation studies is to determine the parameters, which accelerate the simulation process for various systems.

The first two steps, with multifactor regression analysis and correlation matrices, from the study methodology aim to define effective multidimensional space of the study. The study is structured according to type of system in focus. Initial studies are heuristic. The main steps of the methodology for conducting simulation experiments of the RESTART algorithm and various teletraffic systems, are as follows:

- In order to be accurate, for regression analysis it's necessary to be calculated number of required experiments;
- Intuitive experiments are conducted, which are not registered;
- Analytical probabilities for occurring of rare event are calculated in systems, where analytical results can be obtained;
- Experiments with various teletraffic system parameters and RESTART are conducted;
- Experimental and analytical results, along with set parameters should be inputted in advanced analytics software package Statistica v10, in order to obtain:

  - Correlation matrix for optional parameters against blocking probability;
  - Multifactor regression analysis for all variable parameters.

Simulation experiments have to be evaluated statistically with correlation matrix, in order to be verified whether there's linear relation between parameters and the evaluation of the correct application of multifactor regression analysis. Depending on the significance of the experiments, a simulation is conducted, through which the area with optimal parameters for RESTART applying in the investigated system is specified.

Multifactor regression analysis is applied with variation of investigated parameters. The Fisher coefficient along with Steward's criterion are calculated [3].

It's difficult to calculate analytical values for blocking probability for non DTCM teletraffic systems. For that reason, for systems in which this is applicable, experiments with variation of teletraffic system's parameters are calculated and experiments with variation of teletraffic system's parameters and variation of RESTART algorithm's parameters are conducted, to these analytical derived rates for blocking probability.

Teletraffic systems, represented as networks of queues are studied, for which a methodology is developed for accelerated simulation of overflow probability in network of two consecutive queues and blocking probability for whole system. Correlation matrices are determined for consecutive queues of type: M/M/1/N and M/M/1/N; Geo/Geo/1/N and Geo/M/1/N; Pareto/M/1/N and M/M/1/N [11–13]. Working parameters of teletraffic system, along with effective simulation parameters are studied. Multifactor regression analysis estimates the impact of studied parameters for simulation with RESTART method.

4.1 Tandem Queues

The efficiency from accelerated simulation is estimated by analytical results for overflow probability and blocking in consecutive queues with Markov pattern.

A teletraffic system with two consecutive queues with Markov pattern is calculated by limited capacity in queues N (Fig. 4). It is proven, that main parameters affecting simulation overflow probability in queues are:

- System load ρ, intensity of the arrivals λ and intensities of serving μ_1 and μ_2;
- Number of thresholds L.

A teletraffic system is studied with two consecutive queues: Pareto/M/1/N and M/M/1/N. In this system the accelerated simulation can be calculated by varying the parameters, which affect the system, such as:

- Number of customers of two queues N_1 and N_2;
- Number of trajectory samples' branches L.

A teletraffic system is simulated with two consecutive queues. First queue is Geo/Geo/1/N and the second queue is Geo/M/1/N (Fig. 5).

The result data after analysis of statistical experiments define parameters with highest impact as:

- Length of the queues N_1 and N_2;
- Intensity of serving in second queue μ_2.

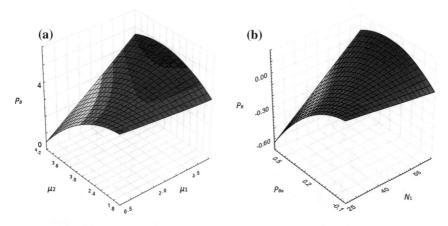

Fig. 4 Graphic correlation between: **a** Poisson incoming flow, exponential distributions for serving first and second queue; **b** N_1 and simulation—and analytical blocking probabilities

(a)   **(b)**

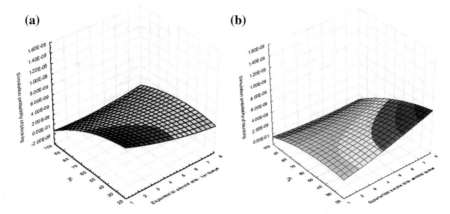

Fig. 5 Graphic correlation between: **a** N_1, serving and simulation blocking probability; **b** N_2, respective serving and simulation blocking probability

4.2 Parallel Queues

Teletraffic systems in net of parallel queues are studied. Initially are analyzed two parallel Markov incoming flows (Fig. 6), as each one of these has certain serving priority.

Acceleration of simulation for reaching rare events is gained by increasing:

- Length of queues N;
- Number of trajectory samples' branches L.

(a)   **(b)**

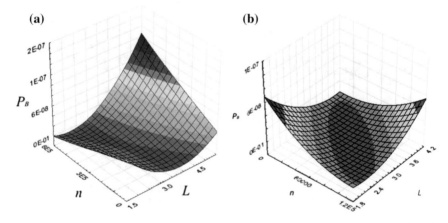

Fig. 6 Graphic correlation between: **a** Simulation blocking probability P_B in terms of 10^{-1}–10^{-7}, with parameters L up to 5 and n up to 50,000 samples; **b** number of levels for trajectory samples' branches L and statistical sampling n, to obtain simulation blocking probability in terms of 10^{-8}

It is used logarithmic geometric distribution as incoming regulation for two parallel incoming flows with assigned priorities for service.

Multiple simulation experiments are conducted and the general acceleration of simulation is gained from:

- Number of trajectory samples' branches L;
- Size of the statistical sample n.

5 Conclusion

The article has reviewed the factors that can affect the efficiency of RESTART/LRE and has focused on the most critical one when the method is applied to basic networks of queue: tandem and parallel, because it is absolutely necessary to verify a simulation method using reference queuing models, whose relevant properties can be described by analytically derived formulas.

The simulation results for teletraffic systems with network of two tandem queues, non-Markov distributions, with RESTART/LRE method show that the significant parameters are the number of customers in N_1 and N_2 [3]. The simulation results for teletraffic systems with network of two tandem queues, non-Markov distributions, with RESTART/LRE method show that the significant parameter is the number of thresholds L.

References

1. L'Ecuyer, P., Demers, V., Tuffin, B.: Rare events, splitting, and quasi-Monte Carlo. J. ACM Trans. Model. Comput. Simul. **17**(2), 1–44, Art. 9 (2007)
2. Görg, C., Schreiber, F.: The RESTART/LRE method for rare event simulation. In: 28th Winter Simulation Conference, pp. 390–397. IEEE Press, Coronado (1996)
3. Ivanova, E.: Modeling of traffic in broadband convergence networks, Ph.D. Thesis, University of Ruse Press, Ruse (2013)
4. Mitkov, A.: Theory of Experiments. University of Ruse Press, Ruse (2010)
5. Radev, D.: Teletraffic Engineering. University of Ruse Press, Ruse (2012)
6. Ivanova, E., Iliev, T., Mihaylov, Gr., Stoyanov, I., Tsvetanov, F., Otsetova, E., Radev, D.: Computer modeling and simulation of teletraffic models for 5G networks. In: 38th International Convention on Information and Telecommunication Technology, Electronics and Microelectronics, pp. 479–482, MIPRO, Opatija, (2015)
7. Radev, D., Iliev, M., Arabadjieva, I.: RESTART Simulation in ATM networks with tandem queue. In: Proceedings of the International Conference Automatics and Informatics, pp. 37–40, UAI, Sofia (2004)
8. Boer, P.: Analysis and efficient simulation of queuing models of telecommunication systems. Ph.D. thesis, University of Twente (2000.)
9. Bucklew, J.A.: Introduction to Rare Event Simulation. Springer Series in Statistics, Springer, New York (2004)

10. Bueno, D., Srinivasan, R., Nicola, V., van Etten, W., Tattje, H.: Adaptive importance sampling for performance evaluation and parameter optimization of communication systems. J. IEEE Trans. Commun. **48**(4), 557–565 (2000)
11. Cerou, F., Guyader, A.: Adaptive multilevel splitting for rare event analysis. Research Report, RR-5710, INRIA (2000)
12. Garvels, M., Kroese, D.P.: A comparison of RESTART implementations. In: 30th 1998 Winter Simulation Conference, pp. 601–609. IEEE Press, Washington (1998)
13. Garvels, M.: The splitting method in rare event simulation. Ph.D. thesis, University of Twente (2000)

The Proposal of the Soft Targets Security

Lucia Duricova Prochazkova and Martin Hromada

Abstract The proposal of the soft targets security contains with technical and management solutions. Management solution is based on legislative safety and security and organization management requirements. Technical solution is based on technical requirements of applied products (for example: detectors, sensors etc.) and interconnection requirements. The aim of the article is the description of the basic functional processes, which will be applied to the software architecture. The proposed solution could connect a methods, technical components and management procedures into system. The methodology exploits a lot of inputs for supporting the operating staffs in relation to smart and quickly reaction. At the end the article proposes software solution, which is realized by fuzzy logic.

Keywords Security · System · Soft targets · Safety · Security

1 Introduction

Soft targets are specific objects with specific facility issues. Every object has different specifications for security and safety requirements. This security and safety requirements are defined by law framework of the Czech Republic and European Union. These aspects must be implemented in the operating processes. It must fulfil these requirements, and this is reason, why directors and objects owners understand the needs of security requirements fulfilment. This proposal solution is based on specific laws requirements. It could be more effective, when we will suggest comprehensive security solutions in the object design. It is only effective system solutions. Owners and managers do not need to pass any special security education.

L. Duricova Prochazkova (✉) · M. Hromada (✉)
Faculty of Applied Informatics, Tomas Bata Univerzity in Zlin,
Zlin, Czech Republic
e-mail: duricova@fai.utb.cz

M. Hromada
e-mail: hromada@fai.utb.cz

© Springer International Publishing Switzerland 2016
R. Silhavy et al. (eds.), *Automation Control Theory Perspectives
in Intelligent Systems*, Advances in Intelligent Systems and Computing 466,
DOI 10.1007/978-3-319-33389-2_32

2 Safety and Security Requirements in Czech Republic

It is necessary to define two groups of security and safety law requirements. The first is known as an OHS, and it means Occupational Health and Safety. The second is known as a Fire protection. Companies need to integrate the ISO standards in processes too. ISO standards describe management systems and quality of management in companies. ISO is International standardization organization. It means companies, which have ISO standardization, are in same level in the international ranking. It is good tool and approach for others companies when they are deciding on mutual cooperation. Our proposal does not find new methods, but we can use current requirements and optimize it in object for system's solving.

3 The Proposal of the Security Solution

The next paragraph will discuss about the proposal of the security solution. It can be seen as a conceptual structure. It is necessary to integrate the current management to one and complex system's solution.

The proposal structure will be represented by following steps:

- Analytical tool for objects classification.
- The proposal of action and security solutions.
- Monitoring object in process.
- The early threats and risks identification and involvement of risks and involvement technical components for efficient decision making.
- Adequate development.

Our solving is based on relevant scientific methods. The first step has to categorize objects in security analytical process. The second has to categorize incidents, which could occur in objects. This comparison could show how object is vulnerable to the incident which could occur in object [3] (Fig. 1).

Requirements for the proposal:

- Users friendly and friendly for owner.
- Fulfil user's requirements.
- System will help owner to fulfil law's requirements.

The system integration consider from next requirements. It is the solving system for monitoring and reaction when incidents occur (Fig 2).

The proposal of the security solution is based on several group of requirements. These requirements are of different type. Requirements are inputs for decision making. Experts make decision in case of incidents according to their own logic and experiences. The aim of proposal is not a review expert's decision and searching mathematical context, but the aim is support security and safety managers in soft targets. The software tool will link every technical components and it can be

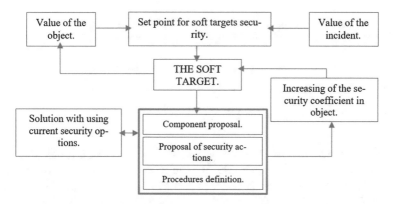

Fig. 1 The proposal of the analytical process [3]

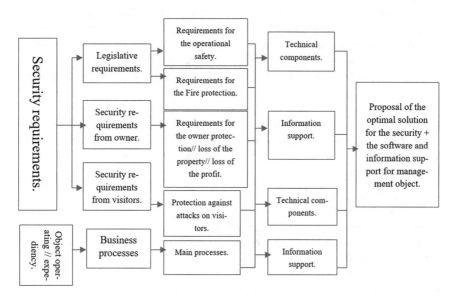

Fig. 2 The proposal of the software tool

programming by fuzzy logic. For this solution was fuzzy logic chosen, and reasons are specific decision-making processes. It takes too long, to examine some specific incidents therefore is necessary to talk about conditions.

For example: "If detector "A" will detected alarm (temperature is more than 25 °C) and detector "B" will detected alarm (moving in room), then camera 1 will start recording and turn on to display (full screen)."

All the process consider from more as one examining. It starts with the first step and it is the analytical part for object and will classify groups of objects and after then classify incidents (Fig. 3).

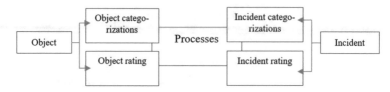

Fig. 3 The schema of analytical part [3]

When it is known, how the security parameter in object is, it is able to evaluate the security level. But in other side, it is necessary to know security parameter which could describe incidents risks. It is obvious that every incident has some other probability and risk too, and it depends on the bond between object and risk. Primary we must classify these input parameters. There is implemented more as one methodology here for analytical process. It considers on binary logic, but mathematic operations are special for two inputs. The base is FMEA analysis. FMEA analysis is Failure Mode and Effects Analysis. It depends on three parameter's evaluation. The main advantages are two groups of acceptable action. The first is immediate actions and second is permanent correction actions.

It is obvious that every process has some special ideal behaviour. In real the process hasn't ideal behaviour, because we do not have ideal conditions. In real we have more external parameters and conditions and it could have influences internal processes.

It can be involved more standards and more solutions but for more effectivity we must integrate it to system. It must be linked to past, present and future. It means for our proposal next:

- Which incidents were in the past?
- How has been reaction?
- How was valuation to this reaction? How would it be better?
- What is in present? How are parameters in object?
- What we can do?
- How is our preparation? How is our training?
- Which incidents could happen in object?
- What is risk?
- What is probability?
- Will be reaction and how?

This questions will be in software and analytical part too, but it is about more examines. In the next flowchart we can see the one of processes (Fig. 4).

The system integration could increase effectivity and usability of the adopted measures. Every object has some special conditions for use and for security too. For development of harmonized instrument built on a part of the current requirements, It could help a lot of parts security and service too. Alternative principle is based on solution, which consider from part of security and safety solutions. The society

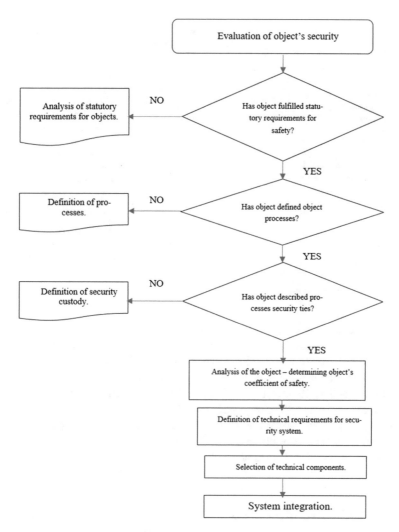

Fig. 4 The flowchart of the assessment

knows methods for determining risk of object individual part. This solution integrates methods in each group, and effectiveness is maximized [6].

3.1 The Process of the Evaluation

Every incident has special conditions, and it can be identified and eliminated by security actions. The next figure describes system's solving in three groups—technical security, guarding of building and protection processes (Fig. 5).

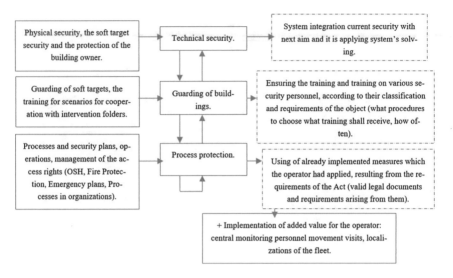

Fig. 5 The system's solving

It is necessary to examine every object, but the aim is the examining and determining the security parameter. The fuzzy logic can be used, because it necessary to know and work with a lot of plans and a lot of experts in one systems. It is a wide range of numerical values too, and a lot of technical components. It is important to define parameters and it is not about two values, but is about a lot of values, and it is reason for fuzzy logic use [5].

This system will be adaptable with other methods, therefore every inputs can be set up and mechanism can be set up to. Then is the implementation of results from other group of security, for support solutions.

4 The Software in Objects

This proposal will be used in computers and it will be connected with other object in network. This software must accept input parameters directly from the object, in real time. On the other side, next group of inputs are requirements from law requirements and other standards, which should company accept. The software will have a lot of subsystems. These subsystems consider of technical components and processes. In the first every experts must know about domain a lot of information. Information about region, building, situation's and after that need to connect it. It can be explained as one big brain, which will make decision which is based on skills of experts. The proposal of system should be user friendly too.

4.1 The Fuzzy Logic in Software

Fuzzy logic is the sophysticate programating logic, which can work with a lot of values. It is not about two parameters, but it is about the examining whole range of values. It is important to implement expert's findings and translate it to fuzzy logic. It means, definition of range for the decision support (Fig. 6).

In the classical theory we can have two values and it is 1 and 0. It means, element D belongs to group 1, or element D does not belong to group 1. In the other side, the fuzzy logic knows about each value in the interval from 0 to 1. Fuzzy Logic has special rules and symbols with operates in mathematical operations. They are called Fuzzy statements, and therefore not a classic strident whose value is either 1 or 0, but their truth value assumes a value of between 0 and the 1. In fuzzy logic we use linguistic variables and it is defining by universum. The value linguistic variable we can mark as element of plurality [7] (Fig. 7).

Inputs are sharp values, which are in fuzzifier process transferred to fuzzy sets. With fuzzy sets we can make special fuzzy operations, which are represented by rule base. In the end is process reverse. We have fuzzy set and we must define sharp

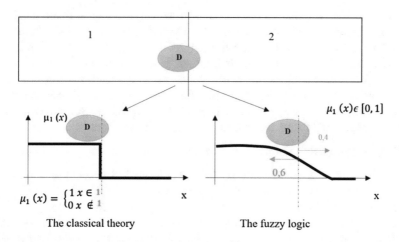

Fig. 6 Basics of fuzzy logic [7]

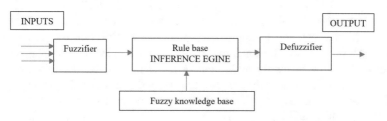

Fig. 7 The diagram of fuzzy system [1]

value. We know more as one method for defuzzifier. For example it is centre of gravity, or the mean of maxima [1].

In the proposal of software, the fuzzifier was used in processes when sharp values will be transformed to linguistic values. Sharp values will be gotten from security devices (for example: firebox, detectors, cameras and others) and from law requirements too. It is process which can define security interval. The defuzzifier is using in process when interval must been transformed to one sharp value, which represents this solution in incidents.

4.2 Algorithmic Examples

Fuzzy rule is in the next text: "if (fuzzy statement) and (fuzzy statement) then (fuzzy statement)" [1].

If (temperature = high) and (oxygen = low) then (ventilation = high).
If (door4 = open) and (mode = secret) then (camera4 = start recording) and (voice prompts = on).

When we are working with the statement we will use logical operators. It will describe the interrelationship between these groups of variables. In security solutions known as a special characteristics. Object has a property with certain parts. It is not about binary system, but it indicates a relation between the object and characteristics. It is obvious that in every object is risk, but we must determine the value, which represents this risk. In other way, when we know about law requirements, we can set up normal interval from legislative. Subprogram can be defined, which will detect limits or some interval before limit [2].

If (staircase = small) and (access = free) then (safety measures = standard).

Small staircase must be defined, and it could be 3–5 stairs. We must define groups of access too, and last but not least safety measures too. Software will detect safety and security measures and it will optimize it for more effectivity.

Detectors have some special function, and it works at technical bases. These technical bases are infrared part, electro part, or Rontgen part, and it works witch numbers and unit of measure, with support special physical phenomena. We can use detectors and linked it in system. It is for example system from company NICE. After, it will be more efficient as every component alone. We must understand physical phenomena and know which status represent normal state or alarm state, and which situations are threatened. We propose set rules, but not only technical rules, but law rules too. The conditions which are in object will be defined, and after we can optimal set security and safety parameters. The impact this solution is system solving situation with technical and lay support. Object's managers could take advantage this expert's knowledge in object and they can make decision quickly and effect will be higher [4].

5 Conclusion

The aim of this paper is to present the development process of new analytical tool for the soft targets assessment. Authors define primary principles, but not special and specific causes. This methodology will be used in soft targets assessment and incidents too. The aim is involving new system's security and safety solution, which will be supported by software and will make support for solving incidents. Values are identified by detectors and we will know object's status in real time and objects will be connection. The advantage of this connecting object is that we can share information and compare incidents between each other objects. It could cause that reaction is quick, effective and smart too. The one of the other benefits is that we can set up inputs according to law requirements and other management requirements too.

Acknowledgments This project is realized as the research with doctoral student and it is the basic input for next research, which we will develop in next term. It was realized with support of the university. This work was supported by Internal Grant Agency of Tomas Bata University under the project No. IGA/FAI/2016/012.

References

1. Dernoncourt, F.: Introduction to Fuzzy Logic. Massachusetts Institute of Technology (2013)
2. Klir, G.J., Yuan, B.: Fuzzy sets and fuzzy logic. In: Theory and Application, p. 574. Prentice Hall, New Jersey (1995)
3. Prochazkova, L.D., Hromada, M.: The proposal system for the safety assessment of soft targets with focus on school facilities. In: Proceeding of 3rd CER Comparative, vol. 2, pp. 30–33. Sciemcee Publishing, London (2015)
4. Rosenberg, F.: Nice Solutions for Critical Facilities. Nidam, NICE (2014)
5. Takagi, T., Sugeno, M.: Fuzzy identification of systems and its application to modeling and control. IEEE Trans. Syst. Man Cybern. **SMC-15**(1), 116–132. IEEE (1985)
6. Valouch, J.: Aggregated coefficient for evaluation of effectiveness of alarm systems. Int. J. Circ. Syst. Signal Process. **9**, 205–2010. Oregon (2015)
7. Zadeh, L.A.: Fuzzy sets. Inf. Control **8**, 338–353 (1965)

Program System for Solving Problems of Monitoring Social and Economic Development of Saint-Petersburg

Oleg Korolev, Vladimir Parfenov and Semyon Potryasaev

Abstract One of the ways of modeling to be used is so called 'visual modeling'. In turn, there are many visual approaches, and only one of them is to apply algorithmic networks. The paper gives a short introduction of the algorithmic network approach and the Cognitron system is based on this approach. Further, the article describes the algorithmic model "Development Evaluation of the Information Society" and the corresponding algorithmic networks, created based on the method of development evaluation of the information society in Saint-Petersburg, developed in St. Petersburg Information and Analytical Center. The program system for working with the model is described, which consists of the Cognitron-Service program and the Cognitron system.

Keywords Decision-making support systems · Visual modeling · Visual programming languages · Simulation · Algorithmic networks · Cognitron

1 Introduction

The city public authorities of Saint-Petersburg request the evaluation of the development of the information society. To perform this task, the Method of Development Evaluation of the Information Society (shortly Method) in Saint-Petersburg is developed in St. Petersburg Information and Analytical Center (SPb IAC).

O. Korolev (✉) · S. Potryasaev
St. Petersburg Institute of Informatics and Automation,
Russian Academy of Sciences (SPIIRAS), St. Petersburg, Russia
e-mail: korolf@rambler.ru

S. Potryasaev
e-mail: spotryasaev@gmail.com

V. Parfenov · S. Potryasaev
Saint Petersburg National Research University of Information Technologies,
Mechanics and Optics (ITMO), St. Petersburg, Russia
e-mail: parfenov@mail.info.ru

© Springer International Publishing Switzerland 2016
R. Silhavy et al. (eds.), *Automation Control Theory Perspectives
in Intelligent Systems*, Advances in Intelligent Systems and Computing 466,
DOI 10.1007/978-3-319-33389-2_33

Initially, the calculations of the method were performed manually or using MS Office tools. The problem there was that the Method is connected with monitoring the fields characterized by a sufficient dynamic of the used index structure. It requires continual modernization of the calculation scheme. The same problem also appears during "maintenance" of the method when new problems emerge which require obtaining a result that is not provided by the original formulation.

To solve the mentioned problem, the algorithmic model "Development Evaluation of the Information Society" based on the Method was created. **The specific of the model** is the necessity to implement the evaluation of index values during calculation using *numerical and qualitative (verbal) scales.*

The basis of the program system implementing the created algorithmic model is the **Methodology of modeling automation and decision-making support on the basis of algorithmic networks**, which is developed in St. Petersburg Institute of Informatics and Automation of Russian Academy of Sciences (SPIIRAS). The program system includes original inventions developed in SPIIRAS: the system of modeling automation—Cognitron—and the secondary program—Cognitron-Service.

2 Algorithmic Networks and Cognitron Modeling System

2.1 Algorithmic Networks Methodology

Currently, the field of decision support systems (DSS) is not homogenous. There are a number of fundamentally different approaches to DSS and each has had a period of popularity in both research and practice. Each of these "DSS types" represents a different philosophy of support, system scale, level of investment, and potential organizational impact [1, 2]. In this paper we propose a new variant of DSS which is based on the algorithmic approach together with knowledge visualization. The tools implementing knowledge visualization include Petri nets, semantic networks, algorithm graph-schemas, informational graphs, restriction networks, case-technology and UML-language. The algorithmic network formalism, introduced in SPIIRAS [3], can be also included and it is intended for describing algorithmic models.

The algorithmic network (further AN) is defined as a finite directed graph, the phenomena characterizing a modeled object (variables) correspond to the arcs of the graph and functional relations between the phenomena (operators) correspond to the nodes.

The algorithmic networks introduce a calculation scheme based on which calculation experiments at models are carried out. The causal relations between the phenomena found during the construction of the modeled process script are mapped to the calculation relations between the operators of the algorithmic network.

Fig. 1 An instance of the
algorithmic network

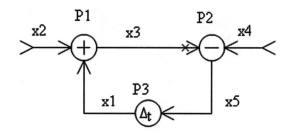

The following kinds of operators are mostly used in AN: arithmetic operations, logarithmic functions, trigonometric functions, condition operators, minimum, maximum, the operator "delay" $\{x(t + 1) = x(t)\}$ allows describing dynamical models.

An algorithmic network introducing the model of some capacity is depicted in Fig. 1. The mathematical equations corresponding to the nodes are shown in formula (1).

$$P1 : x_3(t) = x_1(t) + x_2(t), \; P2 : x_5(t) = x_3(t) - x_4(t), \; P3 : x_1(t+1) = x_5(t) \quad (1)$$

The orientation to the end-user caused the special requirements to provide construction of correct models using AN language: the variable can correspond to the output arc of one node only and there is no cycle without at least one node with the operator "delay".

2.2 Cognitron Modeling System

The Cognitron modeling system [4, 5] supports the full cycle of work with model starting from its graphical representation creation till the model experiment (calculation), which is done by means of three work modes ("Model", "Data", "Experiment") and is intended for the end-users of different categories.

To support intelligent users possessing the AN language and ability to construct new models, there is a graphic editor—the mode "Model". The user can construct a model using several types of building blocks [6]:

- The nodes (to set the operator type);
- The simple arcs (to link nodes);
- The variable labels (to set the variable name corresponding to a given arc);
- The arc signs (to order the arcs corresponding to the node with a non-commutative operator).

The mode provides the syntax check of graphical representation of the model and the computational correctness of the calculation algorithm which is implemented by

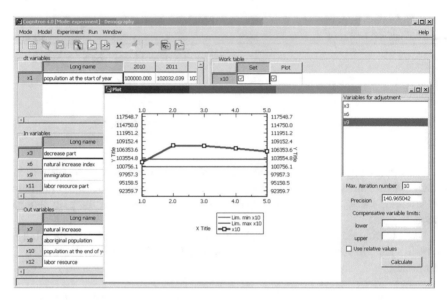

Fig. 2 Decision-making in the Cognitron system

the model. When the model input is finished, the graphical representation is converted to the analytical representation.

The mode "Data" provides the formation of datasets, used in the process of model experiments. The input of variable values is automated thanks to the possibility of the generation of the constants and using of linear interpolation of already entered variable values.

The mode "Experiment" (Fig. 2) allows to carry out the model experiments using the datasets which were created earlier. The mode provides the formation of the decision-making environment (the worktable, the limitations table, the graphs) and the choice of the solution method, if it is required, beginning from the solution methods, using the manual and automatic adjustment of the results, finishing with the methods based on the properties of the inverse function. The use of methods based on inverse function requires some limitations to the structure of algorithmic networks (for instance, the set of used operators).

Since the beginning of the work related to the problem of modeling automation (1982), a number of system versions Sapfir/Cognitron have been developed in SPIIRAS; using such systems, models were created which have successfully solved problems in the macroeconomics [5], the industry and the agriculture. The system was used for the estimation of the ecological state of different objects. Among the applied fields where the Cognitron was used are the chemical industry, the shipbuilding, the military science, and the municipal economy.

3 Model "Development Evaluation of the Information Society"

3.1 General Description

The algorithmic model "Development Evaluation of the Information Society" is created on the basis of the Method of Development Evaluation of the Information Society in Saint-Petersburg (further shortly Method), developed in SPb IAC. In the correspondence with the given method goal, the final goal of the model is the calculation of the **Index of information network readiness of St. Petersburg to join the information society** (Егот)—integral indicator characterizing the degree of development of information and telecommunication technologies both in the city as a whole, and in the most characteristic information and communication "fields" and "groups" in terms of the application of the regulatory impacts of the city public authorities.

In total, the model "Development Evaluation of the Information Society" includes over 600 indices and over 300 operators, including over 200 table functions (scales), numerical and qualitative (verbal).

As for the solution of problems related to the determination of baseline values in certain areas of socio-economic development of St. Petersburg it was decided to use a method of solution based on the inverse function, the algorithmic model "Development Evaluation of the Information Society" was divided into parts, which were implemented using different program tools (because of limitations of the structure of algorithmic networks when the inverse function method is used):

- The network of the model, providing the calculation of indices beginning from the inputs and ending with the outputs of table functions (including them), is created in the Cognitron-Service program (specially developed to solve the task);
- The network of the model, providing the calculation of indices beginning from the outputs of table functions and ending with the Index of information network readiness (Егот), is created in the Cognitron system (extended for solving the task).

3.2 Algorithmic Network of Model in Cognitron-Service

Table 1 shows the instances of the calculation formulas of AN, which are done in the new program Cognitron-Service. The index values received in such a way are estimated further in correspondence with the scales described in the Method (in the program implemented as table functions). The results of the table functions are values from 0 to 1 with a 0.25 step.

Table 1 Calculation formulas of the model in the Cognitron-Service program

№	Index name		Calculation formula
	Long	Short	
1	The length of the primary transport network of telecommunication services, thous. km of all	Сии1	Сии1 = Сии1_1 + Сии1_2 + Сии1_3 + Сии1_4
2	The ratio of used capacity of the automatic telephone exchange to the total used capacity	Сии4	Сии4 = Сии4_2/Сии4_1

Additionally, there are the qualitative input indices in the network, whose values are also estimated in the correspondence with scales given in the Method—the expert has to choose the numerical value on the basis of a given verbal description. The results are again values from 0 to 1 with a step of 0.25.

3.3 Algorithmic Network of Model in Cognitron

Table 2 shows the instances of the calculation formulas of AN in Cognitron.

The AN is realized in the system as the hierarchy of the fragments, which are included in the fragments of the level above. The fragment of the AN calculating the index of information network readiness is shown in Fig. 3. Each operator contains the AN of the underlayer.

Table 2 Calculation formulas of model in the Cognitron system

№	Index name		Calculation formula
	Long	Short	
	...		...
7	The group index "Administration—population"	Сан	Сан = (Сан1т + Сан2т +⋯ + Сан7т)/7
	...		...
11	The field index "Infocommunication urban management"	Дэп	Дэп = (Сан + Саб + Сда + Свэ)/4
	...		...
32	The index of information network readiness	Егот	Егот = (Дди + Дэп + Доит + Дээ + Дио + Дпит)/6

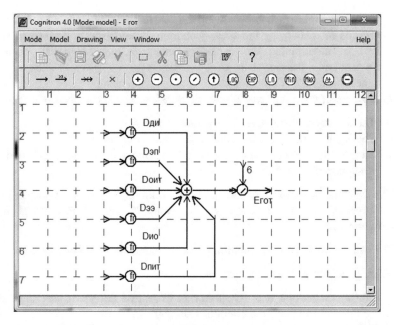

Fig. 3 Fragment of AN for the calculation of the Index of information network readiness

4 Technology of Solving Problems in the System, Implementing Model "Development Evaluation of the Information Society"

4.1 General Description

In the course of working with the program system, the user faces a number of phases, the most important of which are: the maintenance phase of the model software and the modeling software configuration phase.

The work with the program system during the maintenance phase assumes the following actions performed by the user:

1. The formation of the array of input model indices from the data storage of SPb IAC—the *source array of input indices* (performed out of the system by IAC).

2. The evaluation of the values of the input and calculated indicators in the accordance with the scales of values of model indicators and the formation of an array for the calculation of analytical processing of the results of monitoring the socio-economic development of St. Petersburg and determining the values of initial indices for certain areas of socio-economic development of St. Petersburg—*the array of input indices of the task* (performed by Cognitron-Service).

3. The selection of the task to be solved: the calculation of analytical processing of the results of monitoring the socio-economic development of St. Petersburg—task 1 or determining the values of initial indices for certain areas of socio-economic development of St. Petersburg for given values of final parameters—task 2 (Cognitron).
4. In the case of selecting task 2—the clarification of the statement (Cognitron):

   - The selection of the investigated final parameter;
   - The input of the required value of the final parameter;
   - The selection of the source index (indicator), due to which the selected task is supposed to be solved.

5. The calculation of the selected task—determining the value of the Index of information network readiness or the value of the selected initial index (Cognitron).

The work during the configuration phase assumes the following actions:

1. Editing the AN of the model, including:

   - Editing model operators (Cognitron or Cognitron-Service is used depending on the AN to which the operator belongs);
   - Editing model variable names (Cognitron or Cognitron-Service is used depending on the AN to which the variable belongs);
   - Input and editing ambiguity intervals within the indicator scales (Cognitron).

2. Editing the exchange parameters file (Cognitron-Service).

4.2   Cognitron-Service Program

The user can solve the following tasks using Cognitron-Service:

- Editing the AN including editing model operators and names of model variables;
- The evaluation of values of input and calculated indicators in accordance with the scales of values of model indicators;
- The formation of the array for tasks 1 and 2—*the array of input data of the task*, which is used further by the Cognitron system;
- Editing the exchange parameters file, containing the information about the correspondence between codes of model indices in the information storage SPb IAC and short model names in the program system. The file allows providing the information exchange between the program system and information storage SPb IAC.

4.3   Using the Cognitron System

The user can solve following tasks using the system:

- Editing AN of model, including editing model operators and names of model variables and input and editing ambiguity intervals inside of indicator scales;
- The selection of the task to be solved: task 1 or 2;
- In the case task 2 is chosen—the clarification of the statement: the input of the required value of the final parameter and the selection of the source index (indicator), due to which the selected task is supposed to be solved;
- The calculation of the required task.
- Data exchange between Cognitron and Cognitron-Service

The functions listed are performed in Cognitron by modes "Model" (it allows editing operators and names of model variables) and "Experiment" (it allows input ambiguity intervals of model scales and carry out all the steps to solve the task, from setting up to the production of the final result). On loading the system, a form is shown (Fig. 4), providing for the solution of task 1.

Input and Editing Ambiguity Intervals. According to the requirements of maintenance, the Cognitron system should provide the solution of task 2.

To decrease the number of inverse variants which should to be reviewed in the process of task solving, it is suggested to enter the so-called ambiguity intervals for qualitative indicators. *An ambiguity interval* is the range of values of the scale; selection of the value is determined solely by the expert's subjective views. For instance, it is possible to set this range equal to 0.25–0.75 for the index of the availability of On-line projects of urban bills and the possibility of their discussions (Caa6).

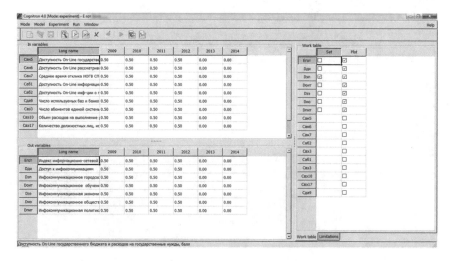

Fig. 4   The initial form of the Cognitron system

Input and editing ambiguity intervals is performed in the mode "Experiment" using standard feature of Cognitron: you need to select the tab page "Limitations" (Fig. 4), add the variable to limitations table and input the ranges of ambiguity interval in the corresponding cell.

Selection of the Task to be Solved. According to the requirements of the specification, the program system is intended to solve two sets of tasks:

1. The calculation of the Index of information network readiness of Saint-Petersburg;
2. The calculation of the required values for task 2 (the set of the tasks based on the use of the "inverse" method).

When task 1 is solved, it is enough to run the Cognitron system, the initial form of the system (Fig. 4) is the beginning of calculating task 1.

For solution of the task from set 2, to select the task, you need to activate the checkbox on the tab page "Work table" in the column "Set" opposite the variable which is used as final parameter.

Clarification of the Statement for Task 2. It takes place in two stages:

1. The input of the value of the final parameter which has to be achieved—it can be done in the table or in the graphical mode. The input of the required value of the field index "Infocommunication urban management" (Dэп = 0.70) in the table mode is shown in Fig. 5;

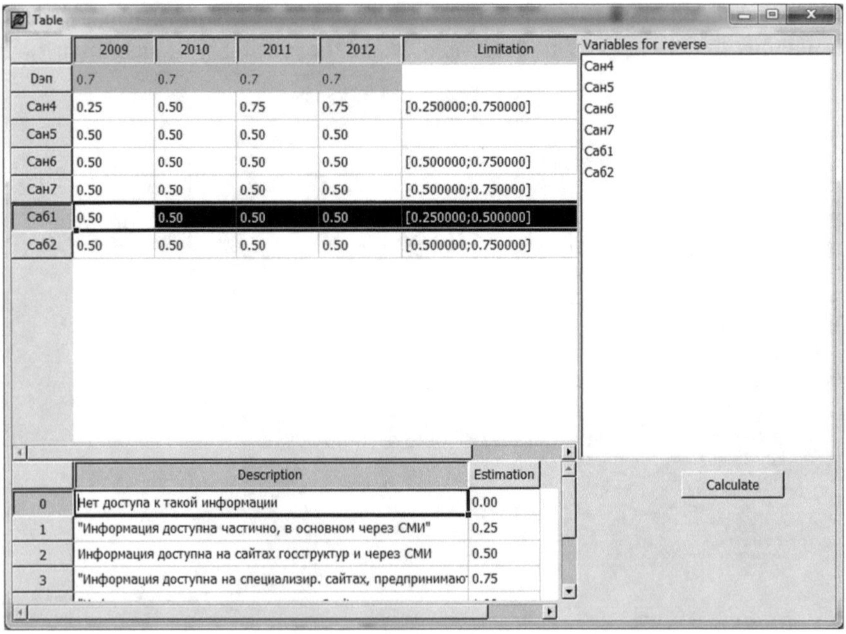

	2009	2010	2011	2012	Limitation	Variables for reverse
Dэп	0.7	0.7	0.7	0.7		Сан4
Сан4	0.25	0.50	0.75	0.75	[0.250000;0.750000]	Сан5
Сан5	0.50	0.50	0.50	0.50		Сан6
Сан6	0.50	0.50	0.50	0.50	[0.500000;0.750000]	Сан7
Сан7	0.50	0.50	0.50	0.50	[0.500000;0.750000]	Са61
Са61	0.50	0.50	0.50	0.50	[0.250000;0.500000]	Са62
Са62	0.50	0.50	0.50	0.50	[0.500000;0.750000]	

	Description	Estimation
0	Нет доступа к такой информации	0.00
1	"Информация доступна частично, в основном через СМИ"	0.25
2	Информация доступна на сайтах госструктур и через СМИ	0.50
3	"Информация доступна на специализир. сайтах, предпринимаю"	0.75

Fig. 5 Setting the value of the field index "Infocommunication urban management" (Dэп)

2. The selection of the source index, due to which the required value of the final parameter is supposed to be achieved. The set of source indices (indicators), which are possible to select, is shown in the "variables for reverse" part (Fig. 5).

The Cognitron system was extended to provide the hints helping to select the "inversed" indicator; it is enough to highlight the string of the analyzed indicator (Fig. 5).

Task Calculation. For task 1, the calculation is performed by the button "Calculate" in the system toolbar. For the task from set 2, the calculation is performed by the button "Calculate", located in the corresponding form. Based on the calculation results, the decision is taken to continue or to finish the work with the system.

Data Exchange between Cognitron and Cognitron-Service. In new menu "Experiment\Exchange settings", you need to specify correspondence (see Sect. 4.2), input and output files. In menu also two commands are added: filling data based on the input file and exporting data to a specified export file.

5 Summary

The newness of the article is the approach used for the creation of the model which requires the evaluation of index values based both on *numerical and qualitative (verbal) scales*. As the Cognitron system using algorithmic network approach only works with numerical variables and selected decision-support method based on inverse function, it was necessary to divide the model into two parts; in addition to Cognitron modifications, the new program Cognitron-Service was developed.

Implementation of the program system in the practice of monitoring the processes of socio-economic development of St. Petersburg allows:

- Saving time/money necessary to get used to the software, reducing the operational cost;
- Reducing the time/cost of resources for the development and "debugging" model;
- Reducing the time of decision-making, increasing the adequacy of solutions to real conditions of tasks by allowing direct work for experts with modeling system, bypassing the stage of interaction with the programmers;
- Reducing the time required to achieve the desired results due to the possible joint use of methods that support the analytical modeling and simulation;
- Improving the validity and quality of decision-making as by bringing in the mathematical apparatus, and by making fuller use of the professional knowledge of the decision-maker.

The important advantages of the Cognitron system are:

- Representation of economists' techniques in the form of AN is accessible and clear for them;
- The ability to create and modify software implementation of models techniques directly by the economists who do not master the programming languages;
- No need to re-enter mathematical expressions, which describe the problems arising during the maintenance of already entered algorithmic model-techniques, including the tasks based on the inverse method.

Acknowledgements The research described in this paper is partially supported by the Russian Foundation for Basic Research (grants 15-07-08391, 15-08-08459, 16-07-00779, 16-08-00510, 16-08-01277, 16-57-00172-Bel_a), grant 074-U01 (ITMO University), project 6.1.1 (Peter the Great St. Petersburg Polytechnic University) supported by Government of Russian Federation, Program STC of Union State "Monitoring-SG" (project 1.4.1-1), The Russian Science Foundation (project 16-19-00199), Department of nanotechnologies and information technologies of the RAS (project 2.11), state research 0073–2014–0009, 0073–2015–0007.

References

1. Arnott, D.: Decision support systems evolution: framework, case study and research agenda. Eur. J. Inf. Syst. **13**(4), 247–259 (2004)
2. Intelligent Systems in Cybernetics and Automation Theory. In: Proceedings of the 4th Computer Science On-line Conference 2015 (CSOC2015), Vol 2: Intelligent Systems in Cybernetics and Automation Theory. Advances in Intelligent Systems and Computing, vol. 348 (2015)
3. Michailov, V.V., Marley, V.E., Korolev, O.F.: The Algorithmic Networks and Their Use: Schoolbook, 2nd edn., ext.—SPb.: SPbSUAI, 136p (2012)
4. Korolev, O., Morozov, V.: The modelling automation system Cognintron. Logistics and supply chain management: German-Russian perspectives. In: Ivanov, D., Lukinskiy, V., Sokolov, B., Kaeschel, J. (eds.) Proceedings of the 5th German-Russian Logistics Workshop DR-LOG 2010 in St. Petersburg, May 2010. pp. 363–368
5. Korolev, O., Raguinia, E.A.D.: The algorithmic networks for total cost ownership (TCO) and return on Investment (ROI) calculation. Logistics and supply chain management: modern trends in Germany and Russia. In: Proceedings of the 4th German-Russian Logistics Workshop DR-LOG 2009 in Cottbus, May 2009. pp. 363–368
6. Korolev, O.F.: The syntax of graphical construction of algorithmic networks. The information technologies and intellectual methods. Issue 3—SPb.: SPIIRAS, pp. 112–131 (1999)

Architecture and Technologies of Knowledge-Based Multi-domain Information Systems for Industrial Purposes

M.G. Shishaev, V.V. Dikovitsky and N.V. Nikulina

Abstract The paper presents an architecture of the information system based on intellectual technologies aimed to serve information needs of heterogeneous users of an industrial enterprise. Technologies for automated creation of semantic model of the subject domain based on the principle "user as an expert" as well as the model of a user's mental stereotypes are discussed. To improve effectiveness of the system in sense of information search speed the authors propose techniques for synthesis of cognitive navigation interface and for information retrieval on the base of subtractive relation. The effectiveness of proposed technologies is confirmed by experiment.

Keywords Knowledge based multi-domain information system · Navigational structure · Cognitive user interface

1 Introduction

Nowadays information systems at enterprises are becoming increasingly smarter. They is collect a lot of information that is aimed to give users a help in various applied tasks. There is a number of large information systems designed to meet the

M.G. Shishaev
Murmansk Arctic State University, Egorova st. 15, Murmansk 183038, Russia
e-mail: shishaev@arcticsu.ru

V.V. Dikovitsky (✉)
Institute for Informatics and Mathematical Modelling of Technological Processes
of the Kola Science Center RAS, 24A, Fersman st., Apatity 184209, Russia
e-mail: dikovitsky@iimm.ru

N.V. Nikulina
Federal State Autonomous Educational Institution of Higher Education "Northern (Arctic)
Federal University Named After M.V. Lomonosov", 17, Severnaya Dvina Emb.,
Arkhangelsk 163002, Russia
e-mail: nikulina_nv15@mail.ru

© Springer International Publishing Switzerland 2016
R. Silhavy et al. (eds.), *Automation Control Theory Perspectives
in Intelligent Systems*, Advances in Intelligent Systems and Computing 466,
DOI 10.1007/978-3-319-33389-2_34

needs of several categories of users simultaneously. To achieve this goal the system should be grounded on the modern intelligent information technologies which allow to adapt the system dynamically for everyone who uses the system. The basis for this adaptation is provided by formalized knowledge on the subject domain and users of the system. Thus, the system's architecture should include adequate means for knowledge acquisition and processing. We call such systems of knowledge based multi-domain information systems (KBMDIS). The first part of the paper briefly describes KBMDIS's architecture and proposes technique for dynamically forming knowledge bases of the system reflecting user's up-to-date view on the subject domain.

With growing of intelligence of information systems the importance of the effectiveness of human-machine interaction is rising. Now, through the front of the user interface is transmitted to a greater degree of knowledge, rather than abstract data. To make this transmission effective we need a common language adapted to the peculiarities of perception between machines and human beings. In the paper we concentrate on adapting a 'speeches' of the machine to the human's perceptive and mental stereotypes. We analyze two types of human-machine interactions— through navigating and through search interfaces. In both cases we construct interfaces according to the user's mental stereotypes. Special techniques for adaptive user interfaces formation and for information retrieval on the basis of subtractive relations are described in the second part of the article.

In the final part of the paper the results of the implementation of the described techniques are given. The implementation was made at the Olenegorsk mining factory. Effectiveness of the techniques was evaluated by using GOMS method and through a series of field experiments.

2   Architecture of the KBMDIS

The structure of KBMDIS is shown in Fig. 1. Its key components are semantic domain model (SDM), user's mental models (UMM), interface unit and index base. The input data for KBMDIS are documents and the user's activity statistics. The output data are the search results and the interface navigation structure. Actual knowledge on the subject domain is stored in the semantic domain model. Initially, SDM is formed by integration of semantic images of documents which are collected in the index bases. Semantic image of the document is the semantic network where vertices denote the document concepts and edges denote the relationships of concepts.

Modern manufacturing processes are dynamic by composition and structure. It reflects on information which is needed to support it: not only the volume and content of the information are changing but also its semantics with changing user's view to the subject domain. To keep knowledge relevance we have to verify the semantic domain model continuously. Traditionally verifying of knowledge bases is made by experts. In our case their role is delegated to actual users of the system. To

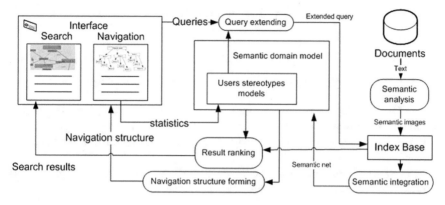

Fig. 1 Structure of knowledge-based multi-domain information system

do so a feedback connecting user's activity within an interface with the SDM is introduced to the system. We name this approach a "user as an expert" principle.

Analyzing user's activity we construct a model of the user's mental stereotypes (UMM) as a part of the SDM. This model reflects user's viewpoint to the subject domain and is used to modify (actually—to expand) search queries and to form cognitive UI of the system. A cognitivity of the UI is evaluated as a degree of matching between UMM and UI structure. UMM is a weighted subnetwork of the SDM where weight of the nodes denotes the value of importance appropriate relation for the user. The model is formed on the basis of users query statistics and statistics of using documents.

3 Automated Formation of Semantic Domain Model Using a "User as an Expert" Principle

The problem of dynamically changing knowledge domain is addressed by many researchers. For example, the paper [1] proposes an algorithms set for dynamically constructing of a subject domain for some innovative task from distributed knowledge collected by different experts. Anyway, for the correction of the knowledge base we have to attract experts.

In our case an actual experts are a priory concentrated on the IS and the task is just to collect their dynamically changing knowledge and to reflect it in the system model of domain. We do that by monitoring users activity within human-machine interactions. Initially the model is formed by analyzing a semantic structure of the documents including to the system's storages. And then it is actualized according to the actual user's view of the domain.

Resulting model is represented as an n-ary weighted semantic network:

$$KB = \{C, L, Tp\},$$

$$Tp = \{synonymOf, HyponymOf, associateWith, subTract\}, L = \{l\}, \tag{1}$$

$$l = <c_i, c_j, tp, \bar{w}>, c_i, c_j \in C, tp \in Tp, \bar{w} = <w_1, \ldots, w_k, \ldots, w_r>$$

where C—a concepts set, L—a relations set, $\bar{w}$—a weights vector, every component of which is the importance of the relation for the respective users category, Tp—a relations types set (synonyms, hyponyms, associations, subtractive relation), r—quantity of users categories.

The SDM forming process is based on documents corpus and expandable thesaurus. This process consists of the following stages:

1. Forming a semantic image of the document. It is presented as a semantic network and it is formed on the base of the statistical and lexical-grammatical document processing methods:

$$D = \{C^D, L^D\}, C^D \subset C, L^D \subset L \tag{2}$$

   where C^D—set of concepts in document, L^D—set of relation.
2. Integrating a semantic images into the model based on complex semantic metric:

   (a) Calculation of estimation of the concept names similarity:

$$\forall c_i : Eq(c_i, c_j) = \frac{x/len}{\max(len_i, len_j)}, c_i \in C^D, c_j \in C^{KB}, i = \overline{1, N_D}, j = \overline{1, N_{KB}}$$

$$\tag{3}$$

   where x—equivalent character string in concepts names, len—concepts names length.

   (b) Calculation of the context similarity:

$$\forall c_i : Syn(c_i, c_j) = \frac{\left| C^D_{syn} \cap C^{KB}_{syn} \right|}{\left| C^{KB}_{syn} \right|}, c_i \in C^D, c_j \in C^{KB}, i = \overline{1, N_D}, j = \overline{1, N_{KB}}$$

$$\tag{4}$$

   where C^D_{Syn}, C^{KB}_{Syn}—concepts synonyms sets.

   (c) Calculation of the structural similarity:

$$\forall c_i : Poseq(c_i, c_j) = \frac{\left| C^D_{Hyp} \cap C^{KB}_{Hyp} \right|}{\left| C^{KB}_{Hyp} \right|}, c_i \in C^D, c_j \in C^{KB}, i = \overline{1, N_D}, j = \overline{1, N_{KB}}$$

$$\tag{5}$$

   where C^D_{Hyp}, C^{KB}_{Hyp}—sets of concepts hyperonyms.

(d) Calculation of the threshold function as average of three estimates:

$$f(c_i, c_j) = \frac{a \cdot Eq(c_i, c_j) + b \cdot Poseq(c_i, c_j) + c \cdot Syn(c_i, c_j)}{3} > z,$$

$$c_i \in C^D, c_j \in C^{KB}, i = \overline{1, N_D}, j = \overline{1, N_{KB}}, \tag{6}$$

where z—threshold function value, a, b, c—expert coefficients.

3. SDM verification by changing weighting coefficients of relations between the concepts. This process is initiated when two concepts are combined in a one user query. Value of changes for weight coefficient is defined in the following way:

$$dw(c_i, c_j) = \frac{\sum\limits_{<c_i,c_j> \in (ACW)} \left(w(c_i, c_j) \right)}{|ACW|} \tag{7}$$

– for weight coefficient between used concepts, and

$$dw(c_i, c_j) = -\frac{\sum\limits_{c_i \in ACW, c_j \in CW} \left(w(c_i, c_j) \right)}{|CW/ACW|} \tag{8}$$

– for weight coefficient between not used concepts, where $w(c_i, c_j)$—weight of relation between concepts c_i and c_j, $dw(c_i, c_j)$—change value for the weight, ACW and CW—relations sets of used and unused concepts.

The resulting model is then used for the synthesis of cognitive navigation interfaces and efficient information retrieval.

4 The Adaptive User Interfaces Formation

Despite the development of effective methods for search, navigation method of information retrieval is still relevant [2, 3]. This approach is justified in cases where the user can not exactly formulate information need that is to describe the object of the search. In this section we consider the technique for synthesis of cognitive navigation interfaces providing structuring information according to the user's mental stereotypes. The technique is based on the following model of a navigating interface. User interface is a pair:

$$UI = \langle I, s \rangle \tag{9}$$

where I—information elements set; s—navigation structure.

Navigation structure determines a hierarchy of information elements (IE) groups. At each hierarchy level set of information elements is divided into subsets in accordance with one or several classification features. SDM concepts attributes are used as classification features. Sets of IE can cross when the several attributes are used at the same level. Let's introduce the following notation:

$\Gamma^l = \{G_i^l\}$—set of information elements on lth level of the navigation structure;
G_i^l—ith group of information elements of lth level of the navigation structure;
$P^l = \{p_i^l\}$—set of classification attributes which used for groups formation of lth level of the navigation structure.

Interaction between a user and KBMDIS includes searching of information elements pertinent to image about the required information in a user's mind. Often this image is inaccurate and only some of classification attributes can be specified. With varying confidence degrees a user can suppose in what of the groups of IE and on what level of navigation structure the required information element is placed. This degree is the higher, the more accurately a user can present the potential content of a group of information elements. Let's introduce function of numerical evaluation of a user's confidence:

$$p^u : \Gamma^l \to (0, 1] \tag{10}$$

Estimation of the time required to access the required information element within the navigation structure for the lth level is equal to:

$$O\left(\frac{\max_i |G_i^l|}{p^u(G_i^l)}\right) \tag{11}$$

With other things being equal, the degree of a user's confidence determines the quality of the interface in terms of information access speed. The supposed navigation structure of the interface has l-layers depth. Then, the following sum can be used as a quantitative evaluation of cognition of the interface for the user u:

$$\sum_{l=1}^{\hat{l}} p^u(l) \tag{12}$$

where $\hat{l}$—navigation structure levels number, $p^u(l)$—function of numeric evaluation of a user's confidence in which group of IE at layer $\hat{l}$ the required element is placed.

This measure can be used for evaluating existing interfaces where $\hat{l}$ is known. To initial structuring of IE set it is required to account additional restrictions. These restrictions are caused by the psychology of human perception. There are limits to the maximum simultaneously perceived objects number, group size of information elements, and navigation structure depth. An optimal structured interface is a solution of the following task with restrictions:

$$\max_{s} \sum_{l=1}^{\hat{l}(s)} p^u(l)$$
$$g(s) \leq K,$$
$$\hat{l}(s) \leq K'. \qquad (13)$$

where $\hat{l}(s)$—number of levels of navigation structure, $g(s)$—maximum size of group of information elements; K, K'—cognitive constants. K determines number of IE simultaneously displayed to user. K' determines maximum number of navigation structure levels. Constants K and K' determined by expert based on psychology of perception [4–6].

The procedure of the navigation interface formation is described bellow. This procedure meets above formulated synthesis problem and it is based on the user's models. The procedure has several stages:

1. Determining of the information needs based on the user's mental model and query:

$$UQ = f_m(Q, UMM_k),$$
$$\forall c_i : \exists l : (c_i \in Q) \wedge (c_j \in UMM_k) \wedge (\overline{w_k} > 0) \qquad (14)$$
$$Q = \{c_i\}, l = <c_i, c_j, tp, \bar{w}>, i = \overline{1, N_Q}, j = \overline{1, N_L}$$

where UMM—a user model; Q—a query, f_m—a function for translation of the query in the user's model, $\bar{w}$—a weights vector, l—the relation between concepts.

2. Defining of the set of elements of navigational structure:

$$G = \{g | \exists c_i = g\}, c_i \in UQ, \qquad (15)$$

where G—a set of the navigation structure elements; c_i—concepts of the user information needs.

3. Dividing of the set of the interface elements into subsets. Ranking of subsets in accordance with the user's model:

$$G_i^d = \{g_k | (\forall g_k, g_m \exists l \in L : w(c_k, c_m) > x) \wedge (\exists g_z : \exists l_{Hyp}(c_k, c_z), l_{Hyp}(c_m, c_z) \in L)\},$$
$$k, m, z = \overline{1, N_G}, z = \overline{1, N_L}, G_i^d \subset G, L \in UM$$
$$\qquad (16)$$

where G_i^d—elements of ith group on dth layer of the navigation structure; $l_{Hyp}(c_k, c_z)$—hyponym-type relations in the user's model; $w(c_k, c_m)$—weight of relation l over concepts c_k, c_m; x—threshold value for the entering of an element in the navigation structure; d—number of navigation structure levels defined based on limits of the maximum number of the information elements.

5 The Information Retrieval Technique Based on Subtractive User Model Relations

This technique aimed to increase the search results relevance by automatically including constraints in queries. The constraints are formed on the basis of subtractive-type relations of UMM. In KBMDIS documents are represented as SDM fragments and as keywords in the index base. The search process consists of the following steps:

1. Query formation in terms of the SDM:

    (a) The formation of an expanded query. The expanded query contains relations and concepts from SDM.

    $$EQ = f_q(Q, KB) = \{C^Q, L^Q | (Eq(c_i^Q, c_j^{KB}) > 1 - \varepsilon)\},$$
    $$C^Q \subset C, L^Q \subset Li = \overline{1, N_Q}, j = \overline{1, N_{KB}} \tag{17}$$

    where KB—a semantic domain model, C^q—a SDM concepts set from the query, L^Q—relation set over C^q, $f_q()$—function that assigns fragment of the SDM to the query, $Eq()$—names similarity evaluation function, ε—concepts similarity estimation error.

    (b) Query expansion is based on the weights of relation and limiting context of query based on subtractive relations:

    $$EQ = \{C^Q, L^Q\} \cup \{C', L' | l : c_i \in C^Q, c_j \in C', |\overline{w_k}| > x\},$$
    $$C' \subset C, L' \subset L, l \in L' \tag{18}$$

    where C'—a SDM concepts set related to C^Q by L' relations, x—a coefficient of the inclusion of SDM relations in an extended query.

2. Receiving the documents corresponding to the extended query:

    $$D = \{d_i | C^{d_i} \cap C^Q \neq \emptyset\}, i = \overline{1, n} \tag{19}$$

    where C^{d_i}—a SDM concepts set from a document d_i, C^Q—a SDM concepts set from EQ.

3. Ranking of the documents:

$$R(d_k) = \sum_{L_{d_k}} (f_u(\bar{w}_k, r)) - \sum_{L'_{d_k}} (f_u(\bar{w}_k, r)),$$

$$L_{d_k} = \{l^d | (c_i, c_j \in d_k) \wedge (tp \in \{synonymOf, HyponymOf, associateWith\})\}$$

$$L'_{d_k} = \{l^d | (c_i, c_j \in d_k) \wedge (tp \in \{subStract\})\}, i, j = \overline{1, n}, k = \overline{1, m}, tp \in Tp$$

$$(20)$$

where $f_u(\bar{w}_k, r)$—the function returns the weights of relations from the set L_{d_k} for the rth category of users, Tp—a relation type set.

Thus, if a UMM has subtractive relations between concepts, then documents containing these concepts will have a lower priority after ranking.

6 Evaluation

The proposed methods have been tested in the information system of the large industrial enterprise—"Olenegorsk mining and processing plant". Initially the Russian-language thesaurus was used as a semantic domain model. Then SDM has been extended by semantic analyzer over the normative-reference documents collection. The effectiveness of the described methods was verified in full-scale experiment involving test users. We proposed users to solve similar information retrieval tasks using the existing search tool and using the adaptive interface. To evaluate adaptive interface formation method we also used GOMS model [7]. Temporal characteristics of the task execution have been compared. The task was to obtain access to requested document. The average time of the operation is shown in Table 1.

Search method was evaluated by searching in pre-indexed collection of 14 thousand documents. The experts were estimated search results by ten queries. The criteria for estimation are search speed—the time taken to satisfy the information needs expressed by one request; accuracy—matching results to the query; and results completeness—coverage of mentioning object in search results.

$$Precision = \frac{|D_{rel} \cap D_{retr}|}{D_{retr}}, Recal = \frac{|D_{rel} \cap D_{retr}|}{D_{retr}} \qquad (21)$$

where D_{rel}—relevant documents, D_{retr}—results.

Table 1 Estimation of the search methods

Estimated characteristics	Existing interface	Adaptive interface
The average time spent on navigation search with sec.	31.8	20.85

Table 2 Evaluations results

Estimated characteristics	Search method	
	Input line interface	Described method
Search speed estimation	0.45	0.82
Results accuracy	0.96	0.91
Results completeness	0.73	0.94
Ratings average value	0.71	0.89

The experts used linguistic scale for evaluation of the alternatives. Evaluation of ith alternative was made by jth expert by the formula:

$$v_{ij} = 1 - \frac{(l-1)}{k},\tag{22}$$

where l—linguistic school index values; k—scale values number.

The experts used this formula for alternatives evaluation:

$$s_i = \sum_{j=1}^{n} v_{ij}\tag{23}$$

The evaluations results are shown in Table 2.

Experimental results are demonstrate the validity of the dynamic automatically formed SDM. Also we show that SDM can be used to implement the adaptive interface and information retrieval method based on subtractive relations.

7 Conclusion

The advanced intelligent technologies can significantly improve the efficiency of information systems for industrial enterprises. The article suggests technologies for automated synthesis of the semantic domain models and the mental stereotypes models, the technologies for the cognitive navigation structures synthesis and contextual search. The technologies demonstrated their effectiveness in solving the actual problems in the information support of processes of an industrial enterprise.

References

1. Karkanitca, A.: Development of dynamic subject domain based on distributed expert knowledge. In: Modeling and Simulation: MS'2012: Proceedings of the International Conference, 2–4 May 2012, Minsk, Belarus, 178p. Publishing Center of BSU, Minsk (2012)

2. Liawa, S.: Information retrieval from the World Wide Web: a user-focused approach based on individual experience with search engines. Comput. Hum. Behav. **22**, 501–517 (2006)
3. Marchionini, G.: Toward Human-Computer Information Retrieval Bulletin, Bulletin of the American Society for Information Science. http://www.asis.org/Bulletin/Jun-06/marchionini. html (2006)
4. Koffka, K.: Principles of Gestalt Psychology, p. 720. Routledge, New York (1935)
5. Miller, G.: The magical number seven, plus or minus two. Psychol. Rev. **63**, 81–97 (1956)
6. Brooke, J., Jordan, P.W., Thomas, B., Weerdmeester, B.A., McClelland, A.L.: SUS: a 'quick and dirty' usability scale. Usability Evaluation in Industry, pp. 189–194. Taylor & Francis, London (1996)
7. John, B.E., Kieras, D.E.: The GOMS family of user interface analysis techniques: comparison and contrast. In: ACM Transactions on Computer-Human Interaction, pp. 320–351 (1996)

Creation of Intelligent Information Flood Forecasting Systems Based on Service Oriented Architecture

Viacheslav A. Zelentsov, Semyon A. Potryasaev, Ilja J. Pimanov and Sergey A. Nemykin

Abstract In this paper a new approach to the creation of short-term forecasting systems of river flooding is being further developed. It provides highly accurate forecasting results due to operative obtaining and integrated processing of the remote sensing and ground-based water flow data in real time. Forecasting of flood areas and depths is performed on a time interval of 12–48 h to be able to perform the necessary steps to alert and evacuate the population. Forecast results are available as web services. The proposed system extends the traditional separate methods based on satellite monitoring or modeling of a river's physical processes, by using an inter-disciplinary approach, integration of different models and technologies, and through intelligent choice of the most suitable models for a flood forecasting.

Keywords Modelling · Flood forecasting system · Service oriented architecture · Interdisciplinary approach · Intelligent interface

1 Introduction

Large cities and small settlements all over the world have experienced more than once the destructive effects of floods. The number of floods is forecast to continue in the foreseeable future, and there is an emerging trend of growth of the total number of large-scale natural hydrological disasters [1–3]. In the current situation, the forecasting of floods caused by spring floods and freshets of various origins is one of the main components of providing for the safety of settlements located in river valleys and it is an important scientific problem [4].

There are a lot of ways to analyze floods.

V.A. Zelentsov (✉) · S.A. Potryasaev · I.J. Pimanov
Russian Academy of Sciences, Saint Petersburg Institute of Informatics
and Automation (SPIIRAS), St. Petersburg, Russia
e-mail: v.a.zelentsov@gmail.com

S.A. Nemykin
Design Bureau "Arsenal" Named M.V.Frunze, Saint-Petersburg, Russia

© Springer International Publishing Switzerland 2016
R. Silhavy et al. (eds.), *Automation Control Theory Perspectives
in Intelligent Systems*, Advances in Intelligent Systems and Computing 466,
DOI 10.1007/978-3-319-33389-2_35

371

Satellite monitoring has been one of the most important methods so far. Although the benefits of this approach are widely known, it is far from being flawless from the decision-making point of view. Most notably, it provides only post-event fixation of the fact and the boundaries of an inundation as well as post-event damage analysis as a rule.

The second group of well-known methods are medium-term and long-term forecasting for a period of several weeks to several months. These methods are based on hydrological modelling and they require large amounts of initial data, such as the characteristics of the snow cover, soil properties, and meteorological parameters, etc. [5].

Naturally, medium-term and long-term forecasting gives valuable information which allows analyzing possible scenarios for probable flooding in the following few months. Nevertheless, the low reliability of the initial data and the large time interval of forecasting do not provide accurate estimates of the flooding time at each specific point of the river valley.

So, *the first basic question is* whether it is possible to find a compromise between the quality of the initial data and the duration of the forecast interval, on the one hand, and the accuracy of forecasting on the other hand. In the investigation presented here we have found this balance by shifting to the short-term forecasting of river floods. In this case the forecast is performed in advance for 12–48 h and initially it starts only when the water level increases.

The *second basic question* is how can we combine, integrate, and synchronize all necessary remote sensing data, other heterogeneous data, information and models to predict the zones and depths of flooding, at the same time perform the necessary verification and validation procedures to ensure the required reliability of the results, and provide the results to users in a timely and convenient form.

In the our previous papers [6, 7], the satellite monitoring and the traditional hydrological forecasting approach based on river physical processes modelling was enhanced by integrating different models and technologies—such as input data clustering and filtering, digital maps of relief and river terrain, data crowdsourcing, hydrodynamic models, visualization tools based on geoinformation technologies, and techniques for flooding scenario generation and comparison (Fig. 1).

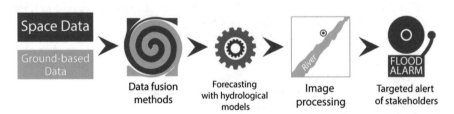

Fig. 1 Short-term forecasting based on the integration of space and in situ data

Furthermore, the principles of software implementation of a flood forecasting system (FFS) based on SOA were considered and the first successful case study using one of the possible hydrodynamic models—LISFLOOD [8], was described.

But it should be noted that high-quality forecasts need fundamentally not one but a complex ensemble of hydrodynamic models to choose model types and parameters depending on the specifics of the riverbed section simulation, the nature of water movement along the riverbed and floodplain, the initial data composition, and a number of other factors. The multimodal approach can significantly improve the forecasting quality but implies the development and implementation of new methods of automatic (or interactive) selection of model types and adjustment of the model parameters.

So the *third basic question* is how can we choose and adjust the model automatically. In this paper, a method and algorithms for the multimodal approach are presented, as well as the results of FFS testing using two types of hydrological models—one-dimensional and two-dimensional are considered.

2 Flood Short-Term Forecasting System

The real-time FFS was created with the approach described above and it is based on the integrated use of ground-based and aerospace data. This system allows operational forecasting of the river water levels, discharges and inundation areas, depths of flooding, and provides prior notification for the citizens in emergency situations at the GeoPortal and/or by using mobile devices (smartphones, tablet computers). The system contains the following main components (Fig. 2):

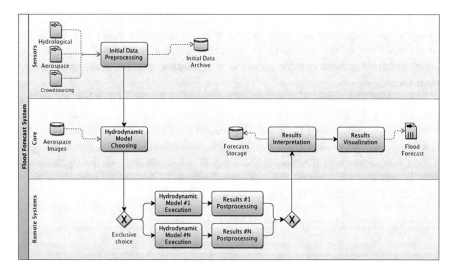

Fig. 2 The multimodal forecasting cycle implementing

(1) The Input Data Collection and Pre-processing component provides input data collection (e.g., satellite images, on-line hydrological data from meteorological stations; a digital terrain model) and pre-processing (i.e., primarily image processing, information filtering and information fusion).

(2) The Modelling and Forecasting component allows forecasting of water levels, calculation and forecasting of the water discharge, depths of flooding, and also automatic modelling of the water spread using hydrological models.

(3) The Post-Processing and Visualization component provides post-processing of modelling results, output data storage in a database, flood map vectorization and visualization of inundation areas.

(4) The Information Distribution component provides dissemination of the modelling results to the external software, publishing short-term forecasts at the public GeoPortal, and automatic notification of the local citizens and organizations using web services and mobile personal devices.

The input data collection and pre-processing component aims to effectively integrate all the available data from various geographically distributed data sources with analytical capabilities to enhance the reliability of the monitoring information and the speed at which it becomes available to decision makers. Another important aspect is to provide the user with the most accurate spatial and temporal resolutions of data, models, and tools, according to the actual tasks to be solved. Consequently, assessment at different geographic scales should take advantage of different data sources. The component implements automated data fusion allowing essential measurements and information to be combined to provide knowledge of sufficient richness and integrity so that decisions may be formulated and executed autonomously. The FFS architecture and functionality are universal regarding the input data type diversity, density, and speed.

The proposed system operates automatically and provides flood forecasts for the following 12–48 h with hourly outlines of the potential flooded zones, objects, and a water depth map, both available via the standard protocols WMS and WCS. The flood forecasting results are provided as a remote web service. Moreover, the users are not required to have specific knowledge in modelling and simulation or to have programming skills.

Two additional useful functions are implemented in the FFS. First, every user can move the timeline slider and see the inundated area 12 h in advance, all the complexity is hidden from the user. The second additional function also allows the user to see the flood area for every possible water level. Finally, automatic generation and analysis of flooding scenarios will allow analysis of the dynamics of river floods and to evaluate their potential effects in the near future to support preventive actions to mitigate the impacts of the floods.

3   Algorithm and Models

3.1   Short-Term Forecasting Algorithm

The short-term flood forecasting algorithm includes the following main steps in the case of the multimodal approach: (1) on-line hydrological data on the river water levels are obtained from a hydrological station; (2) the observed data are processed by their smoothing with trend extraction in order to predict the river water levels in the short term; (3) if necessary, a symbolic regression-based forecasting model is built which provides water flow discharge forecasts; (4) water flow discharge data, accompanied by river terrain information are processed by an adaptive hydrological model which simulates water floating routes and thus provides forecasts on water covered territories in the river basin, choice of the hydrological model and its adaptation on the basis of integrated usage of heterogeneous data comprising remote sensing data and river terrain model, and execution of calculations; (5) visualization software is used for geo-presentation of the river simulation results of water flows and inundation areas; and (6) archival data and images, satellite images of the current situation and data crowdsourcing through a Geo-Portal are used to validate and calibrate the hydrological model.

3.2   Hydrological Models and Their Selection

The basic components of the suggested FFS are hydrological models. Regardless of the specific software implementation, mathematical models of water flow movement are mostly based on the set of Saint-Venant equations [9]. When writing the set of Saint-Venant equations, the water flow can be represented both in the one-dimensional (flow characteristics are averaged by cross section) and in the two-dimensional (averaged by depth) schematizations.

   Thus far, various countries have developed and implemented a large number of such models: the Russian software packages Flood, River, and STREAM_2D, developed by Belikov et al. [9]; Mike 21 of the Danish Hydraulic Institute [10]; Delf 3D of the Deltares Institute in Delft, the Netherlands [11]; HEC_RAS of the United States Army Corps of Engineers [12]; and others. The models mainly differ in the methods of schematization of the computational domain (triangular, tetragonal, and mixed meshes), computational schemes and methods (finite differences and finite elements), and sets of additional blocks (admixture dissemination, load transport, etc.). The design of operational flood forecasting automated systems implies, as was noted above, that no universal model exists to describe flood development processes at river sections that are different in length and configuration. When choosing hydrodynamic models to solve flood forecasting problems, it is advisable to implement the multimodal approach: depending on the length of the monitored section and the presence of initial data it is possible to choose between one-dimensional hydrodynamic models for river valley sections of 100–1000 km long and two-dimensional models for sections shorter than 100 km with wide beds and floodplains, complex configurations, and with various structures located inside them.

Combined (hybrid, complex, multiscale) computing using one-dimensional and two-dimensional models can yield a significant effect in effort and fund economy when investigating and monitoring lengthy river areas.

Therefore, a principal issue is to form a mechanism of choosing the most adequate model for specific conditions.

When building operational flood forecasting systems, this technology is implemented through the creation of an intelligent interface, which unites heterogeneous in situ and space-based data and expert knowledge to be used later for forecasting.

The intelligent interface's main purpose is to choose and adapt a specific model to compute water propagation zones and flood depths on the basis of contextual information (baseline data accuracy, flood dynamics, operational result efficiency, etc.). The methods and algorithms that are currently being developed intensively within the theory of evolutionary modeling as well as the theory of models and multimodal complexes qualimetry have been chosen as the scientific basis for the adaptation of hydrological models [13, 14].

Overall, the creation of an intelligent interface ensures a synergetic effect from the joint processing and use of heterogeneous ground–space data and the implementation of a model complex to assess possible flood zones.

3.3 Application of Remote Sensing Data

Taking into account the vastness of river floodplains and the shortage of the existing in situ data, the implementation of the flood forecasting systems implies the extensive use of Earth remote sensing (ERS) data. In many cases, ERS data are the basis for flood monitoring. Their inclusion in the flood forecasting systems compensates for the shortage of in situ data and allows for additional adaptation of the models used. We may single out the following vectors for implementing ERS data in the FFS design.

Updating cartographic information about the existing infrastructure in river valleys—the boundaries of settlements in the flood zone, roads, line structures that affect the flow structure in floodplains, and riverbed configurations. To this end, high-resolution optical satellite imageries, which are obtained, for example, from satellites Resurs_P, GeoEye, and WorldView, or from drones, during the summer–autumn low water period.

Acquisition of information about river valley terrain. To this end, digital terrain models (DTM) based on highly detailed space and air surveys appear promising. At present, the main possibilities for obtaining more accurate and usable terrain models are related to the programs of radar spacecraft TerraSar and TanDem to create the Earth's DTMs with height resolutions of about 2 m, as well as drone-based surveys or laser scans, which yield a centimeter accuracy during DTM plotting.

The calibration and verification of hydrodynamic models based on information about the boundaries of flooded territories from space images [15] and crowdsourcing results. The calibration parameters of hydrodynamic models are data describing the roughness of the underlying surfaces which determines hydraulic resistance and which is characterized, as a rule, by the roughness coefficient. Setting the fields of roughness coefficients as a first approximation is based on decoding and highlighting areas with various plant covers and microreliefs, as well as on selecting coefficient values for them using the tables in [16]. To this end, both medium and high-resolution images are applicable. Usually, five or six main types of underlying surfaces (beds, plow lands, meadows, forests, and urban development) are singled out. Automated data processing methods and technologies are being developed and implemented successfully to decode these surface types [17, 18].

Note that maximally authentic calculation results require *integrated* processing of the various data types obtained during in situ and aerospace measurements.

4 Short-Term Forecasting System Software

It follows from the previously described FFS composition that the architecture of the system should ensure flexible interaction of many components, including the existing and potential software modules based on hydrological models, input data processing modules, forecast result visualization modules, and control modules, etc. In this respect, it is desirable to build an FFS on the basis of a service-oriented architecture (SOA) [7, 19].

An improved software structure for the practical implementation of the described multimodal approach is presented in Fig. 3.

The service bus with intelligent interface functions is the core of the system. It is implemented as open source software—OpenESB. OpenESB has an imbedded interpreter for the business processes language BPEL that executes the scenarios necessary for the functioning of an FFS. The service bus provides the starting point for the modules intended for the collection, processing, storage and dissemination of information, according to the scenario. Each module is a web service with a standard SOAP interface.

The control module performs starting and stopping of automatic calculation cycles, single calculations, and a number of auxiliary operations.

The calculation module of the intelligent interface synthesizes the particular structure of the FFS on the basis of the initial data of the hydrological conditions.

The initial data collection module enquires of the hydrological sensors and provides the service bus with normalized data. This allows abstraction from the information exchange scheme with specific sensor manufacturers.

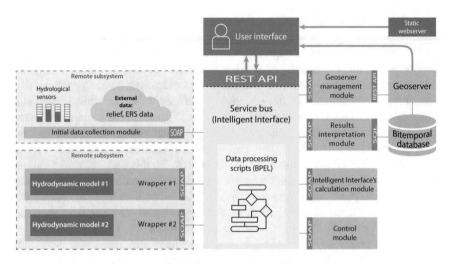

Fig. 3 Structure of the advanced short-term forecasting system software

The Wrappers of the hydrodynamic models provide the formation of files with initial data, starting calculations, performance monitoring, and collecting and sending the calculation results to the Service bus.

The Results interpretation module performs several stages of processing:

1. Creating the depth contours from the matrix of points.
2. Recording the data on the depth intervals and the flooded areas into the attribute Table
3. Simplification of geometry.
4. Generalization and smoothing of contours to restore the natural borders of flooded areas and improve the visualisation of the simulation results.

In addition to the standard python language libraries of data analysis and visualization, such as "scipy", "numpy", and "matplotlib", the Results interpretation module uses the open spatial data libraries: "ogr", "geopandas", "shapely" (vector data processing), "osr" (working with coordinate systems), matplotlib.tri.tricontourf (construction of the Delaunay triangulation elements irregular network on the basis of a field of points), matplotlib.pyplot.tricontourf (calculation of the contours points corresponding to predetermined depth intervals based on triangulation).

The Geoserver management module allows us to put the automatically generated data on a geoserver. A variety of solutions can be used as a geoserver—both open source (Geoserver, Mapserver, QGIS mapserver), and commercial (ArcGIS Server, ArcIMS, GeoMedia, Oracle MapViewer). The developed prototype used Geoserver. Geoserver is managed through a networking software interface (REST API).

In the proposed system every user receives the data on their own personal device– a smartphone, tablet or PC. Powerful searching and visualization tools provide the users with all the necessary data on the areas and depths of inundation.

5   Case Study and Discussions

FFS testing in a multimodal mode was performed on the Russian rivers: r. Oka (near Ryazan city), r. Small Northern Dvina along its 100 km long (from Veliky Ustyug to Kotlas), r.Protva. As a basis for modeling water movement, one-dimensional and two-dimensional models of the Stream-2D software complex (an upgraded version of the software package River) [9] were selected. This complex has demonstrated high efficiency in solving various problems concerned with analysis of water flow dynamics and inundations caused by both natural and technological reasons.

The input data for the simulation were as follows: the results of ground-based measurements (depth, slope of the water surface, water discharge, flow velocity, and observational data from the hydrological stations), different time satellite images for the flood and low water period, and data on the relief.

The FFS (Fig. 3) automatically performed the preparation and preprocessing of the initial data for the hydrodynamic models, namely: spatial information on the territory relief, the distribution of the underlying surface roughness, the time series of the water discharge, and the levels at the input and output boundaries.

Further, in the FFS the visualization of the modeling results was carried out.

It was found during testing that the use of a one-dimensional model is expedient for the river's elongated sections when their length is an order of magnitude or more greater than the width of the potential flood area. A two-dimensional model requires more detailed information on the river valleys morphometry, especially on the relief of potential flood territory. As a result, they allow us to get a forecasted distribution of flow velocities and of water surface levels and depth within the calculation domain.

A significant effect can be achieved by a joint (hybrid, multiscale) calculation of the one-dimensional and two-dimensional models. Here the entire elongated section of the river is modeled by a one-dimensional system of equations, and part of it by a two-dimensional or bilayer two-dimensional system.

The results of multimodal forecasting have demonstrated a good correspondence of the calculated and measured water levels (the difference is not more than 30 cm) and flood area boundaries. Analyses of the modeling results have also shown good correspondence between flood modeled areas and contours with contours derived from satellite imageries. The difference does not exceed 10 % (Fig. 4).

In general, the approach of this proposed and developed FFS provides high accuracy of short-term forecasting due to the integrated use of in situ and aerospace

(a)   **(b)**

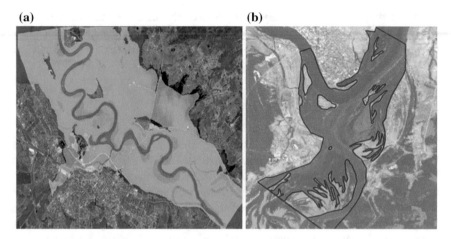

Fig. 4 The results of modeling in comparison with satellite images Landsat-8: the r. Oka in Ryazan (**a**) and the r. Malaya Northern Dvina in Kotlas (**b**), April 2013

Fig. 5 The combination of forecasting results with contours of buildings

data and multimodal forecasting. Convenient visualization of results, notification services, and an automatic mode of operation provide an opportunity for practical application of the described system (Fig. 5).

Acknowledgments The research described in this paper is partially supported by the Russian Foundation for Basic Research (grants 15-07-08391, 15-08-08459, 16-07-00779, 16-08-00510, 16-08-01277), grant 074-U01 (ITMO University), project 6.1.1 (Peter the Great St. Petersburg Polytechnic University) supported by Government of Russian Federation, Program STC of Union State "Monitoring-SG" (project 1.4.1-1), The Russian Science Foundation (project 16-19-00199), Department of nanotechnologies and information technologies of the RAS (project 2.11), state research 0073–2014–0009, 0073–2015–0007.

References

1. Transboundary Flood Risk Management in the UNECE Region. United Nations New York and Geneva (2009)
2. Porfiriev, B.N.: Economic consequences of the 2013 catastrophic flood in the Far East. Herald Russ. Acad. Sci. **85**, 40 (2015)
3. Alekseevskii, N.I., Frolova, N.L., Khristoforov, A.V.: Monitoring Hydrological Processes and Improving Water Management Safety. Izd. Mos. Gos. Univ., Moscow (2011) [in Russian]
4. Vasil'ev, O.F.: Designing systems of operational freshet and high water prediction. Herald Russ. Acad. Sci. **82**, 129 (2012)
5. Sokolov, B.V., Zelentsov, V.A., Mochalov, V.F., Potryasaev, S.A., Brovkina, O.V.: Complex objects remote sensing monitoring and modelling: methodology, technology and practice. In: Proceedings of the 8th EUROSIM Congress on Modelling and Simulation, 10–13 Sept 2013, Cardiff, Wales, United Kingdom, pp. 443–447 (2013)
6. Merkuryev, Y., Merkuryeva, G., Sokolov, B., Zelentsov, V. (eds.): Information Technologies and Tools for Space-Ground Monitoring of Natural and Technological Objects. Riga Technical University, Riga (2014)
7. Sokolov, B.V., Pashchenko, A.Ev., Potryasaev, S.A., Ziuban, A.V., Zelentsov, V.A.: Operational flood forecasting as a web-service. In.: Proceedings of the 29th European Conference on Modelling and Simulation (ECMS 2015), Albena (Varna), Bulgaria. pp. 364–370 (2015)
8. LISPFLOOD-FP, University of Bristol, School of Geographical Sciences, Hydrology Group, http://www.bris.ac.uk/geography/research/hydrology/models/listflood (Accessed: 27 April 2013)
9. Belikov, V.V., Krylenko, I.N., Alabyan, A.M., Sazonov, A.A., Glotko, A.V.: Two-dimensional hydrodynamic flood modelling for populated valley areas of Russian rivers. In: Proceedings International Association of Hydrological Sciences, pp. 69–74 (2015)
10. Skotner, C. et al.: MIKE FLOOD WATCH—managing real-time forecasting, http://dhigroup. com/upload/publications/mike11/Skotner_MIKE_FLOOD_watch.pdf (Accessed: 09 Sept 2015)
11. Delft3D-FLOW Version 3.06 User Manual, http://oss.deltares.nl/web/delft3d (Accessed: 22 Sept 2015)
12. HEC-RAS river analysis system User's Manual, http://www.hec.usace.army.mil/software/hec-ras. (Accessed: 22 Sept 2015)
13. Bukatova, I.L.: Evolutionary Modeling and Its Applications. Nauka, Moscow (1979) [in Russian]
14. Sokolov, B.V., Yusupov, R.M.: Conceptual foundations of quality estimation and analysis for models and multiplemodel systems. J. Comput. Syst. Sci. Int. **43**(6), 831 (2004)
15. Merkuryev, Y., Okhtilev, M., Sokolov, B., Trusina, I., Zelentsov, V.: Intelligent technology for space and ground based monitoring of natural objects in cross-border EU-Russia territory. In: Proceedings of the International Geoscience and Remote Sensing Symposium (IGARSS 2012), Munich, Germany, pp. 2759–2762 (2012)
16. Baryshnikov, N.B.: Hydraulic Resistances of Riverbeds, Izd. Ross. Gos. Gidrometeorol. Univ., St. Petersburg (2003) [in Russian]
17. Sokolov, B.V., Zelentsov, V.A., Brovkina, O. et al.: Complex objects remote sensing forest monitoring and modeling. In: Silhavy, R., Senkerik, R., Oplatkova, Z.K., Silhavy, P., Prokopova, Z. (eds.) Modern Trends and Techniques in Computer Science: Advances in Intelligent Systems and Computing, vol. 285, pp. 445–453. Springer (2014)
18. Sokolov, B.V., Zelentsov, V.A., Yusupov, R.M., Merkuryev, Yu.A.: Multiple models of information fusion processes: quality definition and estimation. J. Comput. Sci. **5**, 380 (2014)
19. Vasiliev, Y.: SOA and WS-BPEL: Composing Service-Oriented Solution with PHP and ActiveBPEL. Packt Publishing (2007)

Possibilities of Use of the Active Knowledge Databases in the Agricultural Sector

Václav Vostrovský, Jan Tyrychtr and Petr Hanzlík

Abstract This contribution deals with the potential of the so called active knowledge database in conjunction with the agricultural resort. The authors of this contribution initially focus attention on the defining the term active databases, their possible roles and expectations associated with them. Agriculture of the Czech Republic certainly has some potential, which can initiate a qualitative change in the functioning of the entire agricultural sector. The authors also deal with the possibility of the creation of such databases and derive their principles from classical relational databases. The crucial matters of this examination will relational database technology (especially the issue of the database triggers) and the theory of measurement and evaluation of the quality of software products. Active database must be created only on the principles of relational database technology and on these principles must also work. The proposed principles of active databases are then verified through a functional prototype and then are generalized. On the basis of these conclusions are then specified the recommendations, which can significantly increase the benefits of this phenomenon.

Keywords Active databases · Knowledge databases · Trigger · Knowledge schema · Agriculture

1 Introduction

The current era of humankind is often labelled as knowledge society. This development stage corresponds tightly with knowledge economy. Agriculture is naturally a part of this economy too. Accordingly, the agriculture in the Czech Republic

V. Vostrovský (✉) · P. Hanzlík
Faculty of Economics and Management, Department of Software Engineering,
Czech University of Life Sciences in Prague, Prague, Czech Republic
e-mail: vostrovsky@pef.czu.cz

J. Tyrychtr
Faculty of Economics and Management, Department of Information Technologies,
Czech University of Life Sciences in Prague, Prague, Czech Republic

© Springer International Publishing Switzerland 2016
R. Silhavy et al. (eds.), *Automation Control Theory Perspectives in Intelligent Systems*, Advances in Intelligent Systems and Computing 466,
DOI 10.1007/978-3-319-33389-2_36

should rely on knowledge, including all of its specific aspects and problems. The crucial principles of knowledge economy are the aspects of knowledge acquiring and storage, as well as its retrieval at the right moment.

The supply of fast, accessible and high-quality knowledge to individual users is the key aspect of information and knowledge strategy of the agricultural sector. This aspect is closely related to the so-called knowledge need. The knowledge need is defined as a basic social need for knowledge of the quality and quantity necessary to solve certain tasks and problems. The real information and knowledge need is based on the task and problem solved by the particular person. Information needs influence the processes of the information systems. The processes and activities in the agricultural sector demand a high level of knowledge. It is therefore important to provide not only the data alone, but also bound with their context, or the knowledge, in accordance with the latest trends of knowledge based society and economy.

The necessary set of information depends not only on the site, where it is used (the decision site), but also on the experience and knowledge of the managing subject [1]. The foundation of information, which is the object of this need, is probably best described by the following definition: "The information is the meaning, which the human subject assigns to the data through the conventions used for its presentation" [2]. Based on this definition, it is possible to derive the relation between information and data and also between information and knowledge as:

DATA—INFORMATION—KNOWLEDGE

The existing information and knowledge need in agriculture is closely related to the issue of so-called transformation sites (TS) [3]. Knowledge need is then the particular set of knowledge required for the operation of the particular TS. The above mentioned variety of agricultural activities affects the number of TS of a regular agricultural company, which determine its information resources. Every individual TS has its own particular knowledge need and the sum of this particular knowledge needs represents the aggregate information need of the agricultural company, which should be covered by the information resources of the company. If this condition is not met, it is necessary to gather the missing knowledge from external sources, which allows to exploit the active databases potential.

The high knowledge need of the agricultural sector is a logical consequence of the variety of activities carried out in a typical agricultural company. This variety emerges from the principles of agriculture and from the effort to diversify the strategy of agricultural business to decrease the potential risk by splitting the activities to different domains. The increased information need of the agricultural sector is also influenced by:

- the existence of a large amount of agricultural activities suffering from the lack of available specialists or other information sources;
- the existence of a large amount of problems accompanying these activities, which require a rapid and highly skilled solution;
- the existence of a real risk of irreversible loss of knowledge required to pursue agricultural activities;

This knowledge need of the agricultural sector is primarily caused by problems related to:

- excessive incidence of diseases and pests,
- adverse weather conditions,
- ecological circumstances,
- failure to comply with agronomic terms.

The knowledge need of each realized activities in agricultural practice relates to following types of knowledge:

- *technological knowledge*—describe mainly the matters of what was done, and how it was done,
- *value-assessing knowledge*—express the financial burden of realized activities,
- *dynamic knowledge*—relates to the time properties of these activities [4].

2 Methods

The goal of this article is to define the principles and purpose of active knowledge databases and specify their role in agricultural practice. This paper also aims to demonstrate the possibilities of active knowledge databases in the form of association rules. These possibilities will be illustrated on the case study of the methodical manual for plant protection published by the Czech Ministry of Agriculture. The usual methods for relational database design and knowledge representation are used.

Knowledge using is a topic that comes up in a lot of discussions of knowledge management. Various methods of knowledge retention are examined. For the management knowledge can be employ knowledge retention in relational databases. In this context the database methods are in the center of interest.

The main term of the relational database theory is a relation [5]. The relation is defined as a set of entities that have the same attributes. An entity usually represents an object and information about that object. A relation is usually described as a table, which is organized into rows and columns All the data referenced by an attribute are in the same domain and conform to the same constraints. The relational model specifies that the entities of a relation have no specific order and that the entities, in turn, impose no order on the attributes [6–8].

For the knowledge retention purpose the relational database technology provides the relative efficient tool—the query language SQL (see [9, 10]) for the definition of tables and manipulation with data. SQL (Structured Query Language) is a standard computer language for accessing and manipulating relational database management systems. It is standardized by ANSI.

A big opportunity for solution or decision support of above-mentioned problems can provide so-called active databases. The principles of traditional active databases have been described and discussed thoroughly. These principles can be utilized as a basis for designing of knowledge-oriented active databases, also known as active knowledge databases.

An active database monitors a database scheme and operations that change it. In principle, a specific action is carried out when a set of conditions is fulfilled. This is hence an example of event-driven architecture. Many authors define active database as a database defined with the help of triggers, assuming the triggers are characterized as active rules [11–15]. In classical relational database technology, triggers are linked with the implementation of integrity constraints that follow the logical sequence event-condition-action. The active databases are based on the logical sequence Event-Condition-Action. In relationship with active databases, the term event means a change in database state. Characteristic examples of events are INSERT, UPDATE, or DELETE, which occur when the corresponding SQL command of same name is executed. In Oracle and DB2 DBMS, the triggers are implemented in accordance with the preliminary version of SQL3 standard. The triggers' actions are defined in PL/SQL language, a relatively robust programming tool.

The syntax of database triggers in Oracle DBMS is:

```
CREATE TRIGGER <trigger name>
{BEFORE|AFTER} <event> [OF <column name>] ON <table name>
[REFERENCING <reference>]
[FOR EACH ROW]
[WHEN <condition>] <PL/SQL code>
```

Within this syntax, Oracle DBMS supports two different trigger levels:

- Row-level—called individually for every row influenced by the triggering event.
- Statement-level—called at once for the whole table.

Database triggers designed in this way have following properties:

- rules are executed immediately,
- action can execute PL/SQL code,
- events can be defined on row-level or statement level,
- rules are evaluated before event and after event (Before a After).

3   Results

Active knowledge database can be described as an information tool for decision support. It is based on knowledge that is represented by a set of trigger rules. The state space of problem domain is given by this set of corresponding triggers. The interaction between users and such application will be conducted through comments to the action carried out. Let us consider an insertion of data describing the problem situation (e.g. the occurrence of disease or pest in the produced crop) into the database as an action. The comment will then have a form of relevant expert recommendation to a problem event (i.e. a proposal for countermeasures—agronomic intervention).

If we compare the general structured of a traditional trigger with the structure of general knowledge rule, we can find great similarities. The clause **AFTER** <event> **[OF** <column name>] **ON** <table name> represents a knowledge rule activation event, the **[WHEN** <condition>] <PL/SQL code> provides a possibility to define a conditional part of the rule, including the appropriate result and comments, like **SIGNAL SQLSTATE**.

The active knowledge databases are based on knowledge independence. Knowledge is extracted from an application and converted to a specific database rule. The rule is consequently embedded into the database scheme. This has significant advantages when a need to modify the knowledge occurs. It is then sufficient to modify the database rule only, instead of laborious modification of all relevant applications.

Classical knowledge applications (and knowledge systems specifically) work on the principle of dialogue between a user and the application. This dialogue is defined by a problem the user is facing, and thus looks for a relevant information (knowledge) support. The core of such applications is formed by a *knowledge base* (Fig. 1).

The inspiration for shaping the principles of active knowledge databases can be some existing relational databases, consisting of the set of relations, representing two-dimensional tables, with rows corresponding to recording entities and columns corresponding to attributes of re-cording entities. The entities recorded as the rows in relations can be often interpreted as simple rules represented a corresponded knowledge. An example from the agriculture area can be the methodical manual for plant protection published in the printed form by the Ministry of Agriculture Czech Republic (Table 1). The records containing in the mentioned manual can be simple transformed into the corresponding relation tables of the following form:

This set of the registered preparations mentioned above contains in the essence the set of production rules of the IF E THEN H types, in which E and H are two propositions in the implication. These propositions can be interpreted as evidences

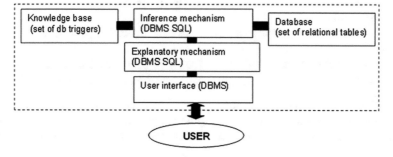

Fig. 1 Schematic representation of the principle of active knowledge databases

Table 1 Example of the set of registered preparations for plant protection included in the methodical manual for plant protection published in the printed form by the Ministry of Agriculture Czech Republic

Growth	Harmful factor	Dose kg/ha	Protection period	Application day
1635-OP TOLKAN FLO				
Sugar-beet	Couch grass	1.5	AT	Postharvest
1640 MARATON				
Spring barley	Hair-grass	0.015	AT	During autumn
1391-1-ARELON 500				
Spring barely	Catchweed	0.25	–	During spring
3885-1 REFINE				
Rye	Catchweed	0.25	–	During spring

or hypothesis from the table. From this table can be for example derived the following rule:

RULE1   IF harmful_factor = couch grass AND growth = sugar-beet THEN use protective_preparation TOLKAN FLO in dose 1.5 kg/ha and application day is postharvest

RULE2   IF harmful_factor = hair-grass AND growth = spring barley THEN use protective_preparation MARATON in dose 0.015 kg/ha and application day is during autumm

RULE3   IF harmful_factor = cathweed AND growth = spring barley THEN use protective_preparation ARELON 500 in dose 0.25 kg/ha and application day is during autumn

   etc.

For example, rule No. 1 can be interpreted as follows: For the couch grass eliminated from sugar-beet can be recommend protective preparation TOLKAN FLO in dose 0.25 kg/ha and application day is during autumn.

   The rules specified above can be represented as a corresponding set of triggers:

```
CREATE TRIGGER rule1
AFTER UPDATE OF harmful_factor ON growth_table
FOR EACH ROW
WHEN   (Growth = 'sugar-beet'  AND  harmful_factor = 'couch
grass')
SIGNAL   SQLSTATE   '8010'  ('Use  protective_preparation
TOLKAN FLO in dose 1.5 kg/ha application after harvest')
```

The set of triggers mentioned above is in fact a knowledge base focusing on problems with harmful agents in crop production. The principle of a knowledge application conceived in this way can be demonstrated on the following example (see Fig. 2).

GROWTH_TABLE:

Growth	harm-ful_factor	date_identification
sugar-beet		
spring barley		
rye		
etc.		

Growth	harmful _factor	date_ identi-fication
sugar-beet	couch grass	'1.6.2015'
spring barley		
rye		
etc.		

event: **UPDATE** Growth_table
SET harmful_factor = 'couch grass'
WHERE Growth = 'sugar-beet' **AND** date_identification = '1.6.2015'

call trigger RULE1 with a conclusion:
SIGNAL SQLSTATE '8010' ('Use use protective_preparation TOLKAN FLO
in dose 1.5 kg/ha application after harvest')

Fig. 2 Schematic representation of the operating principle of the proposed active knowledge databases

3.1 Knowledge Base—Example

inserting (UPDATE) attributes **harmful_factor = 'couch grass'** and **date identi-fication = '1.6.2015'** into record with the **growth = 'sugar-beet'**

4 Discussions and Conclusions

The designed form of the active knowledge database makes easy the knowledge base update. The benefit of this conception of active knowledge database consists in relative independence on high level programming solutions as for artificial intelligence tools. These tools require the active complicity of knowledge engineers and are often isolated from existing program support in related enterprises.

The above mentioned table conception of knowledge base provides:

- transparency of such set of knowledge (knowledge base),
- simplification of this knowledge base,
- optimal knowledge organization,
- checking for consistency can already be performed during the design of the knowledge based system.

The benefits of the active database draft proposed can be:

- timely response to a problem situation,
- easy integration of knowledge application into existing information security policies of agricultural enterprise.

If the active knowledge databases implemented may become relevant information and knowledge to support for tactical and strategic decision-making in the agriculture.

Acknowledgements The results and knowledge included herein have been obtained owing to support from the Internal grant agency of the Faculty of Economics and Management, Czech University of Life Sciences in Prague, grant name: "The potential of open data in the agricultural sector with regard to their correctness and aggregation".

References

1. Říhová, Z.: Informační zabezpečení a organizační změny. Vysoká škola ekonomická, V Praze (1996)
2. Rosický, A.: Informace a Systémy. Oeconomica, V Praze (2009)
3. Vostrovský, V., Tyrychtr, J., Ulman, M.: Potential of Open Data in the Agricultural eGovernment. In: AGRIS on-line Papers in Economics and Informatics, vol. 7, no. 2, p. 103 (2015)
4. Mlejnek, P.: Informační potřeba středního manažera. VŠE Praha (2010)
5. Atzeni, P., De Antonellis, V.: Relational Database Theory. Benjamin/Cummings Publication Co., Redwood City, Calif (1993)
6. Kroenke, D.: Beyond the relational database model. Computer **38**, 89–90 (2005). doi:10.1109/mc.2005.151
7. Paredaens, J.: The Structure of the Relational Database Model. Springer, Berlin (1989)
8. Pokorný, J., Valenta, M.: Databázové Systémy. ČVUT Praha, Praha (2013)
9. Kim, J.: A confirmative knowledge management framework based on RDBMS, Knowledge Map, and SQL inference. Information (Japan) **17**, 89–98 (2014)
10. Telepovska, H., Toth, M.: Support of relational algebra knowledge assessment. In: 6th International Joint Conference on Computer, Information, and Systems Sciences, and Engineering, CISSE (2010)
11. Chezian, R., Devi, T.: Design and development of algorithms for identifying termination of triggers in active databases. Int. J. Inf. Syst. Change Manag. **5**, 251 (2011). doi:10.1504/ijiscm.2011.044527
12. Kasbon, R., Shaharom, S., Mazlam, E., Mahamad, S.: Implementing an active database for maintaining asset data. In: Pacific-Asia Conference on Knowledge Engineering and Software Engineering, KESE (2009)
13. Medina-Marín, J., Pérez-Lechuga, G., Seck-Tuoh-mora, J., Li, X.: ECA rule analysis in a distributed active database. In: International Conference on Computer Technology and Development, ICCTD 2009, pp. 113–116. IEEE, Kota Kinabalu (2009)
14. Rabuzin, K.: Simulating proactive behaviour in active databases. In: 5th International Symposium on Computational Intelligence and Intelligent Informatics, pp. 25–29. IEEE, Floriana (2011)
15. Wang, G.: A rule model in active databases. In: IEEE Workshop on Advanced Research and Technology in Industry Applications, WARTIA (2014)

Conceptual and Formal Modelling of Monitoring Systems Structure-Dynamics Control

Viacheslav A. Zelentsov, Sergey Nemykin and Boris Sokolov

Abstract Elements of the methodological basis of the theory of monitoring automated (computer-aided) systems structure-dynamics control are proposed in the paper. This theory can be widely used in practice. It has an interdisciplinary basis provided by the classic control theory, operations research, artificial intelligence, systems theory and systems analysis. The proposed approaches were implemented in software prototypes. The software prototypes were developed to simulate control processes for space-facilities control system (SF CS). The unified description of various control processes allows synthesizing both technical and functional structures of SF CS simultaneously. The presented multiple-model complex, as compared with known analogues, has several advantages. It simplifies decision-making in SF CS structure dynamics management, for it allows seeking for alternatives in finite dimensional spaces rather than in discrete ones. The complex permits to reduce dimensionality of SF CS structure-functional synthesis problems in a real-time operation mode.

Keywords Monitoring automated (computer-aided) systems · Natural and technological objects · Structure-dynamics control

V.A. Zelentsov (✉) · B. Sokolov
St. Petersburg Institute of Informatics and Automation,
Russian Academy of Sciences (SPIIRAS), St. Petersburg, Russia
e-mail: v.a.zelentsov@gmail.com

B. Sokolov
e-mail: sokol@iias.spb.su

S. Nemykin
The Arsenal Design Bureau Named After M.V. Frunze Federal
State Unitary Enterprise, St. Petersburg, Russia

B. Sokolov
Saint Petersburg National Research University of Information Technologies,
Mechanics and Optics (ITMO), St. Petersburg, Russia

© Springer International Publishing Switzerland 2016
R. Silhavy et al. (eds.), *Automation Control Theory Perspectives
in Intelligent Systems*, Advances in Intelligent Systems and Computing 466,
DOI 10.1007/978-3-319-33389-2_37

391

1 Introduction

The main object of our investigation is monitoring automated (computer-aided) systems (MS). Classic examples of MS are: MS for various classes of moving objects such as surface and air transport, ships, space and launch vehicles, etc., MS for geographically distributed heterogeneous networks, MS for flexible computerized manufacturing. In our paper, MS for the control of natural and technological objects (NTO) under a changing environment has been analysed [1–4].

As applied to MS we distinguish the following main types of structures: the structure of MS goals, functions and tasks; the organization structure; the technical structure; the topological structure; the structure of special software and mathematical tools; the technology structure (the structure of MS control technology) [1–3].

One of the main features of modern MS for NTO is the changeability of their parameters and structures due to objective and subjective causes at different stages of the MS life cycle. In other words, we always come across the MS structure dynamics in practice. Under these conditions, if we want to increase (stabilize) MS potentialities and capacity for work, structure control is to be performed [1].

By structure-dynamics control we mean a process of control inputs production and implementation for the MS transition from the current macro-state to a given one.

Structural and functional control in MS involves a change in system objectives; reallocation of functions, tasks, and control algorithms between different levels of the monitoring system, control of reserves; and transposition of MS elements and subsystems.

The problem of MS structure-dynamics control consists of the following groups of tasks: the tasks of structure-dynamics analysis of MS; the tasks of evaluation (observation) of structural states and MS structural dynamics; the problems of optimal program synthesis for structure-dynamics control in different situations.

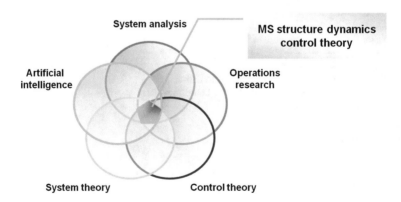

Fig. 1 The place of the theory of MS structure-dynamics control among the interdisciplinary investigations

From our point of view, the theory of structure-dynamics control will be interdisciplinary and will accumulate the results of classical control theory, operations research, artificial intelligence, systems theory, and systems analysis [5–12]. The two last scientific directions will provide a structured definition of the structure-dynamics control problem instead of a weakly structured definition. These ideas are summarized in Fig. 1.

2   Conceptual Modelling of Structural Dynamics of Monitoring Systems for Natural and Technological Objects

Structural dynamics and control problems require innovative solutions and investigation of models to obtain them. The conceptual model represents concepts and relations between them. It is a descriptive model of a system based on qualitative assumptions about its elements and relations between them. The conceptual model is used to understand the system operation in a specific environment using natural language and naïve logic statements [10, 11]. Conceptual models play a key role in software application development. Moreover, the conceptual modelling is an important step in the simulation study.

With regard to NTO monitoring and control problems, conceptual modelling supposes the definition of a clearly formulated target set to various stages of object lifecycle under investigation; setting-up basic concepts in the area of concern and relations between these concepts; as well as definition borders between an investigated system and its environment.

The control system for natural and technological object monitoring may have a multi-level hierarchical structure. It may include control points, control stations, space and ground-based measurement equipment, and a telecommunication system. Each MS subsystem can be considered a control subsystem with respect to its inferior elements, on the one hand, and a controlled element with respect to ranking all subsystems, on the other hand.

For further specification of the structural dynamics and control problems let us introduce a subclass of control objects that represents artificially created or virtual objects (e.g., a set of devices) that could be moved in a real or virtual space and interact with other objects by information, material or energy exchange.

The object structure allows a wide interpretation, so these objects can be used for the description of natural objects, e.g., a part of land, forest area, water basins as well as ground and space facilities. They can be interpreted as different users who want to implement results of space-ground monitoring for their own goals [1, 2, 13–15]. In this case, interactions between objects could be of active or passive character.

The following basic concepts are used for the conceptual modelling of the structural dynamics of monitoring and control systems.

(a) "Operation" defines any action or a system of actions to accomplish a goal. Operations require resources, including information, material and energy exchange. Operation content is formulated by specifying parameters that define the operation results (e.g., volume, quality, and operational time), required resources as well as information, energy and material flows.

(b) Concept "Resources" includes materials, energy, production means, technical equipment, transportation tools, and finances. The operational time and personnel involved are defined as resources as well.

(c) Concept "Task" is used to describe the desirable result of actions for a specified time period. Tasks are derived from the goals and are characterised by quantitative data or parameters of the desirable results.

(d) "Flow" is characterised by current volume (level), transmission intensity, velocity of flow level variations, etc. Different types and kinds of flows are determined in real systems [3], e.g., material, energy and information flows; single-commodity and multi-commodity flows; continuous and discreet flows; homogeneous and heterogeneous flows; synchronous and asynchronous ones.

(e) "Structure" characterises stable links and interactions between system elements. The structure defines the integrity and composition of the system, and its organisation framework. Here, the following basic forms of structures [1–4, 16–18] are defined: structure of goals, functions and tasks; organisational structure; technical structure; topological structure; information support, hardware and software structure; structure of system basic elements and its subsystems at various stages of its lifecycle. Additionally, for various classes of relations between basic elements of MS, multiple constraints (i.e., space-time, technical, technological, energy, material and information) are specified for different application areas.

Furthermore, the main system function in any computer-aided control system belongs to the structure of the control subsystem that communicates with all other types of structures of other types, and each structure is related to MS objectives.

3 Formalisation of Structural Dynamics Control Problems for Monitoring Natural and Technological Objects

The structural dynamics and control problem for monitoring natural and technological objects includes the following main subgroups of tasks [1–4]: (1) control tasks of structural dynamics; (2) investigation tasks for structural dynamics under the condition of zero inputs (neither controlling nor perturbation inputs are considered); (3) investigation tasks for structural dynamics and structural control over nonzero inputs.

The problem is formulated as follows. The following data are assumed to be known: alternative system structures; an initial structural state; inputs affecting system elements and subsystems; space and time technical and technological constraints; a list of system measures to evaluate the quality of the control process, e.g., goal abilities, structural and spatial characteristics [1], and information technology abilities. Multiple criteria are introduced to evaluate structural dynamic states of the MS.

To solve the problem, first, the existence of the solution is analysed. Then controllability and stability of the MS and sensitivity of optimal solutions are investigated. Finally, the analysis, classification and sorting of MS multi-structural states are performed.

Let us introduce the following basic sets of objects and structures:

$\tilde{B} = B \cup \bar{B}$ is a set of objects, where B is a set of objects (subsystems, elements) of the MS, and $\bar{B}$ is a set of external objects interacting with the MS through information, energy or material exchange. $\tilde{C} = C \cup \bar{C}$ is a set of channels (hardware facilities) that are used by objects; D, Φ and P is a set of operations, resources and flows, correspondingly.

$G = \{G_\chi, \chi \in NS\}$ is a set of MS structural types, where the main structures are topologic (spatial) structure, technology (functional) structure, technical structure, software structures and organisational structure. To interconnect these structures, the dynamic alternative multi-graph is introduced:

$$G_\chi^t = \left\langle X_\chi^t, F_\chi^t, Z_\chi^t \right\rangle, \tag{1}$$

where χ denotes a structure type, $\chi \in NS = \{1, 2, 3, 4, 5, 6\}$, where 1 indicates a topologic structure, 2—a functional structure, 3—a technical structure, 4 and 5 indicate math and software structures, and 6 indicates an organisational structure, time point t belongs to a given set T; $X_\chi^t = \{x_{\chi l}^t, l \in L_\chi\}$ is a set of elements of structure G_χ^t that presents multi-graph vertices at time t; $F_\chi^t = \{f_{\langle \chi, l, l' \rangle}^t, l, l' \in L_\chi\}$ is a set of arcs of the multi-graph G_χ^t; the arcs represent relations between the multi-graph elements at time t; $Z_\chi^t = \{f_{\langle \chi, l, l' \rangle}^t, l, l' \in L_\chi\}$ is a set of parameters that numerically characterise these relations.

The graphs of different types are interdependent; thus, for the structural control of each particular task the following mapping should be constructed:

$$M_{\langle \chi, \chi' \rangle}^t : F_\chi^t \to F_{\chi'}^t. \tag{2}$$

The compositions of the mappings can also be constructed at time t:

$$M_{\langle \chi, \chi' \rangle}^t = M_{\langle \chi, \chi_1 \rangle}^t \circ M_{\langle \chi, \chi_2 \rangle}^t \circ \cdots \circ M_{<\chi'', \chi' >}^t. \tag{3}$$

A multi-structural state is defined as follows:

$$S_\delta \subseteq X_1^t \times X_2^t \times X_3^t \times X_4^t \times X_5^t \times X_6^t, \quad \delta = 1, \ldots, K_\Delta \tag{4}$$

Thus, we obtain a set of MS multi-structural states:

$$S = \{S_\delta\} = \{S_1, \ldots, S_{K_\Delta}\}. \tag{5}$$

Feasible transitions from one multi-structural state to another one can be expressed as:

$$\Pi^t_{\langle \delta, \delta' \rangle} : S_\delta \to S_{\delta'}. \tag{6}$$

It is assumed that each multi-structural state at time $t \in T$ is defined by a composition (3). Hence, the problem is defined as the selection of a multi-structural state $S^*_\delta \in \{S_1, S_2, \ldots, S_{K_\Delta}\}$ and transition sequence (composition) $\Pi^{t_1}_{\langle \delta_1, \delta_2 \rangle} \circ \Pi^{t_2}_{\langle \delta_2, \delta_3 \rangle} \circ \Pi^{t_f}_{\langle \delta', \delta \rangle} (t_1 < t_2 < \cdots < t_f)$. The results of the selection can be presented as an optimal programme for MS transition from a given structural state to a specified one.

The interpretation of the object structural dynamics process is given in Fig. 2. Multi-graphs $G^{t_1}_\chi$ and $G^{t_1}_{\chi_1}$ describe dynamics of functional and technical structures, where $\Gamma^{t_1}_\chi$ and $\Gamma^{t_1}_{\chi_1}$ present object functional and technical structures at moment t_1.

Our investigations have shown that the development of the programme for optimal structural dynamics and control in the monitoring systems includes two stages. At the first stage, a set of feasible multi-structural macro-states is generated, so the structural and functional synthesis of a new system is performed in accordance with an actual or forecasted state of the control object. At the second stage, a single macro-state is selected, and an adaptive plan that specifies the transition of the monitoring system to the selected macro-state as well as provides system stable operations in the intermediate macro-states is designed. The construction of the transition programme is formulated as a multi-level multi-stage optimisation problem [1].

Fig. 2 Structural changes in the system

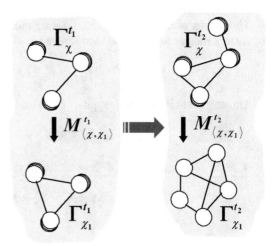

Well-known approaches to solving the problem are based on the PERT description of scheduling and control problems and the process dynamic interpretation [7, 10, 19]. The implementation of these approaches results in algorithmic and computational difficulties caused by high dimension, non-linearity, non-stationary and uncertainty of appropriate models.

4 Modern Optimal Control Theory Application in Monitoring and Control of Natural and Technological Objects

Let us introduce the following modification of the dynamic interpretation of monitoring and control processes. The main idea of the model simplification is to implement non-linear technological constraints in sets of feasible control inputs rather than in the right parts of differential equations. In this case, Lagrangian coefficients, keeping the information about technical and technological constraints, are defined via the local section method [1–4, 14, 15, 20]. Furthermore, interval constraints instead of relay ones can be used. Nevertheless, the control inputs take on Boolean values caused by the linearity of differential equations and convexity of a set of alternatives. The proposed substitution enables the use of fundamental scientific results of the modern control theory in various monitoring and control problems of natural and technological systems.

As provided by the concept of a multiple model description, the proposed general model includes the dynamic models of [1–5]: system motion control (M_g model); channel control (M_k model); operation control (M_o model); flow control (M_n model); resource control (M_p model); operational parameter control (M_e model); structural dynamic control (M_c model); and auxiliary operation control (Mv model).

Procedures of structural dynamics problem solution depend on the variants of transition and output function (operators) implementation. Various approaches, methods, algorithms and procedures of a coordinated choice through complexes of developed heterogeneous models have been developed by now.

The control problem of structural dynamics of monitoring and control systems has some specific features in comparison with classical optimal control problems. The first feature is that the right parts of the differential equations undergo discontinuity at the beginning of interaction zones. The considered problems can be regarded as control problems with intermediate conditions. The second feature is the multi-criteria nature of the problems. The third feature is concerned with the influence of uncertainty factors. The fourth feature is the form of time-spatial, technical, and technological non-linear conditions that are mainly considered in control constraints and boundary conditions. On the whole, the constructed model is a non-linear non-stationary finite-dimensional differential system with a re-configurable structure. Different variants of model aggregation have been studied.

These variants are associated with the tasks of the model quality selection and reduction of its complexity. Decision-makers can select an appropriate level of the model thoroughness in the interactive mode. The level of thoroughness depends on the input data, external conditions, and required level of solution validity.

Classification and analysis of perturbation factors having an influence upon operation of MS were performed. Variants of perturbation-factors descriptions were considered for MS models. In our opinion, a comprehensive simulation of uncertainty factors with all adequate models and forms of description should be used during investigation of MS. Moreover, the abilities of MS management should be estimated both in normal mode of operation and in emergency situations. It is important to estimate destruction "abilities" of perturbation impacts. In this case the investigation of MS functioning should include the following phases [1, 2, 4]:

(a) Determining scenarios for MS environment, particularly determining of extremely situations and impacts that can have catastrophic results.
(b) Analysis of MS operation in a normal mode on the basis of a priori probability information (if any), simulation, and processing of expert information through theory of subjective probabilities and theory of fuzzy sets.
(c) Repetition of item b for the main extremely situations and estimation of guaranteed results of MS operation in these situations.
(d) Computing of general (integral) efficiency measures of MS structure-dynamics control.

Algorithms of parametric and structural adaptation for MS models were proposed. The algorithms were based on the methods of fuzzy clusterization, on the methods of hierarchy analysis, and on the methods of a joint use of analytical and simulation models

The proposed interpretation of MS structural dynamics and its control processes provides advantages of applying the modern optimal control theory for the analysis and synthesis of monitoring and control systems. During the performed investigations, the main classes of structural dynamics problems have been defined. They are as follows: the analysis of structural dynamics of mobile objects and its diagnosis; observation, multi-layer control problems; synthesis of generalised states of monitoring and control systems; generation of optimal transition programmes providing the transition from a given system structural state to feasible (or optimal) one.

Methodology for structural dynamics control developed in [1–4, 13–16, 20] includes the methodology of a general system analysis and the modern optimal control theory for monitoring and control systems with re-configurable structures. As stated above, the solutions are based on principles of goal-programmed control, external complement, multi-scale variety, multiple-models and multi-criteria decision analysis [9–12]. The dynamic interpretation of the structural control allows applying known results from the theory of dynamic systems (in particular, stability and sensitivity analysis in optimal control) to the analysis of monitoring and control systems of natural and technological objects (see Table 1).

Table 1 Practical use of theoretical results in the analysis of monitoring systems

№	Qualitative analysis of space facility control processes	Practical implementation of the results
1	The existence of the solution in a control problem	Adequacy analysis of the system process description in control models
2	Controllability and attainability conditions for a control system	Reliability analysis of the planning time interval (major factors of object goal setting and IT abilities)
3	Conditions for uniqueness of optimal controls in scheduling problems	Analysis of the possibility to obtain optimal schedules for MS functioning
4	Necessary and sufficient optimality conditions for control problems	Preliminary analysis of an optimal control structure and scheduling algorithms
5	Conditions for solution reliability and sensitivity	Reliability and sensitivity analysis of control processes regarding perturbation effects and alterations in a system structure

5   Summary

Methodological and methodical basis of the theory of MS structure-dynamics control has been developed. This theory can be widely used in practice (including the integrated modeling and simulation for structure-dynamic control and monitoring of computer distributed networks). It has an interdisciplinary basis provided by the classic control theory, operations research, artificial intelligence, systems theory and systems analysis. The dynamic interpretation of MS reconfiguration process provides strict mathematical base for complex technical-organizational problems that were never formalized before and have high practical importance. The proposed approaches were implemented in software prototypes. The software prototypes were developed to simulate control processes for space-facilities control system (SF CS). The unified description of various control processes allows synthesizing both technical and functional structures of SF CS simultaneously. The presented multiple-model complex, as compared with known analogues, has several advantages. It simplifies decision-making in SF CS structure dynamics management, for it allows seeking for alternatives in finite dimensional spaces rather than in discrete ones. The complex permits to reduce dimensionality of SF CS structure-functional synthesis problems to be solved in a real-time operation mode. Moreover, it enables us to establish the dependence relation between control technology applied to spacecrafts and the results of their use according to the main goal. It is important that the presented approach extends new scientific and practical results obtained in the modern control theory to the field of space programs. This statement is exemplified by an analysis of SF CS information-technological abilities and goal abilities. All theoretical results and conclusions of this research are explored and validated via simulation on the basis of the prototype software developed by the authors.

The MS structure-dynamics control methodological and methodical basis was implemented in research project 2.1/ELRI -184/2011/14 "Integrated Intelligent Platform for Monitoring the Cross-Border Natural-Technological Systems" (INFROM) as part of "Estonia-Latvia-Russia Cross-Border Cooperation Programme within European Neighborhood and Partnership Instrument 2007-2013".

Acknowledgements The research is supported by the Russian Foundation for Basic Research (grants 15-07-08391, 15-08-08459, 16-57-00172-Бел_a, 16-07-00779, 16-08-01068, 16-07-01277), grant 074-U01 supported by Government of Russian Federation, Program "5-100-2020" supported by Government of RF, Department of nanotechnologies and information technologies of the RAS (project 2.11).

References

1. Okhtilev, M.Y., Sokolov, B.V., Yusupov, R.M.: Intellectual Technologies of Monitoring and Controlling the Dynamics of Complex Technical Objects, 409p. Nauka, Moskva (2006)
2. Ivanov, D., Sokolov, B., Kaeschel, J.: A multi-structural framework for adaptive supply chain planning and operations with structure dynamics considerations. Eur. J. Oper. Res. **200**(2), 409–420 (2010)
3. Ivanov, D.A., Sokolov, B.V.: Adaptive Supply Chain Management, 269p. Springer, London (2010)
4. Sokolov, B., Ivanov, D., Fridman, A.: Situational modelling for structural dynamics control of industry-business processes and supply chains. In: Sgurev, V., Hadjiski, M., Kacprzyk, J. (eds.) Intelligent Systems: From Theory to Practice, pp. 279–308. Springer, London (2010)
5. Norenkov, I.P.: The approaches to designing of automation systems. Inf. Technol. **2**, 2–9 (1998)
6. Ackoff, R.L.: The Art of Problem Solving. Wiley-Interscience, New York (1978)
7. Aframchuk, E.F., Vavilov, A.A., Emel'yanov S.V., et al.: In: Emel'yanov, S.V. (ed.) Technology of System Modelling. Mashinostroenie, Moscow (1998)
8. Casti, J.L.: Connectivity, Complexity and Catastrophe in Large-Scale Systems, 203p. Wiley-Interscience, New York (1979)
9. Gigch, J.: Applied General Systems Theory. Harper and Row, New York (1978)
10. Peregudov, F.I., Tarrasenko, F.P.: Introduction to Systems Analysis. Vysshaya Shkola, Moscow (1989)
11. Peschel, M.: Modellbilding für Signale und Systeme. Verlag Technik, Berlin (1978)
12. Shannon, R.E.: Systems Simulation, 387p. Prentice-Hall, Inc., Englewood Cliffs, New Jersey (1975)
13. Romanovs, A., Lektauers, A., Soshko, O., Zelentsov, V.: Models of the monitoring and control of natural and technological objects. Inf. Technol. Manage. Sci. **16**, 121–130 (2013)
14. Merkuryev, Y., Sokolov, B., Merkuryeva, G.: Integrated intelligent platform for monitoring the cross-border natural-technological systems. In: Proceedings of the 14th International Conference on Harbour Maritime and Multimodal Logistics M&S, HMS 2012. 19–21 Sept 2012, Vienna, Austria, pp. 7–10 (2012)
15. Merkuryeva, G.V., Merkuryev, Y.A., Lectauers, A., Sokolov, B.V., Potryasaev, S.A., Zelentsov, V.A.: Advanced river flood monitoring, modeling and forecasting. J. Comput. Sci. **10**, 77–85 (2015)
16. Sokolov, B.V., Okhtilev, M.Y., Zelentsov, V.A., Maslova, M.A.: The Intelligent monitoring technology based on integrated ground and aerospace data. In: Proceedings of the 14th

International Conference on Harbour Maritime and Multimodal Logistics M&S, HMS 2012. 19–21 Sept 2012, Vienna, Austria, pp. 112–117 (2012)

17. Arnott, D.: Decision support systems evolution: framework, case study and research agenda. Eur. J. Inf. Syst. **13**(4), 247–259 (2004)

18. Intelligent Systems in Cybernetics and Automation Theory. In: Proceedings of the 4th Computer Science On-line Conference 2015 (CSOC2015), Vol 2: Intelligent Systems in Cybernetics and Automation Theory. Advances in Intelligent Systems and Computing, vol. 348 (2015)

19. Skurikhin, V.I., Zabrodskii, V.A., Kopeichenko, Yu.V.: Adaptive Control Systems for Manufacturing. Mashinostroenie, Moscow (1989)

20. Petuhova, J., Lektauers, A., Zelentsov, V.: Classification of natural-technogenic objects in remote sensing applications. In: Proceedings of the 14th International Conference on Harbour Maritime and Multimodal Logistics M&S, HMS 2012. 19–21 Sept 2012, Vienna, Austria, pp. 91–95 (2012)

Development of Event-Driven Models for Operation Data of Some Systems of Small Satellites

Vyacheslav Arhipov, Vadim Skobtsov, Natalia Novoselova, Victor Aliushkevich and Alexander Pavlov

Abstract The paper presents several event-driven models for generation of test operation telemetry data of some systems in small satellites. The models, based on the Poisson flow of events and Gaussian distribution of state duration, are showed to give the unsatisfactory simulation quality. The proposed event-driven model with nonstationary probabilities of transitions between states are more suitable for generation of telemetry data and can be used for the development of the algorithms for revealing the predictors of hardware failures.

Keywords Survivability evaluation · Event-driven models · Telemetry · Satellite board equipment · Hardware failures

V. Arhipov · V. Skobtsov · N. Novoselova (✉) · V. Aliushkevich
United Institute of Informatics Problems, National Academy of
Sciences of Belarus, Minsk, Belarus
e-mail: novos65@mail.ru; runovos65@mail.ru

V. Arhipov
e-mail: arkhipau@gmail.com

V. Skobtsov
e-mail: vasko_vasko@mail.ru

V. Aliushkevich
e-mail: aliushkevich@newman.bas-net.by

A. Pavlov
St. Petersburg Institute of Informatics and Automation, Russian Academy of
Sciences (SPIIRAS), St. Petersburg, Russia
e-mail: pavlov62@list.ru

A. Pavlov
St. Petersburg National Research University of Information Technologies,
Mechanics and Optics (ITMO), St. Petersburg, Russia

© Springer International Publishing Switzerland 2016   403
R. Silhavy et al. (eds.), *Automation Control Theory Perspectives*
in Intelligent Systems, Advances in Intelligent Systems and Computing 466,
DOI 10.1007/978-3-319-33389-2_38

1 Introduction

One of the important tasks of safety evaluation of the complex technical systems such as small satellites is the evaluation of their reliability and survivability.

In order to test the complex of methodologies for the evaluation of reliability and survivability indices of the satellite board equipment (SBE) and individual modules of small satellites (SS) it is necessary to have the representative sample of telemetry data (TM). Especially the sufficient numbers of telemetry data prior to faults in the SBE systems of small satellites are of great demand for the statistical analysis.

In practice the hardware failures in SS SBE systems are insufficiently frequent event in order to develop the system of failure predictor detection. This paper proposes the approach to the generation of test operation telemetry data, which are the result of equipment failure simulation, including the predictable ones.

In our work we consider the models of predictable failures of two types:

1. Chains of failure events, which can be detected using TM data. The failure events are modeled as a non-stationary flow with increasing intensity. In some local neighborhood (window) it can be considered as a Poisson flow [1–3]. The continuous series of events with predetermined length $n \geq 3$ is considered as the final breakdown of the system.

   For example, the switching on failure is possible during the activation of some SS SBE system. Such an event is yet not the breakdown of the system and the switching on instructions are repeated. The intensity of the failure flow is constantly increased with time. The fatal system breakdown is the n-fold repetition of switching on failures, which continuously follow each other.
2. Continuous change of some characteristics of TM signal. Fatal breakdown of the system is modeled as overrunning some of them or going out of the interval of acceptable values.

   For example, some node of SS SBE is in the state X, its signal S must have the value y. When the signal value differs from y more than d the node is considered to be in the breakdown condition. In fact in the state X the signal S has the normal distribution with mean value y and standard deviation much less than d [1]. In the course of time the expectation value of signal S, which corresponds to the node in state X gradually linearly varies and the standard deviation grows exponentially. Therefore the long-term stay in the state X increase the probability of fatal breakdown of the system.

In the first case we have the functional failures, in the second case the parametrical failures [2, 3]. Having the model of failures and their events-predictors the methodology for the evaluation of SS SBE reliability is tested as follows:

1. Modeled TM data is analyzed and the above-mentioned events of the first and second types are revealed.
2. The intensity of the failure events of the first type and the rate of changes of the second type events are dynamically evaluated.

3. The prognosis of the reliability and survivability of the modeled system is constructed.
4. The simulation is continuously proceeded until the destruction of the modeled system.
5. The experiment is repeated many times.
6. The predicted and actual reliability and survivability values are compared.
7. According to the prognosis the conclusion about the operational capability of the tested methodology with regard to the discussed failure models is made.

2 Telemetry Data Generation

Using the real telemetry data of the satellite board equipment, working in the normal mode, we reveal the possible states of some parameter and the law of transition (probability scheme) between these states. Also the statistics (expectation value and dispersion) of the time intervals, corresponding to each state is collected. All these data are the basis for the development of the event-driven model for the TM data generation.

The example of the TM time series, indicating the cyclically repeated six states of the system $x = 1 : 6$ is presented in Fig. 1. For each state the mean values v_x of the some signal, mean values and standard deviations $X : t_x, d_x$ of its duration are known.

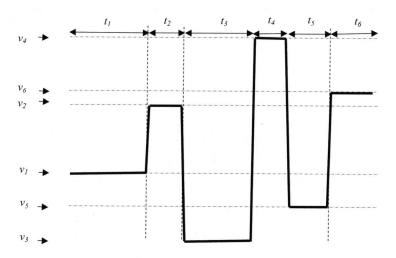

Fig. 1 The signal example for modeling

The following failures are simulated individually or in combination:

1. False transition between the states, which violates the general regularity.
2. The violation of the time of stay in the particular state: preterm transition.
3. The violation of the time of stay in the particular state: delayed transition—"sealing" in the state.
4. The violation of the acceptable range of signal values.

2.1 The Model of Poisson Flow of State Transitions

The simulation process of the state transition flow can be described as follows: the system is in the state X, it can either move to the next state with some probability p_x or can stay in the current state with probability $1 - p_x$. Even not considering the false transitions, which violate the state sequence, the number of difficulties arises.

The mean time t_x of stay in the state X determines the transition probability p_x: $p_x = t_x^{-1}$. At the same time the transitions between the states present the Poisson flow of events with intensity identical to t_x. But in this case the dispersion of the duration in the state X is also identical to t_x.

Using such an approach it is impossible to generate the data with the arbitrary distribution of time durations in the particular state. Figure 2 illustrates the generated data example following the Poisson flow of state transitions.

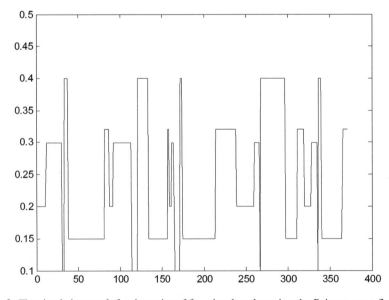

Fig. 2 The simulation result for the series of five signal cycles using the Poisson event flow

2.2 The Model of Gaussian Distribution

The second approach, proposed in the paper, is lacking the high dispersion of time durations in the particular state. For each state X the mean duration t_x and the dispersion d_x are known. Than the duration of each stay of the system in the particular state X is modeled as the random variable, distributed according to Gauss law with predefined expectation value t_x and standard deviation $\sigma_x = \sqrt{d_x}$.

Strictly speaking this distribution is not normal as the time intervals are the strictly positive values. Such a situation can be corrected by the normalization procedure. The probability of the system stay in the particular state within the predefined time interval is normalized using the quantity according to the Gaussian distribution $(1-F(0))$, where $F(x)$—Gaussian probability function with corresponding expectation means and dispersions. If the expectation means are much more than 0 and the standard deviation is relatively small the normalization can be ignored.

Figure 3 illustrates the example of the generated data following the Gaussian model.

2.3 HMM Model for Estimating the Modeled Parameters

In our research we propose to use the Hidden Markov model (HMM) for estimation the state parameters of the multivariate telemetry data together with the law of transition (probability scheme) between these states.

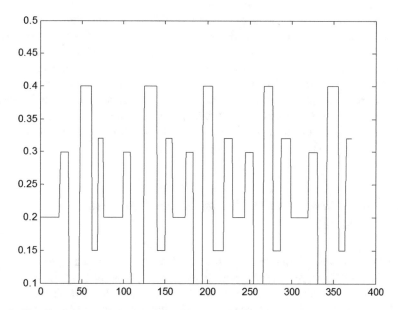

Fig. 3 The simulation result for the series of five signal cycles using the deterministic model with the Gaussian distribution of the state durations

The results of estimation help to collect the statistics of time duration of each state for the development of the event-driven model.

HMM presents the statistical model, which emulates the Markov process with unknown parameters. In contrast to the Markov model, where the object states are visible, the HMM deals only with the series of observations of unknown states. The task is to estimate the parameters of probability distributions, which are called the emission parameters and characterize the states and the probabilities of state transitions. According to [4] at any time point, the observations are distributed as a mixture with n components, and the time dependencies between the observations are due to time-dependencies between the mixture components.

The unknown parameters λ of the HMM model are evaluated using the likelihood function $P(O|\lambda)$ and present its maximum. For estimation the parameters the training set of real observation is used. The likelihood function can be presented as

$$P(O|\lambda) = \sum_{p \in S^T} P[O, p|\lambda] = \sum_{p \in S^T} P[O|p, \lambda] \cdot P[p|\lambda] \qquad (1)$$

where $p = p_0, \ldots, p_{T-1}$ is the sequence of T states. Parameters are estimated using the expectation-maximization algorithm or through the use of a general Newton-Raphson optimizer.

In our research the HMM model considers the multivariate time series of the form $O_{1:T} = (O_1^1, \ldots, O_1^m; O_2^1, \ldots, O_2^m; \ldots; O_T^1, \ldots, O_T^m)$, where m is the number of telemetry variables. The values of variables can be modeled not only using the multinomial probability distribution, but also using the exponential family of distributions, e.g. normal, Poisson, Gamma etc.

Figure 4 illustrates the example of the HMM model with four states for the analyzed multivariate data of the electricity supply system of the small satellite.

Unfortunately the direct application of the HMM model with parameters, estimated from real process cannot be used for the artificial generation of telemetry data with failures. The reason for this is the HMM inability to model the time interval distribution of the particular states. Additionally the application of the first-order HMM is unable to take into account the complex dependences of state transitions. To overcome some of the HMM deficiencies it is possible to take into account the nonstationary transition probabilities between the states.

2.4 The Model with the Nonstationary Transition Probabilities

The proposed event-driven model of the TM data, which is based on the transition probabilities between the system states is more appropriate for the joint application with the other statistical approaches such as for example Markov chains, Hidden Markov models and finite automata [5–7]. In this case the transition to the new

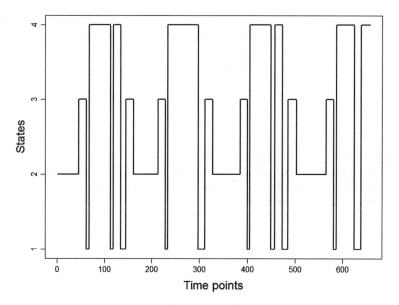

Fig. 4 The optimal sequence of states of the electricity supply system estimated by HMM modeling

state at each time point is regulated by the probabilistic mechanism. The variants with stationary probabilities and classical Poisson flow of events give the unsatisfactory results in simulation quality. The proposed approach, described in Sect. 2.1, is based on a priory determination of state duration of the system and is applicable only for modeling the simple sequence of events as it regulates the distribution of state durations for the whole ensemble of states. At the level of separate time points of system operation such an approach has the deterministic and not the dynamic nature.

In the paper we propose the model of system behavior, which generates the TM data and for which the state transition at each time point is determined through the flow of events (with memory, unlike the Poisson flow). Meanwhile for the ensemble of states the correspondence to the predefined statistical distribution of state durations is maintained.

Without loss of generality suppose that for some state X the duration of the system stay in this state as in the previous examples is normally distributed with expectation mean t_x and standard deviation σ_x. Than the probability $P_x(T)$ to meet the time interval T of the continuous stay of the system in state X is the following:

$$P_x(T) = \frac{e^{-\frac{(T-t_x)^2}{2\sigma_x^2}}}{\sigma_x\sqrt{2\pi}}. \tag{2}$$

The exit of the system from the state X at the time point T is possible if the system didn't move to the next state up to this time point. If the transition probability $p_x(t)$ from the state X is known for each time point t, than the probability $\hat{p}_x(t)$ to stay in the state X up to the time point T is as follows:

$$\hat{p}_x(T) = \prod_{t=1}^{T-1} (1 - p_x(t)). \tag{3}$$

The probability of the system exit from the state X at the time point T forms the distribution of the system stay durations in the state X and is determined jointly by the probability of system stay in the state X up to the time point T and the transition to the new state at this time point:

$$P_x(T) = \hat{p}_x(T)p_x(T) \tag{4}$$

from which $p_x(T)$ can be expressed:

$$p_x(T) = \frac{P_x(T)}{\hat{p}_x(T)} \tag{5}$$

where $P_x(T)$ and $\hat{p}_x(T)$ are obtained from the expressions 2 and 3 correspondingly.

It must be stressed, that $P_x(T)$ can have any distribution, the algorithm to calculate the probability of transition to the new state will be the same. But there is one important remark: if for the objective distribution $\exists T_0 : T \geq T_0 \rightarrow P_x(T) \equiv 0$ (e.g. when the distribution is uniform in the interval $(a, b) : T_0 = b$), than the exit from the state X must be not later than the time point T_0, therefore $p_x(T_0) = 1$.

Figure 5 illustrates the example of the generated signal using the nonstationary transitions.

3 Generation of TM Data with Failures

To simulate the data, which will be used in training algorithm for detection the failure predictors and evaluation the reliability and survivability of SS SBE the probabilities of single failures and the deviations of the signals from the norm are considered in advance corresponding to normal mode of equipment operation and increase as the algorithm proceeds, simulating the system "wear". As a result the increasing fault number in unit time can serve as a predictor of the fatal breakdown (if the model of fatal breakdown is specified as the n-fold failure).

Figure 6 illustrates the example of the generated TM signal with possible preterm transitions and "sealings" using the nonstationary transitions. The signal values are normally distributed.

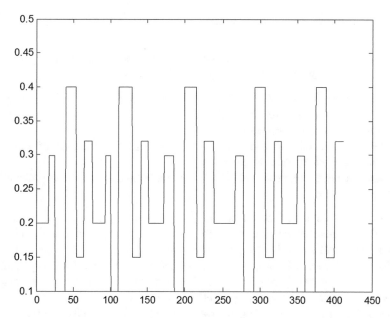

Fig. 5 The simulation result for the series of five signal cycles using the event flow with nonstationary transition probabilities

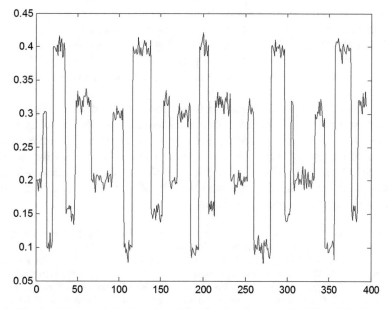

Fig. 6 The simulation result for the series of five signal cycles using the TM model with the time intervals violation of system stay in the particular states and deviation of the signal values from the mean

4 Conclusion and Discussion

In the paper we consider four approaches to the simulation and artificial generation of telemetry data of SS SBE operation:

- based on the Poisson flow model of state transitions;
- based on modeling time intervals of the system stay in each state with predefined Gaussian distribution;
- based on the Hidden Markov model together with the law of transition (probability scheme) between states;
- based on the event flow model with the nonstationary transition probabilities between states.

Using the first approach, based on the Poisson flow model, does not give us a possibility to generate the data with the arbitrary distribution of time durations in the particular state. And in context of given problem it is the worst method.

The second approach, proposed in the paper, is lacking the high dispersion of time durations in the particular state. Than the duration of the system stay in the particular state is modeled as the random variable, distributed according to some predefined distribution, like Gauss law in our case. This model allows to more adequately generate the telemetry data sequences of SS SBE compared to the first approach.

The Hidden Markov model is more accurate mathematical tool to simulate the multivariate telemetry data together with the probability scheme of transition between these states. But unfortunately the direct application of the HMM model is not applicable for the generation of telemetry data with failures. The reason of that is the HMM inability to accurately emulate the time interval distribution of the particular states in case of the data generation. To overcome the HMM deficiencies it is possible to take into account the nonstationary transition probabilities between states.

And the final approach for the generation of the TM data, described in the paper, is based on the event-driven model with nonstationary probabilities of transitions between states and allows modeling the data with specified distribution of time intervals of the system stay in different states. It is able to be used for modeling failure conditions of the SS SBE. And this method gives the most adequate way for the simulation and artificial generation of the SS SBE telemetry data with failures among all suggested in paper approaches.

Acknowledgments The research described is partially supported by the Russian Foundation for Basic Research (grants 15-07-08391, 15-08-08459, 16-57-00172-Бел_а, 16-07-00779, 16-08-01068, 16-07-01277), grant 074-U01 supported by Government of Russian Federation, Program "5-100-2020" supported by Government of RF, Department of nanotechnologies and information technologies of the RAS (project 2.11). This investigation was executed in framework of a project of the scientific program "Monitoring-SG" of the Union State of Russia and Belarus also.

References

1. Wentzel, E.S.: Probability Theory. Science, Moscow (1969). (in Russian)
2. Kaliavin, V.P.: Reliability and Diagnostics. Elmor, St. Petersburg (1998). (in Russian)
3. Druzhinin, G.V.: Reliability of Automated Manufacturing Systems. Energoatomizdat, Moscow (1986). (in Russian)
4. Visser, I., Speekenbrink, M.: DepmixS4: an R package for hidden Markov models. J. Stat. Softw. **36**(7), 1–21 (2010)
5. Rabiner, L.R.: A tutorial on hidden Markov models and selected applications in speech recognition. Proc. IEEE. **77**(2), 257–286 (1989)
6. Bishop, C.: Pattern Recognition and Machine Learning. Springer, Heidelberg (2006)
7. Gill, A.: Introduction to the Theory of Finite-States Machines. McGraw-Hill Book Company, New-York (1962)

The Method of Lossless 3D Point Cloud Compression Based on Space Filling Curve Implementation

Victor V. Alexandrov, Sergey V. Kuleshov, Alexey J. Aksenov
and Alexandra A. Zaytseva

Abstract This paper proposes a method of reordering point cloud into a bit stream which has ordered structure, and features the ability to save local space features, which allows to use the methods of run-length encoding to reduce the bit volume of the scanned 3D data. The proposed method can be implemented in software designed to work with 3D scanned data, raw 3D data transmission for further post-processing of complex spatial forms.

Keywords 3D-modeling · 3D scanning · Point cloud · 3D model · 3D compression · Space-filling curve

1 Introduction

Since the works of R. Descartes and G. Monge, a paradigm of Cartesian coordinate system in space and ways of describing objects by means of descriptive geometry have been stated.

Later, at the beginning of the dominance of digital technology over analogue one the "pixel representation" of different objects has become the industry standard, unlike analog "vector representation". Nowadays, 3D technology in various fields of human activity has become common.

V.V. Alexandrov · S.V. Kuleshov · A.J. Aksenov · A.A. Zaytseva (✉)
St.-Petersburg Institution for Informatics and Automation of RAS,
St.-Petersburg, Russia
e-mail: cher@iias.spb.su

V.V. Alexandrov
e-mail: alexandr@iias.spb.su

S.V. Kuleshov
e-mail: kuleshov@iias.spb.su

A.J. Aksenov
e-mail: a_aksenov@iias.spb.su

© Springer International Publishing Switzerland 2016
R. Silhavy et al. (eds.), *Automation Control Theory Perspectives
in Intelligent Systems*, Advances in Intelligent Systems and Computing 466,
DOI 10.1007/978-3-319-33389-2_39

The development of digital programmed technology has led to the possibility of direct digitization of three-dimensional objects, computer 3D-modelling and object replication.

This paper represents the continuation of a series of studies on the 3D technology [1–3] as well as the development of methods of spatial data compression [4]. As described in [1, 3], the development of digital programmed technology has led to the possibility of fast and high-quality scanning, i.e. immediate digitizing of complex artistic shapes and other spatial three-dimensional objects.

3D data can be obtained in various ways: modeling, scanning, application of tomography, 3D data recovery from stereo image pairs. This paper considers the data obtained via 3D scanning process.

The rest of the paper is organized as follows: Sect. 2 presents an overview of the 3D-scanning process, Sect. 3 presents the data used to estimate efficiency of the proposed method in Sects. 4 and 5 presents the results and comparison of the different methods exposed in this paper, and finally Sect. 6 summarizes the possibilities that can be achieved with proposed method.

2   3D Scanning

3D scanning (three-dimensional scanning) is a process of translating of the physical form of a real object to digital form obtaining a computer model of the object.

Three-dimensional scanning is widely used for solving the problems of reverse engineering; developing equipment, producing spare parts in the absence of original documentation; digitizing surfaces of complex shape, including artistic forms and molds. 3D scanning also plays an important role in health problem solving as in prosthesis design and fabrication [5].

Non-contact 3D scanners mainly use dual system of object coordinates obtaining. Many devices combine laser sensors and a digital photo camera that allows to model objects with texture mapping.

In some cases, instead of laser sensor more complex systems can be used. For example, there are 3D scanning systems based on ultrasonic devices which have the advantage over its competitors such as the ability to scan objects with the internal structure immersed in a homogeneous medium.

It is known that the development of the magnetic scanners that are used to determine the spatial coordinates of the object by measuring changes in its spatial magnetic field is being carried out. The variety of such complex devices means that without good software and human interaction, data received by 3D scanners, will still remain as an unusable set of digits [6]. Currently, there are numerous methods of representing three-dimensional objects and related imaging techniques.

Depending on the method of obtaining 3D data its volume will be different, as shown in Table 1.

Table 1 shows the comparative amounts of the initial (uncompressed) representations of various 3D objects. It can be seen that the efficient storage of 3D

Table 1 Approximate bit volume comparison of various 3D objects representations

Type of object	Method of data obtaining	Size (bytes)
Bass-relief	3D scanning	108 265 982
Helicopter (model)	CAD	5 505 893
Stereo pair in HD-quality (film)	Camera	12 441 600

objects requires representation formats of these objects in a compressed format, targeted to a specific presentation class.

3 Initial Data for Experiments

For experiment performing the point cloud obtained by 3D scanner Artec Spider was used (see Figs. 1 and 2). The point cloud means the set of points in 3D coordinate system, which are used for representation of the object surface.

Parameters of scanning: the process of scanning was performed in real time using the Artec Studio software, version 9.2.3.15. Point cloud was obtained during continuous scanning mode.

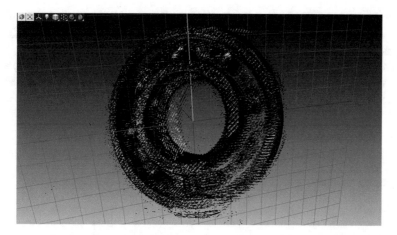

Fig. 1 Initial dot cloud for a 3D object

Fig. 2 Initial dot cloud for a
3D object

4   Method Description

While functioning, 3D scanner obtains sets of images from onboard cameras, which
then are transformed into a set of points with calculated distances from them to
scanner with the mutual coordinate system using special software.

Traditional representations of scanned data are the following formats: ply, STL,
xyz, WRML, 3D dots cloud (asc), aop, obj, ptx, xyzrgb, e57. They contain the list
of the points coordinates in text or binary form. For the efficient representation of
the coordinate list the algorithm represented in Fig. 3 is proposed. We consider the

Fig. 3 General scheme of
compressing algorithm

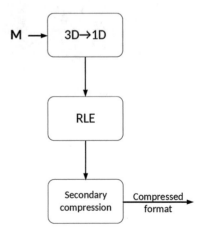

fact that the point cloud is a sparse 3D bit array with finite resolution, provided by scanner, and thus it can be effectively compressed.

Moreover the points forming scanned object surfaces are occupying compact areas of scanned space. Thus the main stage of compression (Fig. 3) is to convert the three-dimensional array in a linear binary sequence by traversal algorithm that preserves local space features such as a space-filling curve (SFC) generation (in the terminology of [7] this step is also referred as the normalization oriented on data semantics).

The method of constructing SFCs to be used in the encoding procedure is described in [7]. The SFC visualization for the case of 8 × 8 × 8 points is shown in Fig. 4.

Run-Length Encoding. After the transformation of the 3D coordinates array into 1D sequence, where 1 indicates that the current point of space belongs to the scanned object's surface and 0 if it does not, the long bit sequence can be efficiently compressed by using the run-length encoding algorithm (RLE).

RLE algorithm performs run-length encoding of long sequences of zeros which are emerging due to the sparse array of point coordinates. The RLE algorithm creates the pairs of the following form <skip, number>, where "skip" is counter of skipped zeros, and "number"—the value you want to put in the next cell. Thus, vector 42 3 0 0 0 −2 0 0 0 1... will be converted into pairs (0,42) (0,3) (3, −2), (4,1)....

Secondary compression. As a final stage (the secondary compression) one of the entropy compression methods, for example, Huffman compression [8] or arithmetic compression [9] can be used. These compression algorithms can reduce the average length of a code word for the characters of the alphabet.

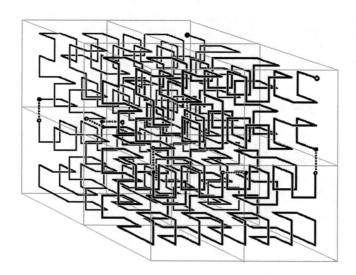

Fig. 4 Space-filling curve for the three-dimensional case with the dimensions of 8 × 8 × 8

5 Results

Table 2 shows the results of applying the proposed method for various complex spatial forms obtained via 3D scanning. Scanned space was divided into independently compressed three-dimensional $256 \times 256 \times 256$ cells. The compression results then were combined into single output stream. The Table 2 shows the average results of compression of a $256 \times 256 \times 256$ cell for different scanned objects. The comparison shows the uncompressed bitmap point cloud volume defined as 3D coordinates (RAW-file), volume of RAW-file compressed by Deflate algorithm, as well as the results of applying of the proposed $3D \rightarrow 1D$ compression algorithm using a space filling curve.

For empirical verification of the proposed method a cross-platform software application written in Java was used which implements a coder/decoder for the compressed 3D data stream (see Fig. 5). In the encoder mode the initial format (cloud of points or a set of polygons) is converted into an internal representation as an array of point coordinates. Then the SFC software module generates route for traversing of the scanned object volume for the subsequent run-length encoding and secondary compression. If necessary, the output stream can be stored in a standard container (XML, RIFF, etc.) or be the basis of the network protocol. Subsequently, the encoder/decoder may be configured as a module (plug-in) in order to work directly with given 3D data format. The decoder works in the opposite direction and using the same SFC generator.

Figure 6 shows the compression level (in bit per point) in dependence on the number of points in the given object for the proposed method based on SFC in comparison to known lossless compression methods based on octree and spanning tree [10, 11]. It is shown that for the proposed method the level of compression increases with the growth of number of points in the point cloud resulting in increased efficiency in the case of 3D scanner accuracy improvement.

Table 2 Results of the used method

Object	Horse (bass-relief)	Bearing
Number of points in the cloud	21312	11198
Size of 3D cube elements	256	256
Size of the file containing cloud before compression (byte)	127872	67188
Size after compression using deflate method (byte)	44152	21991
Size after compression using the proposed methods (byte)	10891	5519

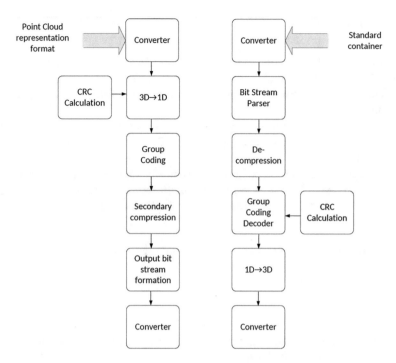

Fig. 5 Block diagram of a software implementation of the proposed 3D data compression algorithm

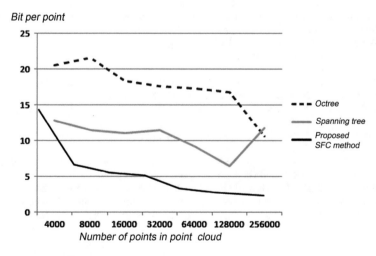

Fig. 6 Proposed method based on SFC in comparison to known lossless compression methods based on octree and spanning tree

6 Conclusion

This paper proposes a method of reordering point cloud into a bit stream which has ordered structure, and features the ability to save local space features, which allows to use the methods of run-length encoding to reduce the bit volume of the scanned 3D data.

The proposed algorithm is using space-filling curve to reorder point clouds into bit stream and does not require the preliminary 3D data simplification or scanning artifact reduction.

The proposed method can be implemented in software designed to work with 3D scanned data, raw 3D data transmission for further post-processing of complex spatial forms and for more efficient usage of 3D data storage systems and the libraries of 3D objects.

Acknowledgments This work is supported by state research № 0073-2014-0005.

References

1. Alexandrov, V.V., Sarychev, V.A.: Digital programming technologies. J. Inf.-Meas. Control Syst. **11**(8), 3–9 (2010) (in Russian)
2. Alexandrova, V.V., Zaytseva, A.A.: 3D technology and cognitive programming. J. Inf.-Meas. Control Syst. **5**(10), 61–64 (2012) (in Russian)
3. Alexandrova, V.V., Zaytseva, A.A., Tyzhnenko, D.A.: The scanning and editing of the 3D object for rapid prototyping. J. Inf.-Meas. Control Syst. **9**(11), 53–57 (2013) (in Russian)
4. Kuleshov, S.V.: The real 3D scene representation format for three dimensional television (True3D Vision). J. Inf.-Meas. Control Syst. **7**(4), 49–52 (2009) (in Russian)
5. 3D scanning, http://reprap.org/wiki/3D_scanning
6. 3D scanning in the Interests of 3D modelling, http://www.comprice.ru/articles/detail.php?ID= 40134 (in Russian)
7. Kuleshov, S.V.: Criterion for evaluating the energy efficiency of video data compression process. J. Inf.-Meas. Control Syst. **8**(11), 16–18 (2010) (in Russian)
8. Huffman, D.A.: A method for the construction of minimum redundancy codes. Proc. IRE **40**, 1098–1101 (1962)
9. Witten, I., Neal, R.M., Cleary, G.: Arithmetic coding for data compression. J. Comm. ACM **30**(6), 520–540 (1987)
10. Schnabel, R., Klein, R.: Octree-based point-cloud compression. In: Proceedings of Symposium on Point-Based Graphics 2006, pp. 111–120 (2006)
11. Merry, B., Marais, P., Gain, J.: Compression of dense and regular point clouds. In: Proceedings of the 4th International Conference on Virtual Reality, Computer Graphics, Visualization and Interaction in Africa, pp. 15–20. ACM Press, New York, NY, USA (2006)

Control of the Air Transportation System with Flight Safety as the Criterion

**Alexander Rezchikov, Vadim Kushnikov, Vladimir Ivaschenko,
Aleksey Bogomolov, Leonid Filimonyuk and Konstantin Kachur**

Abstract In this article an approach is suggested to ensuring safe functioning of air transportation systems. This approach allows calculation of the probability of complicated situations appearing out of combining of events, which would not be dangerous if taken separately. Mathematical support of an information/management system is worked out, which allows suggestion of actions at different time periods to minimize the probability of appearance of a critical combination of events in the process of functioning of an air transportation systems.

Keywords Air transportation system · Probabilistic analysis of safety · Emergency situation · Combination of events · Differential equations

1 Introduction

The development of domestic aircraft industry, as well as entering the international market of air traffic and achieving reliability in the country's defence is impossible without increasing the level of flight safety and reducing the number and the severity of accidents and catastrophes happening to aircraft (AC). One of the promising ways of solving this problem is based on utilizing modern means of processing information that improves mathematical support of air transportation systems (ATS), which is used to prevent critical combinations of events affecting flight safety.

Despite the considerable success in the building of aircraft control systems, the level of automation, computerization and equipment in general of the ATS's in

A. Rezchikov (✉) · V. Kushnikov · V. Ivaschenko · A. Bogomolov · L. Filimonyuk
Institute of Precision Mechanics and Control of Russian Academy of Sciences,
Saratov, Russia
e-mail: rezchikov1939@mail.ru

K. Kachur
Yuri Gagarin State Technical University of Saratov, Saratov, Russia
e-mail: konstantin.kachur@gmail.com

© Springer International Publishing Switzerland 2016
R. Silhavy et al. (eds.), *Automation Control Theory Perspectives
in Intelligent Systems*, Advances in Intelligent Systems and Computing 466,
DOI 10.1007/978-3-319-33389-2_40

general remains inadequate for emergency prevention when the emergency arises because of a combination of unfavourable factors, each of which taken separately wouldn't affect safety in a critical way. Nowadays, there is practically no information about automated systems, which would allow control of an ATS with flight safety as the main criterion, as well as calculation of probability of emergencies related to adverse combinations of failures and errors for various periods of time.

The above mentioned considerations stipulate the urgency and practical importance of this research by means of developing new problem statements, models and algorithms of their control with flight safety taken as the main criterion [1–7].

2 Air Transportation System as Control Object

Let's look at an ATS as an object of control with flight safety as the main criterion. The elements comprising this system are: the aircraft; the crew; the system that controls motion in the air (AMCS); airport personnel that prepare and control the flight.

Complex interactions between the ATS's subsystems and their processes can cause critical combinations of factors and events, which, as a rule, lead to accidents.

Flight safety is defined as an ATS's property that describes the level of danger to people and material objects onboard the aircraft in the process of its operation. Flight safety failure can occur even if there are no failures in the aircraft or other parts of the ATS. The reason for that can be errors in equipment design (not all conditions and requirements are met), errors in regulatory and operational documentation, or occurrence of unpredictable or not accountable for operating conditions (force majeure). To prevent the emergencies, this article suggests analysis, identification and elimination of possible dangerous combinations of events.

3 Problem Statement

Situations are described as accidents if they cause destruction of the aircraft without deaths of human beings; if there are victims among human beings, the situation is described as catastrophic [1].

Let us assume that we have a list of accidents and catastrophic situations $\{A_1, A_2, \ldots, A_n\}$, each of which is a consequence of a combination of events possible for the given type of an ATS, and for every $A_i \in \{A_1, A_2, \ldots, A_n\}, i = \overline{1, n}$ one or several event trees are built: [2–4] $\{D_1, D_2, \ldots, D_m\}, m \geq n$, which describe the process of appearance and development of critical combinations of these events for every $A_i \in \{A_1, A_2, \ldots, A_n\}, i = \overline{1, n}$.

To solve the problem, not only models of Markov's processes can be used, but other processes, as well. In this case, calculation of probability is carried out using neural network apparatus, fuzzy logic, etc.…

Let us also accept that we know the current state of the ATS, including the intensity of the streams of appearance and restoration of failures of the ATS elements $\lambda_i(t), \mu_j(t), i = \overline{1,n}, j = \overline{1,m}$, which lead to or prevent, respectively, accidents or catastrophic situations $\{A_1, A_2, \ldots, A_n\}$.

In view of the above, the problem statement is as follows:

1. $\forall t \in [t_H, t_K]$ calculate probabilities $P_i(\lambda_1(t), \lambda_2(t), \ldots, \lambda_k(t), \mu_1(t), \mu_2(t), \ldots, \mu_k(t), \overrightarrow{x}(t), t), i = \overline{1,n}$ describing the possibility of accidents or catastrophes, critical combinations of events included in the list $\{A_1, A_2, \ldots, A_n\}$;

2. $\forall A_i \in \{A_1, A_2, \ldots, A_n\}, i = \overline{1,n}$ determine the vector of controlling factors $\overrightarrow{\mu^*}(t) \in \{\overrightarrow{M}(t)\}$, which allow for the time interval $\Delta T = t_H - t_K$ and all acceptable environmental conditions achievement of the maximum for the safety criterion of ATS operation for accidents and catastrophes included in the following list: $\{A_1, A_2, \ldots, A_n\}$. This criterion looks as follows:

$$\int_{t_K}^{t_H} \left(\sum_{i=1}^{n} (\eta_i(1 - P_i(\lambda_1(t), \lambda_2(t), \ldots, \lambda_k(t), \mu_1(t), \mu_2(t), \ldots, \mu_k(t), \overrightarrow{x}(t), t)) \right) dt. \rightarrow \max$$

(1)

The limitations are:

$$P_i = P_i(\lambda_1(t), \lambda_2(t), \ldots, \lambda_k(t), \mu_1(t), \mu_2(t), \ldots, \mu_k(t), \overrightarrow{x}(t), t) > A, i = \overline{1,n}, \quad (2)$$

$$P_i = P_i(\lambda_1(t), \lambda_2(t), \ldots, \lambda_k(t), \mu_1(t), \mu_2(t), \ldots, \mu_k(t), \overrightarrow{x}(t), t) < B, i = \overline{1,n}. \quad (3)$$

And boundary conditions are:

$$\begin{aligned} F_i^{(t_H)}(\overrightarrow{x}(t), \overrightarrow{\mu}(t), \overrightarrow{\lambda}(t)) = 0, \quad i = \overline{n_1 + 1, n_2}, \\ F_i^{(t_K)}(\overrightarrow{x}(t), \overrightarrow{\mu}(t), \overrightarrow{\lambda}(t)) = 0, \quad i = \overline{n_3 + 1, n_4}, \end{aligned}$$

(4)

$\{\overrightarrow{X}(t)\}, \{\overrightarrow{M}(t)\}, \{\overrightarrow{\Lambda}(t)\}$ are the sets of the acceptable changes of the vectors $\overrightarrow{x}(t), \overrightarrow{\mu}(t)$ and $\overrightarrow{\lambda}(t)$, respectively; t is the time; $\eta_i, i = \overline{1,n}$ are weight coefficients ordering the following list: $\{A_1, A_2, \ldots, A_n\}$ by the severity of the accidents and; $k, m, n_i, i = \overline{1,4}, A$ and B are known constants).

In other words, to solve the problem, it is necessary for every $A_i \in \{A_1, A_2, \ldots, A_n\}, i = \overline{1,n}$ to choose such intensity of failure restoration $\mu_i^*(t).i = \overline{1,k}$. The pilot can increase the intensity of the ATS's elements, which will allow achievement of the maximum value for criterion (1) on the preset time interval $[t_H t_k]$ with limitations (2), (3) and initial conditions (4).

The method of solving the problem being developed here is founded on the idea of distributing the principles of analysis of the safety of aircraft, most well worked in the models of reliability and safety [1], to the sphere of safe operation of the whole air transportation system.

According to this approach, the article suggests calculating the probability of critical combinations of events causing emergency situations as the probability of the realization of corresponding sections of the event tree.

Therefore, in order to solve the stated problem, it's necessary:

- to work out mathematical models and algorithms of formal tree synthesis for ATS's events $\{D_1, D_2, \ldots, D_m\}$, including those based on reduction and completion methods [1];
- to build an event tree D, which combines trees $\{D_1, D_2, \ldots, D_m\}$ and allows calculation of minimal sections corresponding to critical combinations of events in the ATS;
- to work out an algorithm allowing numerical calculation of probabilities $P_i(\lambda_1(t), \lambda_2(t), \ldots, \lambda_k(t), \mu_1(t), \mu_2(t), \ldots, \mu_k(t), \overrightarrow{x}(t), t)$, $i = \overline{1, n}$ for different critical combinations, which are also called minimal sections of the event tree D.

To solve the problem, the following approach is used. An event tree D is built, the minimal sections of that tree are determined, which correspond to possible critical combinations of events. The minimal sections are categorized by the numbers of elements included in them, as two-element ones, three-element ones, etc.... For each class of sections a graph of states is built, for which a system of Kolmogorov-Chapman's differential equations for Markov's processes is created. From their solution, the probability is calculated of an adverse combination of circumstances formalized by the corresponding minimal section. In the process of ATS's operation, constant changes of the event tree take place and, consequently, minimal sections are changed too. An automated system for each adverse combination of circumstances determines its probability and suggests a list of actions, which will eliminate the reasons for such situations to appear. For the solution of the problem the list of actions is chosen, which provides for the minimal probability of the emergency situation.

In this article it is suggested to calculate dependencies $P_i(\lambda_1(t), \lambda_2(t), \ldots, \lambda_k(t),$ $\mu_1(t), \mu_2(t), \ldots, \mu_k(t), \overrightarrow{x}(t), t)$, $i = \overline{1, n}$ with the help of the apparatus of Markov's processes. If the processes under consideration can't be described by this formal apparatus, using statistical analysis methods is also possible, as well as the theory of neural networks, expert systems and other methods. In case we use Markov's processes, which are often used to calculate the reliability of aircraft, the sought for probabilities of the realization of critical combinations of events are calculated by solving the Kolmogorov-Chapman's equation system, which is built from the state graph [8]. Its solution can be gained analytically, if the number of section elements is small, or using numerical methods. In case when the law of failure distribution is unknown, expert methods can be used to calculate probabilities, and they can be combined with formal methods.

Kolmogorov-Chapman's system of differential equations for n-element minimal section consists of 2^n equations for probabilities $P_0(t)$, ..., $P_{2-1}^n(t)$ of event combinations, preceding the given n-element combination. The equations look as follows:

4   Mathematical Model

To solve the problem a mathematical model has been worked out, including:

- a set of dynamic event trees describing the reasons and the development paths of accidents and catastrophic situations caused by critical combinations of events;
- a set of graphs of minimal sections of the event tree, used to build a system of differential equations;
- logical and mathematical models allowing calculating the probability of critical combinations of events when the law of time distribution between events is not exponential;
- systems of Kolmogorov-Chapman's differential equations allowing calculating the probability of accidents and catastrophic situations.

Nowadays, a large number of mathematical models and algorithms have been worked out and tested practically. They allow building of event trees to analyze modes of complex system functioning, including reduction and completion methods necessary to design aircraft [1]. For example, for an ATS based on a forward-looking twin-engine airplane a fragment of such an event tree will look like this (Fig. 1).

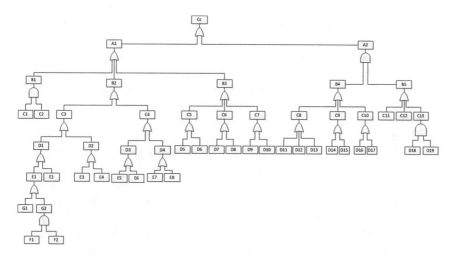

Fig. 1 Fragment of an event tree D describing the appearance of an emergency situation for a forward-looking twin-engine airplane

The following designations are assumed for Fig. 1: Cc—emergency situation; A1—functional failure (FF) of an aircraft creating an emergency situation at landing controlled by steering-control; A2—combinations of FF of an aircraft, erroneous actions of the crew (EAC) and parameters of expected operating conditions (EOC) creating an emergency situation at the landing of the; B1—loss of 50 % the required thrust of the propulsion system; B2—reduction of the efficiency of longitudinal and transversal control by half; B3—FF of flight equipment; B4—FF of an aircraft; B5—EAC and parameters of external conditions; C1—failure of one of the aircraft's engines; C2—failure of the second engine (after the failure of the first one); C3—Noticeable reduction of the efficiency of pitch control; C4—Noticeable reduction of the efficiency of list control; C5—loss of flight parameter indication by the pilot at the controls; C6—faulty indication of one of flight parameters for the pilot at the controls; C7—not signaled autopilot failure at the landing assisted by flight director; C8—loss of effectiveness by the rudder; C9—loss of flight parameter indication by the co-pilot; C10—faulty indication of one of flight parameters for the co-pilot; C11—EAC; C12—adverse external conditions; C13—combination of pilotage inaccuracies and adverse weather conditions; D1—hydrosupply failures; D2—FF in the system of pitch control; D3—FF in the system of list control; D4—hydrosupply failures; D5—loss of attitude indication; D6—loss of indication of altitude and speed parameters; D7—faulty indication of one of attitude parameters; D8—faulty indication of one of altitude and speed parameters; D9—faulty indication of Yagi arrows; D10—disappearance of Yagi arrows; D11—failures on the rudder control system; D12—failures of hydrosully of rudder control; D13—failure of the rudder correction mechanism in the min position; D14—loss of indication of attitude parameters; D15—loss of faulty indication of one of altitude and speed parameters; D16—faulty indication of one of attitude parameters; D17—faulty indication of one of altitude and speed parameters; D18—incorrect approach descent; D19—lateral wind over 8 m/s; E1—loss of pressure in hydraulic system; E2—depressurization of hydraulic system; E3—open-circuit in the wiring of control of one of the sections of elevation rudder (ER); E4—jamming of the booster rod of an ER section; E5—open-circuit in the wiring of control of one of the sections an aileron; E6—jamming of the booster rod of an aileron section; E7—depressurization of hydraulic system; E8—loss of pressure in hydraulic system; F1—failure of hydraulic pump; F2—pumping plant failure; G1—destruction of hydraulic reservoir; G2—failure of pressure sources.

In this article it is suggested to calculate dependencies $P_i(\lambda_1(t), \lambda_2(t), \ldots, \lambda_k(t), \mu_1(t), \mu_2(t), \ldots, \mu_k(t), \overrightarrow{x}(t), t)$, $i = \overline{1, n}$ with the help of the apparatus of Markov's processes. If the processes under consideration can't be described by this formal apparatus, using statistical analysis methods is also possible, as well as the theory of neural networks, expert systems and other methods. In case we use Markov's processes, which are often used to calculate the reliability of aircraft, the sought for probabilities of the realization of critical combinations of events are calculated by solving the Kolmogorov-Chapman's equation system, which is built from the state graph [8]. Its solution can be gained analytically, if the number of section elements is small, or using numerical methods. In case when the law of failure distribution is unknown, expert methods can be used to calculate probabilities, and they can be combined with formal methods.

Kolmogorov-Chapman's system of differential equations for n-element minimal section consists of 2^n equations for probabilities $P_0(t)$, ..., $P_{2-1}^n(t)$ of event combinations, preceding the given n-element combination. The equations look as follows:

$$\frac{dP_v(t)}{dt} = \sum_{w=0}^{2^n-1} \pi_{v,w}^+ P_w(t) - P_v(t)\pi_v^-,\qquad(5)$$

$$\pi_{v,w}^+ = \begin{cases} \lambda, \text{if the arc of the state graph marked } \lambda \text{ goes from state } w \text{ to } v, \\ \mu, \text{if the arc marked } \mu, \text{ goes from state } w \text{ to } v, \\ 0, \text{if there is no arc going from state } w \text{ to } v \text{ in the graph,} \end{cases}$$

π_v^-—is the sum of marks of all arcs going from arc v to other nodes of the graph, $\lambda \in \{\lambda_1, \lambda_2, \ldots, \lambda_n\}, \mu \in \{\mu_1, \mu_2, \ldots, \mu_n\}, v, w \in \{0, \ldots, 2^n - 1\}$.

From the above, let's formulate the common algorithm of solving the problem.

1. Beginning of the algorithm.
2. Determining of the set $\{A_1, A_2, \ldots, A_n\}$ of accidents and catastrophic situations.
3. Building the set of event trees $\{D_1, D_2, \ldots, D_m\}$, each of which corresponds to accidents and catastrophes from the set $\{A_1, A_2, \ldots, A_n\}$.
4. Determining and classification of minimal sections corresponding to critical event combinations for each of the event trees.
5. For the chosen minimal section the terminal nodes are determined, which correspond to events that trigger an accident or a catastrophe described by the given minimal section.
6. For each event a list of actions is determined, realization of which prevents it.
7. For each list of actions the μ_i value is calculated and a system of differential Eq. (5) is solved, which calculates the $P_i(t)$ probability of an accident or a catastrophe; if the process of accident appearance is not Markov's, the probability is calculated using the apparatus of neural networks, fuzzy logic, etc....
8. The minimal probability value $P_i^*(t)$ is selected; using it, a corresponding list of actions is retrieved from the database and conveyed to the crew, the flying control officer and other decision making personnel.
9. End of the algorithm.

Analytical solution of the system (5) for a 3-element section gives us the following expression to calculate the probability of its realization:

$$P_7(t) = \frac{1}{\mu_2\mu_3\mu_1}\left(\mu_2\mu_3\mu_1 e^{-(\mu_2+\mu_1+\mu_3+\lambda_1+\lambda_2+\lambda_3)t}C_8\right.$$
$$- \mu_1\mu_2\lambda_3 e^{-(\mu_1+\mu_2+\lambda_1+\lambda_2)t}C_6 - \mu_3\lambda_2\mu_1 e^{-(\mu_3+\mu_1+\lambda_1+\lambda_3)t}C_7$$
$$- \mu_2\mu_3\lambda_1 e^{-(\mu_3+\mu_2+\lambda_2+\lambda_3)t}C_5 + \mu_1\lambda_2\lambda_3 e^{-(\mu_1+\lambda_1)t}C_3$$
$$+ \lambda_1\mu_3\lambda_2 e^{-(\mu_3+\lambda_3)t}C_2 + \lambda_1\lambda_3\mu_2 e^{-(\lambda_2+\mu_2)t}C_4 - \left.\lambda_1\lambda_3\lambda_2 C_1\right), \tag{6}$$

$C_1, C_2,..., C_8$ are constants.

The numerical solution of the system (5) is presented on Fig. 2, where the graphic dependencies are given illustrating the solution for Kolmogorov-Chapman's differential equation system for $\lambda_1 = 10$, $\lambda_2 = 1$, $\lambda_3 = 1$, $\lambda_4 = 1$, $\lambda_5 = 1$, $\mu_1 = 0$, $\mu_2 = 5$, $\mu_3 = 5$, $\mu_4 = 5$, $\mu_5 = 5$, where $P_1(t)$—is the probability of the fact that all the ATS's subsystems work; $P_2(t)$ is the probability of the fact that the autopilot failed at the landing assisted by the flight director; $P_7(t)$ is the probability of the fact that the autopilot failed at the landing assisted by the flight director and the efficiency of the rudder is lost; $P_{17}(t)$ is the probability of a non-signaled autopilot failure at the landing assisted by the flight director accompanied by the loss of the efficiency of the rudder and the loss of the indication of flight parameters by the co-pilot.

Figure 2 shows that the probability of the fully functional state is the highest, but gets reduced in the process of functioning, while the probability of the failure of all elements is the lowest, i.e. the critical combination of the events doesn't cause an emergency situation.

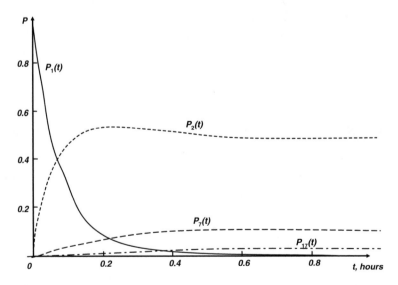

Fig. 2 Solution of the differential equation system for a five-element minimal section

Each value of the μ_i parameters corresponds to certain actions of automated systems, crew and flying control officers directed at the elimination of possible reasons of an emergency.

5 Conclusion

An approach is suggested to increasing the safety of air transportation systems, which is based on analysis of critical combinations of separately non-dangerous failures and errors. A formal problem statement is worked out for air transportation system control with safety as the criterion.

To solve the problem, a mathematical model is developed, which allows calculation of probabilities of accidents and catastrophes related to aircraft when certain events are combined. In particular, formal methods are proposed to allow building and promptly correcting event trees for various operation conditions of aircraft.

Common principles of creating and operation of an informational and logical system are worked out. This system allows numeric estimation of the probability of emergency situations caused by critical combinations of events on different time intervals. It also recommends preventive measures for such situations. The software that has been developed is partly used as a component of the model of reliability and safety of air transportation systems in Open Joint Stock Company "Ilyushin Aviation Complex".

Acknowledgments The research is supported by Russian foundation for basic research, grant number 16-01-00536 a.

References

1. Novozhilov, G.V., Neymark, M.S., Tsesarsky, L.G.: Safety of the flight of an airplane: Conception and technology. In: Publishing house of Moscow Aviation Institute, 196 p. (2007) (in Russian)
2. Novozhilov, G.V., Rezchikov, A.F., Neymark, M.S., Tverdokhlebov, V.A., Tsesarsky, L.G., Filimonyuk, L.Yu.: Causal-investigatory approach to analyzing air transportation systems. In: All-Russian Scientific and Technical Journal "Flight", vol. 7, pp. 3–8 (2011) (in Russian)
3. Neymark, M.S., Tsesarsky, L.G., Filimonyuk, L.Yu.: Model of support of decision-making when aircraft enter the zone of the ariposrt's responsibility. In: All-Russian Scientific and Technical Journal "Flight", vol. 3, pp. 31–37 (2013) (in Russian)
4. Rezchikov, A.F., Bogomolov, A.S., Ivashchenko, V.A., Filimonyuk, LYu.: Approach to providing for and maintaining safety of complex systems based on automat models. Large-Scale Syst. Control **54**, 179–194 (2015). (in Russian)
5. Hansman, R.J., Histon, J.M.: Mitigating complexity in air traffic control: the role of structure-based abstractions. In: ICAT-Reports and Papers (5) (2008)
6. Vaaben, B., Larsen, J.: Mitigation of airspace congestion impact on airline networks. J. Air Transp. Manag. **47**, 54–65 (2015)

7. Alderson, J.C.: Air safety, language assessment policy and policy implementation: the case of aviation English. Annu. Rev. Appl. Linguist. **29**, 168–187 (2009)
8. Ostreykovsky, V.A, Shvyryaev, U.V.: Safety of atomic plants. Probabilistic analysis. In: Fizmatlit, 352 p. (2008) (in Russian)

RFID Technology for Adaptation of Complex Systems Scheduling and Execution Control Models

Boris Sokolov, Karim Benyamna and Oleg Korolev

Abstract In this paper, we investigate the issues of establishing adaptive feedbacks between complex systems (CSs) scheduling and execution from the perspectives of modern control theory. In using optimum control for the scheduling stage, feedback adaptive control for the execution stage, and attainable sets for the analysis of the achievement of the planned performance in a real execution environment, we provide a mathematically unified framework for CSs scheduling and execution control. The proposed framework makes it possible to analyze the correspondence of RFID (Radio Frequency Identification) functionalities and costs to the actual needs of execution control and support problem-oriented CSs adaptation for the achievement of the desired performance. The developed framework can be applied as an analysis tool for the decision support regarding the designing and applying RFID infrastructures in supply chains.

Keywords Complex systems · Scheduling and planning · RFID technologies · Integrated modeling · Multi-agents modeling

B. Sokolov (✉) · K. Benyamna
Saint Petersburg National Research University of Information Technologies, Mechanics and Optics (ITMO), St. Petersburg, Russia
e-mail: sokol@iias.spb.su

K. Benyamna
e-mail: benyamna.karim@gmail.com

B. Sokolov · O. Korolev
St. Petersburg Institute of Informatics and Automation, Russian Academy of Sciences (SPIIRAS), St. Petersburg, Russia
e-mail: korolf@rambler.ru

© Springer International Publishing Switzerland 2016
R. Silhavy et al. (eds.), *Automation Control Theory Perspectives in Intelligent Systems*, Advances in Intelligent Systems and Computing 466,
DOI 10.1007/978-3-319-33389-2_41

433

1 Introduction

The main objects of our investigation are complex systems (CSs). By CSs we mean systems that should be studied through polytypic models and combined methods [1–5]. In some instances, investigations of complex systems require multiple methodological approaches, many theories and disciplines, and carrying out interdisciplinary studies. Different aspects of complexity can be considered to distinguish between a complex system and a simple one, for example: structure complexity, operational complexity, complexity of behavior choice, complexity of development. Classic examples of complex systems are: control systems for various classes of moving objects such as surface and air transport, ships, space and launch vehicles, supply chain, geographically distributed heterogeneous networks, flexible computerized manufacturing [6–18].

Modern developments in information technologies (IT) such as RFID (Radio Frequency Identification), CSEM (Complex System Event Management) and mobile business provide a constructive basis to incorporate the stages of CSs planning and execution. However, the IT service in CSs as organizational systems fulfils the decision-support role (and not automatic decision execution role). Hence, the analysis frameworks for the decision support regarding the designing and applying IT, incl. RFID, infrastructures in CSs are practically needed [19–21].

RFID also does not propose and control actions that should be taken to adapt CSs in the case of changes or disruptions at the execution stage. Quality of adjustment adaptive actions efficiency at the execution stage depends on two factors: (1) control actions that are taken in operations execution dynamics and (2) control actions that have been taken at the planning stage. Hence, the planning and execution models are to be inter-reflected, which means, in both of the models, that the decision making principles of the other model are to be reflected. The preferable way to ensure such integration is to apply the same modeling methods [1–3, 22–26]. In these settings, the extensive development of approaches and models to dynamic CSs scheduling under the attracting adaptation methods is becoming a timely and crucial topic in CSs.

Our investigations are based on the results of the CSs adaptive control theory which is being developed now by Professor Skurihin V.I. in Ukraine [4]. The analysis of known investigations on the subject [5–13] confirms that the traditional tasks of CSs control should be supplemented with procedures of structural and parametric adaptation of special control software (SCS) (see Fig. 1, blocks 3, 7). In this paper further CSs we will consider complex control systems consists of control complex objects (COs) and control subsystems (CnSs).

Here the adaptive control should include the following main phases: parametric and structural adaptation of structure-dynamics control (SDC) models and algorithms to previous and current states of objects-in-service (SO), of CO, of CnS, and of the environment (see Fig. 1, blocks 1, 2, 3); integrated planning and scheduling of CnS operation (construction of SDC programs) (blocks 4, 5); simulation of CS operation, according to the schedules, for different variants of control decisions in

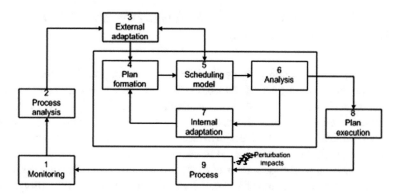

Fig. 1 Conceptual model of parametric and structural adaptation [4]

real situations and analysis of planning and scheduling simulation (block 6); structural and parametric adaptation of the schedule, control inputs, models, algorithms, and SDC programs to possible (predicted by simulation) states of SO, CO, CnS, and of the environment (block 7), realization of CS structure-dynamics control processes (block 8).

To implement the proposed concept of adaptive control let us consider two groups of parameters [4, 5, 7] for CS SDC models and algorithms: parameters that can be evaluated on the basis of real data available in CS; parameters that can be evaluated via simulation models for different scenarios of future events.

The adaptation procedures can be organized in two blocks (models) [1, 4]: external adapter of planning and scheduling models; internal adapter of planning and scheduling models.

When the parametric adaptation of SCS does not provide simulation adequacy, the structural transformations can be needed. Two main approaches to structural model adaptation are usually distinguished [6, 7, 15–18]. The first approach lies in the selection of a model from a given set. The model must be the most adequate to SO and CS. The second approach consists in CS SDC model construction of elementary models (modules) in compliance with given requirements. The second approach provides more flexible adjustment of SO and CS for particular functioning conditions. However, the first one is faster and can be effective if the application knowledge base is sufficiently large.

Both approaches need active participation of system analysts and decision-makers who interact with special control software of simulation system (SIS) and consider hard-formalizing factors and dependences within the general procedure of CS SDC program selection.

2 Integrated Analysis of the CS Scheduling and RFID Based Execution Control

The main question in the proposed execution adaptation approach is to define rules to establish which control actions should be taken by which deviations to maintain the planned performance. In addition, we emphasize that these actions are strongly interconnected with the RFID infrastructure of CSs. Hence, a method is required to consider simultaneously: (1) CS performance and adaptation and (2) control actions (RFID functionality) and perturbation impacts (see Fig. 1, blocks 3, 7, 8).

To approach these complex interrelations, we propose using the concept of CS global stability and the method of attainable sets. Global stability is the ability of a CS to maintain such an execution trajectory which allows for the achievement of the planned output performance despite perturbations in a real execution environment with the help of corresponding control adjustment actions. This approach commits to principles that are laid down in the global stability by Lyapunov, hyperstability by Popov, which allows uncertainty in dynamics, the systems parameters, and control actions [7, 8, 11, 13]. In the approach, stability is considered a dynamic property that emerges through feedback loops. Hence, stability can be considered a system behaviour property that should be maintained despite perturbation influences by means of corresponding adaptive control actions. This approach commits to stabilizing adaptive systems [5–8, 11–14]. As such, stability becomes interconnected with adaptability within the so-called stabilizing feedback control. Let us consider the general logic scheme of decision making on the elimination of disturbances in CS.

Step 1. Conformity analysis of current and planned goals. At the given stage, a comparison of the current values of parameters and the goals of CS execution with the planned values is carried out. If the arisen deviations in aggregate do not lead to a loss of stability and the CS maintains a stable state, necessities for correcting control influences are not present (see Fig. 1, blocks 4, 5, 6). Otherwise, a transition to step 2 is necessary.

Step 2. Alerting managers about the necessity to take regulating decisions. In the case where perturbation influences lead to a loss of stability and the CS loses its stable state. On the basis of the actual stability, analysis data and the planned scenarios of recovering the CS operability, a certain set of actions on the restoration of a planned (or wished for) course of events is proposed to managers (see Fig. 1, blocks 2, 6, 8).

Step 3. Decision making for the CS adaptation. Taking operative decisions is based on a system comparison of various kinds of control influences with various levels of parameter deviations of the CS gained on the basis of the stability analysis (see Fig. 1, blocks 3, 7).

The model global stability analysis is based on the control theory-based dynamic interpretation of the CSs functioning process and attainable sets. The results of the stability analysis can be brought into correspondence with different CS adjustment measures with regard to CS schedules. In addition, the usage of attainable sets makes it possible to quantify the stability in the form of a stability index. The essence of a stability index calculation is based on the construction and comparison of two attainable sets (the area of admissible values of CS performance metrics and the approximated area of CS attainability under the influence of perturbation factors). The stability index is expressed as the area of intersection of these two rectangles. Based on the quantification of stability with the help of attainable sets (AS) [1, 3, 14], it becomes possible to compare different CS and RFID structures with different scope and scale of the resource consumption regarding both the performance indicators and stability. This analysis can be performed with regard to different execution scenarios and different areas of control impacts in order to achieve the planned output performance in a real perturbed execution environment. Hence, the results of the stability analysis can be brought into correspondence with different CS performance and resource consumption scenarios. To summarize, we emphasize that the proposed approach makes it possible to address the issues of both (1) analysis of the mutual correspondence of the investments into RFID infrastructures and the CS performance and (2) analysis of the selection of appropriate CS adaptation actions responding to changes in CS execution environment.

3 Experimental Environment

To investigate the RFID-based feedbacks within the developed modeling framework, a simulation program (Fig. 2) and an experimental stand (Fig. 3) with a transport network (for example, railroad) and some production and warehouse facilities is currently under development (the transportation network with an RFID infrastructure is already elaborated).

We note that the RFID experimental environment is not intended (at least, in its current version) for a full implementation of the developed models. It is much simpler as the modeling framework and serves to gather experimental data for the modeling complex. The modeling complex itself is implemented in a special software environment, which contains a simulation and optimization engine of CS planning and execution control, a Web platform, an ERP system, and APS system, and a SCEM system. The kernel of the computational framework is the decision modeling component, that is, the simulation and optimization engine. The schedule optimization is based on an optimal control algorithm that is launched by a heuristic solution, the so-called first approach. The seeking for the optimality and the CS scheduling level is enhanced by simultaneous optimizing and balancing interrelated CS functional, organizational and information structures. The schedules can be analyzed with regard to performance indicators and different execution scenarios with different perturbations. Subsequently, parameters of the CS structures and the environment can be tuned if the decision-maker is not satisfied with the values of performance indicators. In analyzing the impact of the scale and location of the adaptation steps on the CS performance, it becomes possible to justify methodically the requirements for the RFID functionalities, the stages of a CS for the RFID element locations, and the

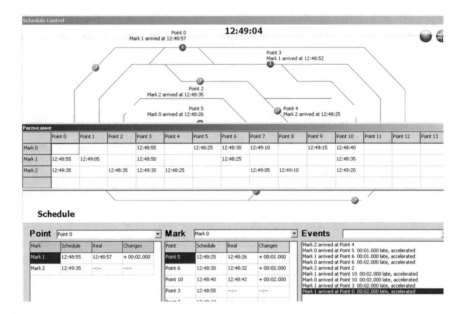

Fig. 2 Screenshot from the transport network simulation program

Fig. 3 Example railroad prototype with RFID readers

processing information from RFID. In particular, possible discrepancies between actual needs for wireless solution of CS control problems and the total costs of ownership regarding RFID can be analyzed. In addition, processing information from RFID can be subordinated to different management and operation decision-making levels (according to the developed multi-loop adaptation framework). Pilot RFID devices with reconfigurable functional structure are developed [27].

A word about simulation models of CS execution control model. In order to simulate the RFID-based transport network, we have created a prototype of a simu-lation model that reproduces a real railway network. To do this, we could use different approaches, but the most suitable is the multi-agent system modelling [28]. But first, let us describe what a multi-agent system is. Research in the multi-agent systems (M.A.S.) [29] intends to apprehend the way that a group of multiple autonomous entities, called agents, can interact and organize themselves in order to solve a problem [30]. In contrast to a classical artificial intelligence (AI) system, the aim is not to reproduce an intelligent human behavior, but to reproduce a social behavior, whether it is a reactive M.A.S. like social insects, or like in the case of human society.

In our case, the system purpose is to ensure the realization of schedule in spite of different events that can disturb it and our agents are the locomotives that carry different amounts of wagons.

To realize the simulation model we could use different tools made for modeling a M.A.S. [31, 32] but most of them have a restrictive or a paid licensing policy. So we managed to develop from the bottom a prototype of a simulation environment based on the C ++ programing language and the M.A.S. approaches. The first step in creating a simulation model is to define the time framework, implemented in the form of a main event loop. In the loop, we have two functions; the first may or may not create a random disturbance in the locomotive speed and the maximum speed. The second function simulates the agent behavior. Our agents are described by a structure; they are able to move ahead in their section based on their speed. The speed itself can vary from 0 to the maximum speed, which is defined by the topography of the section, the locomotive characteristic, and the wagons weight and length. When the locomotive arrives at the end of the section, it moves to the next one according to the schedule. It is also able to check if it is on time, cooperate with other locomotives in the same section and adjust speed in order to meet the schedule. In the function responsible for the agents' behaviors, we calculate the new position of the locomotive, and then we correct it if an RFID-based update has arrived meanwhile. Then we check if we are still on time, if it is the case, we stop there and move to the next agent, if not, first we try to cooperate with other locomotives in the section; the cooperation means that, for example, if the first locomotive is late and there is a locomotive in front of it and it is ahead of the schedule or has less priority, it will try to use a secondary route to let the first train go ahead. Then the first train will adjust its speed in order to meet the schedule.

This system enables us to simulate a basic railroad network, and test the RFID infrastructure on a realistic model. Of course, as any system, this one has its advantages and disadvantages. As regards advantages, they include: the technolo-gies used to create this model (C++ language), it provides us with a great flexibility in terms of functionality, allowing for modification and implementation of any kind of logic we want; agent based modeling is a powerful method allowing a large number of enhancements in the behavior of the system. Additionally, it enables us to define a logic of each individual locomotive, which is close to how decisions are made in a real system; define the system behavior by an independent entity allows great scalability, as the complexity of the system is linear to the number of entities.

In terms of disadvantages, they include: the direct use of a low level programing language, even if it allows us a great flexibility, requirement to create everything from scratch, which is costly as regards time and effort; a model based on time simulation can rapidly become costly in terms of calculation, and thus will require a lot of optimization in order to stay usable; a pure agent based model disallow a system optimization as a whole. But in our case, we delegate this task to the scheduler.

As we can see, the overall advantages and disadvantages are a matter of choice. We have chosen the way of integration results of modern control theory and M.A.S. theory in a perspective of scalability and flexibility in order to allow further developments.

4   Summary

We have implemented the RFID technology of parametric and structural adaptation of models describing CS scheduling and execution control processes via original simulation system (SIS) which is based on original analytical-simulation models, and an experimental stand. RFID technology makes it possible to realize feedback in a complex technical-organization system (for example a supply chain). This simulation system consists of the following elements: (a) simulation models (the hierarchy of models); (b) analytical models (the hierarchy of models) for a simplified (aggregated) description of objects being studied; (c) informational subsystem that is a system of data bases (knowledge bases); (d) control-and-coordination system for interrelation and joint use of the previous elements and interaction with the user (decision-maker).

Acknowledgements The research is supported by Russian Science Foundation (Project No. 16-19-00199).

References

1. Ohtilev, M.Yu., Sokolov, B.V., Yusupov, R.M.: Intellectual Technologies for Monitoring and Control of Structure-Dynamics of Complex Technical Objects, p. 410. Nauka, Moscow (2006) (in Russian)
2. Zaychik, E., Sokolov, B., Verzilin, D.: Integrated modeling of structure-dynamics control in complex technical systems. In: 19th European Conference on Modeling and Simulation ESMS 2005, "Simulation in Wider Europe", 1–4 June 2005, pp. 341–346. Riga Technical University, Riga, Latvia (2005)

3. Ivanov, D., Sokolov, B., Arkhipov, A.: Stability analysis in the framework of decision making under risk and uncertainty. In: Camarinha-Matos, L.M., Afsarmanesh, H., Ollus, M. (eds.) Network—Centric Collaboration and Supporting Frameworks, IFIP TC5WG 5.5 Seventh IFIP Working Conference on Virtual Enterprises, 25–27 Sept 2006, pp. 211–218. Springer, Helsinki, Finland (2006)

4. Skurihin, V.I., Zabrodsky, V.A., Kopeychenko, Yu.V.: Adaptive Control Systems In Machine-Building Industry. Mashinostroenie (1989) (in Russian)

5. Rastrigin, L.A.: Modern Principles of Control for Complicated Objects. Sovetscoe Radio (1980) (in Russian)

6. Bellmann, R.: Adaptive Control Processes: A Guided Tour. Princeton University Press, Princeton, New Jersey (1972)

7. Rastrigin L.A.: Adaptation of complex systems. Zinatne, Riga (1981) (in Russian)

8. Fleming, W.H., Richel, R.W.: Deterministic and Stochastic Optimal Control. Springer, Berlin (1975)

9. Moiseev, N.N.: Element of the Optimal Systems Theory. Nauka (1974) (in Russian)

10. Sowa, J.: Architecture for intelligent system. IBM Syst. J. **41**(3) (2002)

11. Zypkin, Ya.Z.: Adaptation and Teachning in Automatic Systems. Nauka (1969) (in Russian)

12. Bryson, A.E., Ho, Yo-Chi: Applied Optimal Control: Optimization Estimation and Control. Waltham, Massachusetts (1969)

13. Singh, M., Titli, A.: Systems: Decomposition, Optimization and Control. Pergamon Press, Oxford (1978)

14. Petrosjan, L.A., Zenkevich, N.A.: Game Theory. World Scientific Publications, Singapore (1996)

15. Roy, B.: Multi-criteria Methodology for Decision Aiding. Kluwer Academic Pulisher, Dordreeht (1996)

16. Nilsson, F., Darley, V.: On complex adaptive systems and agent-based modeling for improving decision-making in manufacturing and logistics settings. Int. J. Oper. Prod. Manag. **26**(12), 1351–1373 (2006)

17. Rabelo, R.J., Klen, A.A.P., Klen, E.R.: Multi-agent system for smart coordination of dynamic supply chains. In: Proceedings of the 3rd International Conference on Virtual Enterprises, PRO-VE'2002. pp. 379–387 (2002)

18. Wu, N., Su, P.: Selection of partners in virtual enterprise paradigm. Robot. Comput.-Integr. Manuf. **21**, 119–131 (2005)

19. Angeles, R.: RFID technologies: supply chain applications and implementation issues. Inf. Syst. Manag. (Winter), 51–65 (2005)

20. Chalasani, S., Boppana, R.V.: Data architectures for RFID transactions. IEEE Trans. Ind. Inf. **3**(3), 246–257 (2007)

21. Huber, S., Michael, K., McCathie, L.: Barriers to RFID adoption in the supply chain Barriers to RFID adoption in the supply chain. In: IEEE RFID Eurasia, pp. 1–6. 5–6 September, Istanbul, Turkey (2007)

22. Henseler, M., Rossberg, M., Schaefer, G.: Credential management for automatic identification solutions in supply chain management. IEEE Trans. Ind. Inf. **4**(4), 303–314 (2008)

23. Rong, C., Cayirci, E.: RFID security. Computer and Information Security Handbook, pp. 205–221 (2009)

24. Lee, H., Oezer, Oe.: Unlocking the value of RFID. Prod. Oper. Manag. **16**(1), 40–64 (2007)

25. Chuang, M.L., Shaw, W.H.: RFID: integration stages in supply chain management. IEEE Eng. Manag. Rev. **35**(2), 80–87 (2007)

26. Li, S., Visich, J.K.: Radio frequency identification: supply chain impact and implementations challenges. Int. J. Int. Supply Manag. **2**(4), 407–424 (2006)

27. Dashevsky, V., Sokolov, B. (2010). New concept of RFID reader networks structure: hardware and software architecture. In: Proceedings of International Conference on Ultra Modern Telecommunications ICUMT-2009, Saint-Petersburg, Russia

28. Bhardwaj, A., Singh, V.K., Kumar, P.: Multi-agent based train passing in railway system with minimum system delay. In: 2014 IEEE International Advance Computing Conference (IACC) (2014)
29. Niazi, M., Hussain, A.: Agent-based computing from multi-agent systems to agent-based Models: a visual survey. Scientometrics (2011)
30. Müller, J.P.: Des systèmes autonomes aux systèmes multi-agents: Interaction, émergence et systems complexes. Mémoire d'habilitation (2002)
31. Kurve, A., Kotobi, K., Kesidis, G.: An agent-based framework for performance modeling of an optimistic parallel discrete event simulator. Complex Adapt. Syst. Model. **1**, 12 (2013). doi:10.1186/2194-3206-1-12
32. Salamon, T.: Design of Agent-Based Models. Repin: Bruckner Publishing. p. 22 (2011). ISBN 978-80-904661-1-1

Electromagnetic Interference of Components of Intrusion and Hold-up Alarm Systems

Hana Urbancokova, Stanislav Kovar, Jan Valouch and Milan Adamek

Abstract The components of intrusion and hold-up alarm systems are not only the receiver of the electromagnetic interference that is around them but also these components are the source of the electromagnetic interference. If the components generate excessively high interference, their activity could have an adverse effect on other electronic devices in their vicinity or it can cause errors in the functional properties of the entire security alarm system. For this reason, each component must be tested for electromagnetic immunity and also their electromagnetic interference must be measured in accredited laboratories for electromagnetic compatibility.

Keywords Electromagnetic compatibility · Electromagnetic interference · Intrusion and hold-up alarm systems · Electronic device · Level of interference signals

1 Introduction

Electromagnetic interference (EMI) is defined as the process by which a signal is generated by the source of interference. This signal is transmitted through electromagnetic coupling into the disturbed systems. The basis of EMI is the

H. Urbancokova (✉) · S. Kovar · J. Valouch · M. Adamek
Faculty of Applied Informatics, Tomas Bata University in Zlin,
nam. T.G. Masaryka 5555, 760 01 Zlin, Czech Republic
e-mail: urbancokova@fai.utb.cz

S. Kovar
e-mail: skovar@fai.utb.cz

J. Valouch
e-mail: valouch@fai.utb.cz

M. Adamek
e-mail: adamek@fai.utb.cz

© Springer International Publishing Switzerland 2016
R. Silhavy et al. (eds.), *Automation Control Theory Perspectives
in Intelligent Systems*, Advances in Intelligent Systems and Computing 466,
DOI 10.1007/978-3-319-33389-2_42

identification of sources of interference, parasitic transmission paths and measuring the level of interference signals. It is not just about the analysis of the causes of interference but also about their removal on the side themselves sources and their transmission paths [1].

Nowadays, every electronic equipment, system or device is the source and also the receiver of the electromagnetic interference. Because these devices occur more and more in our surroundings, manufacturers of new electronic devices have an increasingly difficult to produce such product, which would resist the highest levels of electromagnetic interference and at the same time it would not disturb the operation of other devices.

The development of electronics and increase of interference signals radically changed not only the requirements for measuring devices but also requirements for measuring workplace and the methods used to measurement of electromagnetic compatibility. The interference environments, that are normally found in our surroundings, can cause unwanted bonds, background noises, resonance or transient phenomena, whether the distortion or depreciation of the measured data in the measurement of electromagnetic radiated of electronic devices in open space. In the implementation of our measurements, the elimination of these effects was achieved through the semi-anechoic chamber, which is located at the Faculty of Applied Informatics of Tomas Bata University in Zlin [2, 3].

All electrical and electronic equipment must be designed in accordance with the standards for electromagnetic compatibility (EMC). In the field of electromagnetic interference, the components of intrusion and hold-up alarm systems (I&HAS) are tested in accordance with the international standard CSN EN 55022 ed. 3. This technical standard determines uniform requirements for the high-frequency interference level of the information technology equipment, defines limits on the levels of the EMI and the methods of measurement [4, 5].

This paper describes the levels of electromagnetic interference that were generated by the basic set of the alarm security system. This basic set consisted of a control panel with an accumulator stored in a plastic box, the keypad, PIR detector and siren. The level of interfering signals that were generated by the components of intrusion and hold-up alarm systems were dependent on the current state of components. In our case, two basic states have been investigated. It was state of turned on and state of alarm.

2 Set of I&HAS and Measuring Devices

This basic set of intrusion and hold-up alarm systems, on which the electromagnetic radiation was measured, was powered from the mains 240 V/50 Hz. The control panel, accumulator and mains power module were closed in the plastic box which is usually supplied with the control panel. All components belong to the product lines Oasis. Figure 1 shows the location of the components of I&HAS on the table which

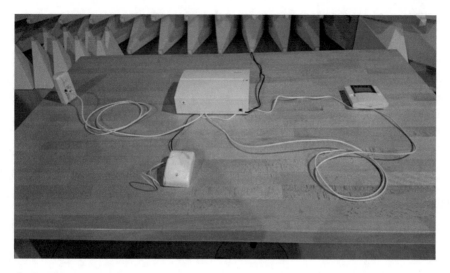

Fig. 1 The set of I&HAS

was placed on the turntable in the semi-anechoic chamber. The set included the following components:

- control panel JA 82-K,
- accumulator 12 V, 2.4Ah,
- mains power module,
- keypad JA-81E,
- PIR detector JS-20,
- siren SA-913TM.

The measurements of electromagnetic interference were performed in the semi-anechoic chamber in the EMC laboratory at the Tomas Bata University in Zlin. The semi-anechoic chamber creates a shielded space in which all inappropriate electromagnetic interferences from the environment that could distort the measurement results are eliminated [5, 6].

Figure 2 shows the BiLog antenna (CBL 6112) which is placed in the semi-anechoic chamber. This broadband biologaritmic-periodic antenna operates with range 30 MHz to 2 GHz in the horizontal and vertical polarization at a height from 1 to 4 m above the ground plane. The distance of the antenna from the equipment under test (EUT) is adjustable according to the requirements of standards. The BiLog antenna is connected on the EMI test receiver (ESU8) with a range of 20 Hz to 8 GHz by using the switching and control units (OSP130 and OSP150). The whole set is controlled by a computer with EMC Software (EMC32) for simplifying control of the antenna, the setting of limits and higher quality display of measured data.

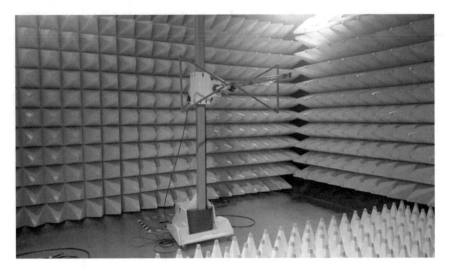

Fig. 2 The BiLog antenna (CBL 6112)

3   EMI Measurement Method in Semi-Anechoic Chamber

According to the international technical standard CSN EN 55022 ed. 3 the radiated interference is measured using measuring devices with a quasi-peak detector or the detector with a peak value can be used when there is a shortening of the measurement time. The frequency range used in the measurements is from 30 to 1000 MHz and measuring antenna should be at a distance 10 m from the EUT. However, if this distance can not be met, the measurement can be performed from a distance of 3 m [4].

In this case, our antenna was located at a distance 3 m from the EUT because our semi-anechoic chamber has a smaller size as semi-anechoic chambers in the big accredited centers. Therefore, the limits of EMI were increased by 10 dB (μV/m). EUT was located on a wooden table which was placed on the turntable. The location of the components of I&HAS on the table is shown in Fig. 3.

The first measurements were focused on identifying the most appropriate the angle of rotation (azimuth) of the antenna to the EUT and the antenna height above the ground plane. The following measurements were performed in the settings when the antenna recorded the strongest intensity of the radiated of electromagnetic interference which was generated by the set of I&HAS. The polarization of antenna (horizontal and vertical) was changed during the measurements by reason of we recorded the maximum radiated interference.

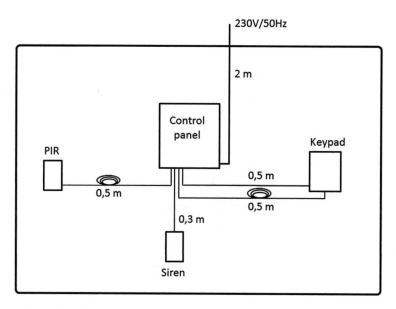

Fig. 3 The location of the components of I&HAS on the table

4 The Results of Selected Measurements

The electromagnetic interference which was generated by the basic set of I&HAS was measured in the mode where the whole set was in the ON state (state of guarding) or when the alarm was induced. The selected measurements are shown in the following figures. The x-axis shows the frequency in Hz (30 MHz to 1 GHz) and the y-axis shows the measured level of electromagnetic interference. The red line shows the maximum level of electromagnetic interference, which electronic devices can generate at the respective frequencies. This maximum level is defined by the standard CSN EN 55022 ed. 3. If the measured interference of the EUT has exceeded this red line, it would mean that the device generates the interference which endangers functionality of electronic devices in its surrounding.

After the first measurements, we identified an ideal setting of the antenna when the highest level of electromagnetic interference of the EUT was recorded. The EUT was placed directly opposite the measuring antenna with 0° rotation and antenna was at the height of 250 cm above the ground plane. The peak detector was used in the measurement.

In Fig. 4, the blue color shows the electromagnetic interference generated by the EUT in the ON state. The measuring antenna was in the horizontal polarization. The green color shows the level of interference of the EUT recorded by the antenna in the vertical polarization. The biggest differences in the levels of the electromagnetic interference of the EUT, when the polarity of the antenna was changed, have been recorded in the frequency range from 40 to 80 MHz and then from 120 to 160 MHz.

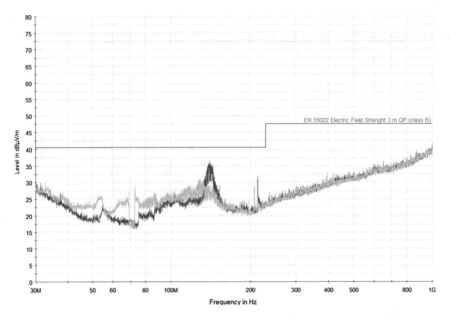

Fig. 4 The EMI of the EUT in the ON state—the antenna in the horizontal (*blue line*) and vertical (*green line*) polarization

In the next figure (Fig. 5) the EUT was in the state of alarm. The antenna height was maintained at 250 cm above the ground plane. The green color shows the level of EMI recorded by the antenna in the horizontal polarization and the EMI recorded by the antenna in the vertical polarization is shown the violet line.

A clearer view changes in the level of EMI generated by the EUT in the ON state and the state of alarm is shown in Figs. 6 and 7. How we can observe, the bigger changes of EMI were recorded in the horizontal polarization of the antenna (Fig. 6)—the blue color (EUT in the ON state), violet color (EUT in the state of alarm). The EMI recorded at frequencies approximately from 180 to 250 MHz (Fig. 6—violet color) was caused by the active siren which indicated the alarm.

The level of change of EMI recorded in the vertical polarization of the antenna (Fig. 7) was the minimum—the blue color (EUT in the ON state), violet color (EUT in the state of alarm).

Another interesting example of measuring was the level of interference generated by EUT in the ON state when the main power supply 230 V/50 Hz was disconnected (Fig. 8). The EUT was powered by the accumulator and the antenna was in the horizontal polarization. The blue color shows the level of interference of the EUT when the main power supply 230 V/50 Hz was connected, the violet color shows the EMI of the EUT when it was powered by the accumulator. As can be seen on the Fig. 8, the level of interference of the EUT, when it was powered by the accumulator, was almost negligible.

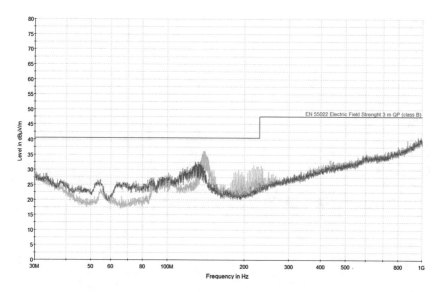

Fig. 5 The EMI of the EUT in the state of alarm—the antenna in the horizontal (*green line*) and vertical (*violet line*) polarization

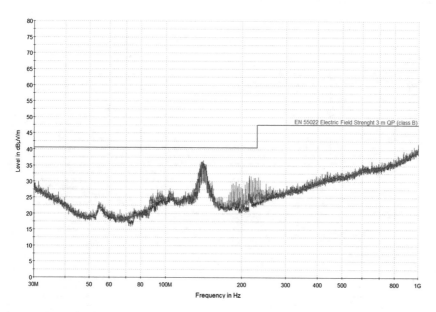

Fig. 6 The EMI measurement of antenna in the horizontal polarization—EUT in the ON state (*blue line*) and the state of alarm (*violet line*)

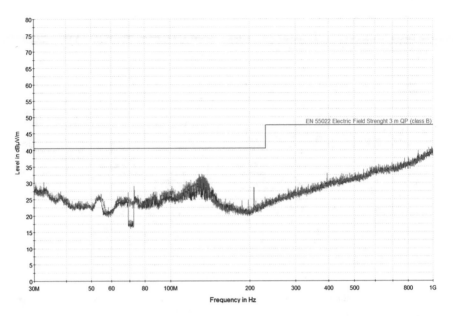

Fig. 7 The EMI measurement of antenna in the vertical polarization—EUT in the ON state (*blue line*) and the state of alarm (*violet line*)

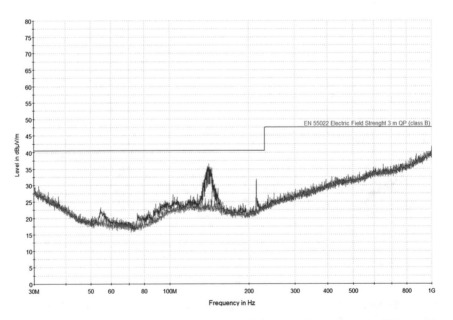

Fig. 8 The EMI measurement of antenna in the horizontal polarization—EUT in the ON state, the main power supply 230 V/50 Hz was connected (*blue line*) and disconnected (*violet line*)

5   Conclusion

The measurement of electromagnetic interference which is generated by the electronic devices is an important part in their development and production. Each electronic device that is sold on the European market must have a certificate of the performed tests of the EMC. The excessively high level of the EMI of devices could have a negative impact on the functionality of other devices in its vicinity, it could even cause damage to or destruction these devices.

The level of electromagnetic interference generated by the components of I&HAS was lower than the maximum level of EMI permitted according to internationally standard CSN EN 55022 ed. 3—Information technology equipment—Characteristics of high-frequency disturbance—Limits and methods of measurement. The tested set met the criteria in this standard and its activity has no relevant influence to the electronic devices in its surrounding. From our measurements, it shows that the maximum EMI of the EUT was generated when the siren was in a state of alarm (provides the audible warning of the alarm) and the main power supply from 230 V/50 Hz was connected to the EUT. The EMI of the power supply has the highest level of 37 dBμV/m at the frequency of 150 MHz. This EMI was the most observed with the positioning of the measuring antenna at a height of 250 cm above the ground plane and the antenna was setting in the horizontal polarization.

Acknowledgments This work was supported by the Ministry of Education, Youth and Sports of the Czech Republic within the National Sustainability Programme project No. LO1303 (MSMT-7778/2014) and also by the European Regional Development Fund under the project CEBIA-Tech No. CZ.1.05/2.1.00/03.0089 and Internal Grant Agency of Tomas Bata University under the project No. IGA/CebiaTech/2016/005

References

1. Urbancokova, H., Valouch J., Adamek M.: Testing of an intrusion and hold-up systems for electromagnetic susceptibility—EFT/B. In: International Journal of Circuits, Systems and Signal Processing, vol. 9, pp. 40–46. North Atlantic University Union, Oregon, USA (2015). ISSN: 1998-4464. 7 p
2. Vaculik, E., Vaculikova, P.: Electromagnetic compatibility of electrotechnical systems: A practical guide to technology limitations HF electromagnetic interference, 1st edn, p. 487. Grada Publishing, Prague (1998). ISBN 80-716-9568-8 (in Czech)
3. Svacina, J.: Electromagnetic compatibility: principles and notes. Issue No. 1. Brno: University of Technology (2001), 156 p. ISBN 8021418737 (in Czech)
4. CSN EN 55022 ed. 3. Information technology equipment—Characteristics of high-frequency disturbance—Limits and methods of measurement. Czech office for standards, metrology and testing, Prag (2011) (in Czech)

5. Valouch, J.: Technical requirements for electromagnetic compatibility of alarm systems. In: International Journal of Circuits, Systems and Signal Processing, vol. 9, pp. 186–191. North Atlantic University Union, Oregon, USA (2015). ISSN: 1998-4464. p. 6
6. Valouch, J.: Electromagnetic Compatibility of Alarm Systems—Legislative and Technical Requirements. In: Security Magazin. Issue No 106, 2/2012, pp. 32–36. Security Media, Praha (2012). ISSN 1210- 8273

Application of Object-Oriented Simulation in Evolutionary Algorithms

Yuriy Skobtsov, Alexander Sekirin, Svetlana Zemlyanskaya,
Olga Chengar, Vadim Skobtsov and Semyon Potryasaev

Abstract In paper two-level genetic and ant algorithms are proposed to optimize the functioning of the automated technological machining complex. For suggested genetic and unt algorithms it is designed the object-oriented simulation model, which allows to calculate the fitness function and evaluate potential solutions. The problem-oriented crossover, mutation and reproduction operators for two-level genetic algorithm are developed. The transition and calculation of the concentration for synthetic pheromone rules are determined for suggested ant algorithms.

Keywords Genetic algorithms · Ant colony optimization · Object-oriented simulation model · Production schedule

Y. Skobtsov
Peter the Great St. Petersburg Polytechnic University, St. Petersburg, Russia
e-mail: ya_skobtsov@list.ru

A. Sekirin · S. Zemlyanskaya
Donetsk National Technical University, Donetsk, Ukraine
e-mail: alx09@list.ru

S. Zemlyanskaya
e-mail: zsaa@yandex.ua

O. Chengar · V. Skobtsov (✉)
United Institute of Informatics Problems, National Academy of Sciences of Belarus,
Minsk, Belarus
e-mail: vasko_vasko@mail.ru

O. Chengar
e-mail: olga.chengar@gmail.com

S. Potryasaev
St. Petersburg Institute of Informatics and Automation,
Russian Academy of Sciences (SPIIRAS), St. Petersburg, Russia
e-mail: spotryasaev@gmail.com

S. Potryasaev
St. Petersburg National Research University of Information Technologies,
Mechanics and Optics (ITMO), St. Petersburg, Russia

© Springer International Publishing Switzerland 2016
R. Silhavy et al. (eds.), *Automation Control Theory Perspectives
in Intelligent Systems*, Advances in Intelligent Systems and Computing 466,
DOI 10.1007/978-3-319-33389-2_43

453

1   Introduction

To increase efficiency and intelligence of CAD systems, and expand their functions are widely used methods of artificial intelligence (AI). Evolutionary Computing (EC) is one of the most promising approaches of AI [1].

EC—is actively investigated new direction in the theory and practice of AI. EC term is applied generally to methods of search, optimization and learning, which are based on certain formal principles of natural evolutionary selection. It is necessary to identify individuals, populations, evolutionary operators and fitness function for solving specific problem with using evolutionary algorithm (EA). A potential solution is represented by a chromosome—some code, consisting of the elements—genes. Typically, EA operates with encoded chromosomes (genotype) and not direct solutions (phenotypes).

2   The Interaction of Genetic Algorithm
and the Simulation Model

In the process of artificial evolution each individual of the population is estimated with calculating of the fitness function value, which determines the quality of a potential solution. This operation consumes more than 90 % of computing resources to solve the problem with genetic algorithm (GA) using. It should be noted that in the general case, the objective function and fitness function may be different. The objective function is used to evaluate the performance of individuals relatively to the global ultimate goal (for example, extremum). Fitness function is used primarily in the selection of individuals for further evolution and here quality characteristics of one individual relative to another individuals are important. After decoding the chromosome where converting genotype to phenotype is executed (e.g., binary code is converted to a real number) then the resulting value is used as an argument for the fitness function. Next fitness function values are calculated to every individual of population. On the basis of obtained fitness function values the individuals are ranked relatively to each other in terms of building good solution.

Definition of fitness function for solving a specific problem with EA is critical to its effectiveness. In particular, the form of fitness-function may depend on the restrictions imposed in solving optimization problems. For example, the genetic operators of crossover and mutation may not consider the solution correctness, whether new constructed individuals—offsprings are in the area of feasible solutions that is caused by imposed limitations.

The definition of fitness function is influenced by the following factors: type of task—maximization or minimization; content of ambient noise in the fitness function; the ability to dynamically change the fitness function in the process of

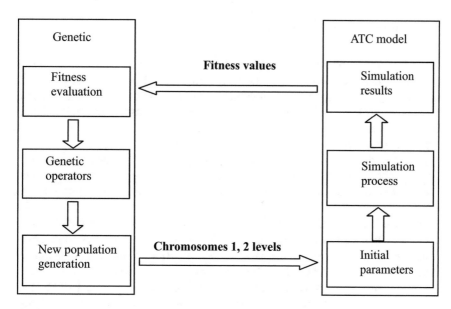

Fig. 1 Interaction of genetic algorithm and the simulation model

solving the problem; the amount of available computing resources—is it allowed to use more accurate methods and significant resources or possible only close approximations that do not require large resources; how different values for individuals should be given by fitness function to facilitate the selection of the parent individuals; whether it should take into account the constraints of the considered problem; whether it can combine different subgoals (e.g., multicriteria problems), etc.

The GA fitness function is often used as a black box: for a given chromosome it calculates the value indicating the quality of a given individual. Internally, it can be implemented in different ways: in the form of a mathematical function, modeling program (including simulation), a neural network or even expert evaluation. In this paper, according to [2] estimation of the fitness function value is performed using an object-oriented simulation model. The interaction of this model with GA shown in Fig. 1.

It should be noted that the high computational complexity and required huge computational resources for real optimization problems often does not allow for its solution the classical optimization methods.

Therefore, it is proposed to use genetic algorithm together with the object model as a new approach to the optimization of complex systems. Designed object models describe the most important characteristics of the system and allow with reasonable certainty to carry out simulation of their operation.

3 Object-Oriented Model

The use of object-oriented models in the EC has been successfully tested by the authors in optimizing following complex systems: (1) distributed database and data warehouse [2]; (2) corporate information systems [2]; (3) simulation and testing of digital systems [3]. Further, as an example, the following object-oriented model of the automated technological complex (ATC) machining [3, 4], which was used in evolutionary algorithms to optimize production schedules.

ATC combines in various combinations of equipment with computer numerical control, robots, flexible production modules (FPM), the single units of equipment and systems to ensure their operation in automatic mode for a predetermined time interval.

ATC must simultaneously produce parts of different types and respond quickly to changes in the production schedule needed, as well as—for all possible deviations (disturbance) such as: the lack of raw materials, equipment, workpieces; deficit; rejects; equipment failure; additional production orders.

On the basis of the developed classes following objects were created: FPM—a flexible production module (basic function—timing of machining completion, failure and recovery), TM—transport module (basic function—timing for transportation, failure and recovery, synchronization with the warehouse) WH—warehouse (basic function—timing the end of the warehouse operations, failure and recovery, synchronization with transport), EM—Event Manager (basic function—detection, fixation and transfer commands and events to destinations, the schedule formation), CS—control system (the purpose of CS ATC as part of the model is reduced to the control of technological equipment, transport and storage equipment by the commands and control of their implementation). The generalized flowchart of the components interaction in the ATC model is shown in Fig. 2.

As a result of the system analysis, based on developed graphical analytical model it is constructed the object model of organizational and technological process of loading equipment, that represents a system of interacting classes of typical components.

In actual production conditions on the functioning of the ATC machining is influenced by various disturbing effects (V_k), which lead to deviations of parts production from the production program.

The control objective of the automated technological complexes machining is to ensure production according to the production program (P_i) in quantity and in a time-bound manner, with the effective use of resources (R_l) under the action of disturbance (V_k). Providing high efficiency of resource using (R_l) and the APC operation in general is achieved by optimizing the scheduling of the equipment.

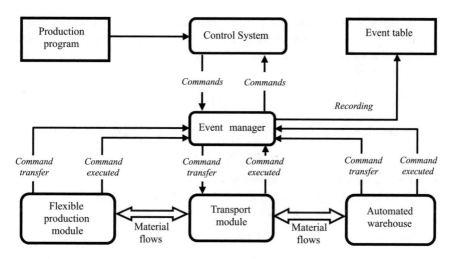

Fig. 2 The generalized functional block diagram of components interaction model ATC

4 The Objective Function

Let us consider the ATC consists of L technological equipment units $Q_l, (l = 1, 2, \ldots, L)$. The process of parts manufacturing $D_i, (i = 1, 2, \ldots, n)$ is split into technological operations $O_{ij}, (i = 1, \ldots, n; j = 1, \ldots, m)$. The parts of one type are combined into transport batches and in the context of scheduling are considered as a technological unit. Each operation can be expressed as:

$$O_{ij} = \langle H_{ij}, T_{ij} \rangle, \tag{1}$$

where H_{ij}—group number of the technological equipment;

T_{ij}—the technological operation length.

The process route is a sequence of operations performed, which passes the i-th part in the machining process:

$$M_i = \langle O_{i1}, O_{i2}, \ldots, O_{im} \rangle. \tag{2}$$

Here operation O_{ij} must be performed without interruption from the its start. If we denote by t_{ij}—the start time of the operation O_{ij}, and $\bar{t}_{ij}$—the finish time of the technological operation, it must have the equality: $\bar{t}_{ij} = t_{ij} + T_{ij}$. Obviously, the start time of operation depends on the time of the previous operations and always have the inequality $t_{ij} \leq t_{ij} + 1$. Then the set $\{t_{ij}\}, (i = 1, \ldots, n; j = 1, \ldots, m)$, satisfying all the technological and time limits, is a schedule (G) operation of the automated technological complex.

The schedule optimization problem for ATC with given machining part routes is in following. It is necessary to determine the startup sequence of parts in manufacture and the sizes of batches run in which the quality criterion schedules $F(G)$ sought to extremum

$$F(G) \rightarrow extrF(G). \tag{3}$$

This criterion should be considered togather with the following restrictions:

1. Limitations on the volume of production: $N_{fi} = P_i$,

   where N_{fi}—the actual part number of i-type $(i = 1, \ldots, K)$,
   P_i—the number of parts of i-th type in the production program,
   K—ATC nomenclature of manufactured parts.

2. Restrictions on the production length:

$$T_{prod.i} \leq T_{dir.i}, i = 1, 2, \ldots, k, \tag{4}$$

   where $T_{prod.i}$—the actual production time of i-th component,
   $T_{dir.i}$—the directive time of manufacture i-th part.

3. The time limit on the technological equipment operation:

$$\sum_{i=1}^{n} P_i \sum_{j=1}^{m} T_{ij} \leq R_l, \tag{5}$$

   where T_{ij}—the duration of technological operations;
   R_l—resource of l-th equipment group.
   Here are the main criteria for evaluating effectiveness of the resulting schedule:

1. Minimizing time cycle of manufacturing parts

$$T_{c.opt.} = T_c \rightarrow \min \tag{6}$$

   where T_c—general time of production time.

$$T_c = \sum_{i=1}^{n} \left(\sum_{j=1}^{m} T_{c.ij} + \sum_{j=1}^{m} \alpha_{ij} \right), \tag{7}$$

   where $T_{c.ij}$—cycle time of the j-th technological operation for i-th part batch,
   α_{ij}—downtime before starting the j-th technological step of the i-th part batch.

2. Maximizing the average load factor K_{lf} of ATC technological equipment

$$K_{lf.opt} = \sum_{i=1}^{l} K_{lf.i}/l \rightarrow \max, \qquad (8)$$

where l—number of flexible manufacturing cell (FMC) in ATC,
$K_{lf.i}$—loading factor for i-th FPM.

The loading factor for i-th FPM is determined as follows:

$$K_{lf.i} = \frac{\sum\limits_{j=1}^{m} T_{c.ij}}{(\sum\limits_{j=1}^{m} T_{c.ij} + \sum\limits_{j=1}^{m} \alpha_{ij})}. \qquad (9)$$

In this case the dataset of time $\{t_{ij}\}_{opt}$, corresponding to the extremum of criteria $F(G)$ for noted restrictions, would be optimal schedule (G) of ATC job.

To solve the stated above problem of scheduling optimization of technological equipment it is necessary to choose a new method for the solution, as all previously used methods did not provide extreme value for the specified performance criteria.

Typically, imitation, a production, network, object-production models are mainly used in ATC control systems of mechanical machining. The disadvantage of these methods is the inability to obtain optimal solutions in acceptable time limits established by production and technological constraints. Optimal solutions, obtained only for the simplest cases, have a purely theoretical value. To obtain acceptable solutions within acceptable time limits there are applied mainly heuristics, based upon use of the various rules for machining parts run. The dis-advantage of heuristics is the inefficiency of the solutions obtained by the error of the rules, which are compiled using the experience of a human expert. The full search method can not be applied, since the time of the solution search exceeds the permissible limits and require significant computing resources.

The way of overcoming these shortcomings is the application of evolutionary methods. To optimize the scheduling of work considered ATC it is suggested to apply genetic algorithms (GAs). The mechanism of evolution and inheritance allows consistently improve the chromosomes in each new population that yields suboptimal (close to optimal) solutions.

5   Two-Level Genetic Algorithm

Object model describes the structure of the classes that make up the system of the production process, their attributes, operations, relationships with other classes. On the basis of specialized problem oriented genetic operators of crossover and mutation, it was developed following two-level genetic algorithm to find the optimal or suboptimal operation schedules of ATC [2]:

1. Input. In this step the initial setting of algorithm parameters is performed. These parameters are the number of individuals in populations at first and second level —N_1 and N_2, respectively, probability values of crossover and mutation operators for both levels—$(P_{oc1}, P_{om1}, P_{oc2}, P_{om2})$, selection the target function F_{ts}— one of the criteria for the effectiveness of the ATC).
2. The initial population generation of the first and second levels. The initial population is represented by chromosomes $\{Chr_i^1\}$, formed with a strict compliance with noted the above restrictions. The unique code (natural number) is assigned to every type of part. The sequence of these codes in the chromosome defines the startup sequence of parts in manufacture.
3. Each of the chromosomes of the first level is associated with a population of chromosomes of the second level $\{Chr_j^2\}$. All chromosomes of the lower level are also being built within the constraints and reflect the size of the transport batch in the run batch by types of parts.
4. Genetic algorithm of second level. Once the initial populations of both levels were generated, you should select the first chromosome of the first level and work with the corresponding population of the second level. The evaluation of chromosome pairs is performed using the object model ATC. Then proportional individuals selection for the crossover operator is executed. Further with a given probability (P_{oc2}) the crossover operator is fulfilled for selected chromosomes. The mutation operator with a given probability (P_{om2}) changes a randomly selected element of individual. After applying the crossover and mutation the size of the intermediate population increases. For reducing population to its original size reduction operator is used that works on the principle of "elite group" [1].
5. The genetic algorithm of the first level. The population of the first level is processed similarly to the second level, but has its own characteristics. To evaluate the objective function on the object model ATC there are used the chromosome of the first level with the best chromosome in the corresponding population of the second level. Then apply with a given probability (P_{oc1}) above problem-oriented crossover operator and mutation (P_{om1}). For reduction of intermediate population size is applied the "elite group" principle too.
6. Check the stopping criterion of genetic algorithm. The counter value (M) is compared to the number of formed generations. If equality is not achieved then the process is repeated starting from the second step. When stop criterion achieved, the top three (in the sense of the objective function) pairs of chromosomes are selected as the sub-optimal solution solution.

A modified GA is designed with a two-level representation of the chromosomes, which allows to vary the sequence of tooling for machining parts at the 1-st level, and the size of the parts batches at the second level. The characteristics of ATC machining problem-oriented crossover and mutation operators are deteremined for the high and lower level of chromosomes, selection and reduction strategies are defined for the new population of chromosomes.

For the modified genetic algorithm the values of its parameters are determined: population power, the number of generations, the probability of crossover and mutation, providing the sub-optimal values of selected criteria. For the one-level genetic algorithm obtained deviation from the optimum solutions does not exceed 5 %, and two-level GA provides an improvement of 27.7 % at the criterion of the production cycle length relatively to the one-level one.

As a result of experiments with genetic algorithm the following parameter values are defined: population power of the first level $n_1 = 8$ and the generation number of the genetic algorithm $T_1 = 10$. Also suboptimal parameters are identified for probability of crossover and mutation: $P_c = 0.7$ and $P_m = 0.01$ (according to the criterion of the of the production cycle length).

In addition to the GA for solving this problem ant algorithm [4] is developed. On the basis of the constructed models the task of modification of ant colony method is set to optimize the load schedules of production site equipment. For the first time it was proposed the "directed" ant algorithm to optimize schedule of production site. For suggested algorithm the preferred choice of a graph vertex is defined which is based on the "directionally proportional" transition rule. The "global rules" are suggested to calculate the concentration of the pheromone in the ant transition to the next graph vertex, which contribute to the directed search.

The population size of artificial ants, corresponding to the amount of techno-logical equipment used in the production (FPM and transport), is defined and validated. In this case for all agents, except for the "elite", the list of forbidden vertices (tabu list) is defined.

6 Conclusion

In paper the object models of typical components are developed for the automated technological complex: a flexible production module, automated warehouse, automated transport management system. For each type of objects are defined their main properties and methods. A generalized model of automated technological complex machining is designed.

A modified genetic algorithm with two-level representation of the chromosomes is proposed. It allows to vary the sequence of machinig part batches at the first level, and the sizes of the part batches on the second level. With taking into account the features of ATC machining there are developed problem-oriented genetic operators of crossover and mutation for the chromosomes at the upper and lower levels. it is defined the strategy of chromosomes selection and reduction for the new population.

To optimize the ATC functioning, together with a modified genetic algorithm are used the object models, that calculate the fitness-function, forming the optimal schedule of ATC equipment in real time for the following key performance criteria:

the minimization of the production cycle length; maximization of the average load factor of technological equipment.

The simulation of actual ATC machining parts such as bodies of rotation established that the "bottleneck" of considered ATC is an automated transport module. It has the highest load, its intensity of the input stream applications exceeds its throughput. The FPM downtime accounts about 27 % of the time, their main causes are: the expectation of transport module service (12 %), lack of billets (10.5 %), equipment failure (4.5 %). To eliminate the "bottleneck" it is proposed to increase the size of the transport batches up to 50–70 parts and the number of seats in the storage for FPM to 4–6, which increases loading production equipment by 11.2 % without additional cost

For the modified genetic algorithm suboptimal values of its parameters are defined: the power of the population, the number of generations, the probability of crossover and mutation, providing the definition of near-optimal values of selected performance criteria of ATK. For the one-level genetic algorithm the deviation of optimum solutions, obtained with full search, does not exceed 5 %. Two-level provides an improvement on the 27.7 % according to the criterion of the duration of the production cycle length.

The structure of the IP system to support decision-making in the management of ATC is developed. It is built on the basis of using object model, genetic and ant algorithms, which make in real-time analysis of the production situation, forecasting and the formation of sub-optimal schedules of equipment. It provides high efficiency operation ATK in general.

Acknowledgments The research is supported by the Russian Science Foundation (project № 16-19-00199).

References

1. Skobtsov, Y.A., Speransky, D.V.: Evolutionary computation: hand book. The National Open University "INTUIT", Moscow (2015).(in Russian)
2. Skobtsov, Y.A., Lazdyn, S.V., Telyatnikov, A.O., Petrov, A.V., Zemlyanskaya, S.Yu.: Simulation and Optimization of Distributed Information Systems. Publishing house "Noulidge", Donetsk (2012). (in Russian)
3. Skobtsov, Yu.A., Skobtsov, V.Yu.: Evolutionary test generation methods for digital devices. In: Adamski. M., et al. (eds.) Design of Digital Systems and Devices. LNEE., vol. 79, pp.331–361. Springer, Heidelberg (2011)
4. Chengar, O.V.: Development of the "directed" ant algorithm to optimize production schedules. Bull. Kherson Nat. Tech. Univ. **1**(46), 212–217 (2013). (in Russian)

Research into Structural Reliability and Survivability of Complex Objects

Anton E. Paschenko, Alexander N. Pavlov, Alexey A. Pavlov,
Alexey A. Slin'ko and Alexander A. Masalkin

Abstract The paper describes the scientific and methodical foundations of study into the central issues of structural modeling of monotone and non-monotone complex objects such as determining their structural reliability and survivability. The studies are based on the original concept of the genome of the system structural construction. Particular attention is paid to the monotone and non-monotone structures of the second type using the example of systems "at least k of n" and "exactly k of n". Features inherent in them are revealed.

Keywords Monotone and non-monotone structure of the second type · Structural reliability · Structural survivability

1 Introduction

Currently, the development and methodological foundations of the complex automation of processes of adaptive evolution of complex objects (CO) are very relevant, which are based on the results of the modern theory of controlled self-organization of complex organizational and technical systems, including the combined models, methods and algorithms for the analysis and management of the structural dynamics of CO in conditions of incompleteness, indeterminateness, inaccuracy, and inconsistency of information about the evolving situation and the

A.E. Paschenko (✉)
SPIIRAS, Saint Petersburg Institute of Automation and Informatics,
Russian Academy of Sciences, St. Petersburg, Russia
e-mail: AEP@iias.spb.su

A.N. Pavlov · A.A. Pavlov · A.A. Slin'ko · A.A. Masalkin
Mozhaisky Military Space Academy, St. Petersburg, Russia
e-mail: Pavlov62@list.ru

A.N. Pavlov
Saint Petersburg National Research University of Information Technologies,
Mechanics and Optics (ITMO), St. Petersburg, Russia

© Springer International Publishing Switzerland 2016
R. Silhavy et al. (eds.), *Automation Control Theory Perspectives
in Intelligent Systems*, Advances in Intelligent Systems and Computing 466,
DOI 10.1007/978-3-319-33389-2_44

presence of a catastrophic threshold time limit for the cycle of formation and implementation of solutions for prevention of possible critical, extraordinary, and emergency situations.

In general, the problem of structural dynamics analysis of the CO includes three major subclasses of problems:

- the problem of the CO structural analysis;
- problems of research into CO structural dynamics in the absence of input (control actions and disturbances);
- problems of research into CO structural dynamics in the presence of input actions.

Traditionally, when using many structural modeling techniques for CO, monotony condition of their functioning was considered mandatory or default [1–3]. As a rule, this is caused by the properties of natural monotony of many real-life complex organizational-technical systems, such as the absence of elements of the system, failure or restoration of which increases or decreases, respectively, the reliability of the system as a whole.

However, in recent years, a lot of new research trends [4–6] have emerged, in which a problem of constructing structural models of the functioning of non-monotonic CO becomes important. For example, study of the systems in a targeted, unknown environment requires inclusion of components and subsystems that reflect the processes of destruction, defeat, penetration, countering the counter-party, the presence of force majeure, in the structural model.

In work [7], a classification of the CO structures was proposed shown in Fig. 1.

According to this classification, all monotone and non-monotone CO structures are divided into two types. The description of the models of the structures of the monotone CO of the first type uses graphs connectivity, two or multi-terminal networks. In these graphs, logical connections between the elements are represented by two logical operators, "AND", "OR". To summarize the mutual influence of CO elements on each other, for monotone systems of the first type, fuzzy graphs are introduced, i.e. graphs with weights from the [0,1] interval. The monotone system of the second type includes such systems, a structure of interaction of elements of which can be described by monotone logical functions, but cannot be presented by graphs connectivity.

To remove restriction on the monotony and allow for the construction of logical conditions for functioning of elements of monotone systems of the first and second type, and any non-monotone complex system objects, a common logical-probabilistic method (CLPM) [6, 8] was designed, which is the development of classical methods of logical-probabilistic reliability calculation, allowing research into the monotone systems of the first type [1, 6, 9].

The main feature and fundamental difference of CLPM from classical monotone LPMs is a fact, that CLPM includes the new means of graphical representation of structures of the systems (the scheme of functional integrity—SFI) including graphic means of a functionally complete set of logical operations "AND", "OR", and "NOT".

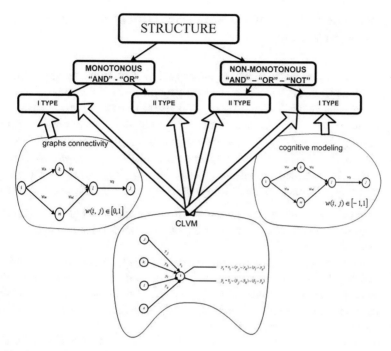

Fig. 1 Classification of the CO structures

Besides, the universal graphic-analytical method of solution of the logical equations was developed, which allows constructing the logical-probabilistic model of CO functioning considering independent, disjoint events, multiple states of elements, multipurpose elements and subsystems.

Also, to "NOT" operation introduction in structural modeling, use methods of cognitive modeling can belong, in which the most widespread classes of models are the weighed (sign) orgrafs, pulse processes, and indistinct cognitive maps [10–13]. The indistinct cognitive map represents a causal network (or an indistinct positive and negative semantic network) in which the weights of communications lie in an interval of $[-1,1]$. Positive and negative communications can be interpreted in terms of fuzzy logic. In a cognitive map, positive communications of "x influences on y" type, in logical interpretation, can be considered an implication $x \Rightarrow y$ or $\bar{x} \Rightarrow \bar{y}$. Negative communications can be represented as $\bar{x} \Rightarrow y$ or $x \Rightarrow \bar{y}$.

To conduct the "NOT" operation, works [10–13] entered negative weights to orgraph arches. Though "NOT" operation is taken into account in the considered approach, it is not possible to describe the structural model of functioning of any non-monotone system. Structures of non-monotone CO, which can be presented by means of an indistinct cognitive map belong to non-monotone structures of the first type. Structure interactions of which elements can be described by non-monotone logical functions but it is not possible to present them by indistinct cognitive maps, belong to non-monotone structures of the second type.

2 Theoretical Bases of the Complex Object Structure Genome Concept

As a result of the conducted research studies [4, 5, 7] into structures (monotone and non-monotone CO), the concept of a direct genome and dual genome of structure, which represents vector $\vec{\chi} = (\chi_0, \chi_1, \chi_2, \ldots, \chi_n)$ and $\vec{\eta} = (\eta_0, \eta_1, \eta_2, \ldots, \eta_n)$, whose components are coefficients of a failure polynomial $T(Q) = \chi_0 + \chi_1 Q + \chi_2 Q^2 + \cdots + \chi_n Q^n$ and a reliability polynomial $R(P) = \eta_0 + \eta_1 P + \eta_2 P^2 + \cdots + \eta_n P^n$ of CO structure made of uniform elements, is introduced. The obtained results allow us to take up the position that the object structure genome stores a complete image of CO structure in the concentrated form:

- topological properties of monotone structures of the first type—parallel-serial structures (P-structures), complex "bridged" structures (H-structures);
- properties of structural monotony;
- allows quantitative estimation of structural reliability, the importance and contributions of elements into structural reliability of uniform and non-uniform, monotone and non-monotone structures of CO both in the case of probabilistic description of failures (reliability) of structure elements and their fuzzy-possibility interpretation;
- besides, the integrated representation of indicators of structural failure (or structural reliability) for various scenarios of CO degradation allows calculation of values of the CO structural survivability indicator.

We will get acquainted with some topological properties of structure reflected in a genome in the following section. As for property of structural monotony,

- if $\chi_0 = 0$ and the sum of vector components is equal to 1, then the polynomial of the structural failure function $T(Q) = \chi_0 + \chi_1 Q + \chi_2 Q^2 + \cdots + \chi_n Q^n$ describes monotone structure of CO;
- if $\chi_0 = 0$ and the sum of vector components is equal to 0, then the CO structure is non-monotone and polynomial of the structural failure function does not remain "1" (i.e. $T(1) = 0$);
- if $\chi_0 = 1$ and the sum of vector components is equal to 1, then the CO structure is non-monotone and polynomial of the structural failure function does not remain "0" (i.e. $T(0) = 1$);
- if $\chi_0 = 1$ and the sum of vector components is equal 0, the CO structure is non-monotone and the polynomial of the structural failure function does not remain "0" and "1" (i.e. $T(0) = 1$, $T(1) = 0$).

To calculate values of indicator of the structural failure of monotone and non-monotone CO taking into account the elements, uniform $(F_{odnor}(\vec{\chi}))$ or non-uniform $(F_{neodnor}(\vec{\chi}))$ by the failure, as well as in the case of fuzzy-possibility

$(F_{vozm}(\vec{\chi}))$ interpretation of element failures, we can use the following formulas [4, 5, 7]:

$$F_{odnor}(\vec{\chi}) = \vec{\chi} \cdot (1, \frac{1}{2}, \frac{1}{3}, \ldots, \frac{1}{n+1})^T, \quad F_{neodnor}(\vec{\chi}) = \vec{\chi} \cdot (1, \frac{1}{2}, \frac{1}{2^2}, \ldots, \frac{1}{2^n})^T,$$

$$F_{vozm}(\vec{\chi}) = \sup_{\mu \in [0,1]} \min\{\vec{\chi} \cdot (1, \mu, \mu^2, \ldots, \mu^n)^T, g(\mu)\}.$$

One of the central problems of structural modeling of CO, especially in the conditions of destructive influences, consists in the determination of structural survivability of objects. The offered ways of estimation of structural survivability [14], such as cornerstone procedures of identification of the minimum sections of network structures, are applicable only to monotone structures of the first type and belong to the solution of combinatory problems of major computing complexity. However, the developed combined method of the occasionally directed search of evolutionary type [15, 16] allowed for the creation of scenarios (trajectories) of structural degradation of CO in the course of failures of the prescribed set or all sets of elements of the structure. For optimistic, pessimistic, or arbitrary scenarios of structural changes of both monotone and non-monotone, uniform, and non-uniform COs constructed with use of the specified method, the indicator of structural survivability $J^k = S_0^k / S^k$ of some k-th scenario of degradation, as ratio of the respective areas presented in Fig. 2, is introduced.

In Fig. 2, parameter $\alpha_j (j \in \{0, 1, \ldots, N\})$ is the number of an intermediate structural state of CO in the considered degradation trajectory. Used as an indicator of structural failure $F_{otkaz}(\vec{\chi}_{\alpha_j}) \in \{F_{odnor}(\vec{\chi}_{\alpha_j}), F_{neodnor}(\vec{\chi}_{\alpha_j}), F_{vozm}(\vec{\chi}_{\alpha_j})\}$ can be one of three indicators on the assumption that the CO structure consists only of elements, uniform by failure, only from elements, non-uniform by failure, and, finally, that there are possibility failures of elements.

Fig. 2 Graphic interpretation of structural survivability of k-th scenario of CO degradation

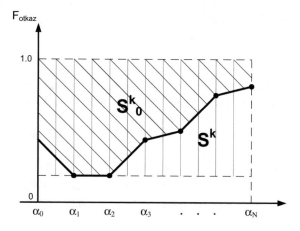

For each of these three cases, indistinct values of survivability indicator $(a^{\hat{i}}, \alpha^{\hat{i}}, \beta^{\hat{i}})$, (a^n, α^n, β^n), (a^w, α^w, β^w), where, say, $a^{\hat{i}} - \alpha^{\hat{i}}$—a pessimistic assessment of structural survivability of CO made of elements, uniform by failure, (the pessimistic scenario), $a^{\hat{i}} + \beta^{\hat{i}}$—an optimistic assessment of structural survivability (the optimistic scenario), $a^{\hat{i}}$—the most expected assessment of structural survivability, are defined (averaged processing of scenarios selected randomly). Then, we will assume the average size of the received results $J_{SG} = \frac{(a^{\hat{i}},\alpha^{\hat{i}},\beta^{\hat{i}}) + (a^n,\alpha^n,\beta^n) + (a^w,\alpha^w,\beta^w)}{3}$ as an integrated value of an indicator of structural survivability of CO J_{SG}.

3   Research into Monotone and Non-monotone Structures of the Second Type

While monotone structures of the first type, within which parallel-serial (P-structures) and complex "bridged" (N-structures), and non-monotone structures of the first type are known and often used in the solution of various practical problems of estimation of reliability, survivability, safety, the risk of structurally complex objects, monotone and non-monotone structures of the second type are not so common. Therefore, it may seem that the structures of the second type cannot exist at all. Indeed, if the interaction of the CO elements can be described by the monotone logic function using logical operators "AND", "OR", it seems that such a structure should be presented with a graph. It turns out it is not so. To confirm this fact, consider a CO in which there are subsystems of "at least k of n" or "exactly k of n" type.

The reliability polynomial of "exactly k of n" structure (the structure consists of *n* uniform elements and it is serviceable, if exactly *k* elements are serviceable) has form $R_{=k,n}(P) = C_n^k P^k (1 - P)^{n-k}$ where P is a probability of failure-free operation of the CO element.

The reliability polynomial of "at least k of n" structure (the structure consists of *n* uniform elements and it is serviceable, if at least *k* elements are serviceable) has a form $R_{\geq k,n}(P) = \sum_{i=k}^{n} C_n^i P^i (1 - P)^{n-i}$.

Without losing the generality of reasoning, we will conduct a structural analysis of these systems for the case when the number of elements n is equal to 7. The dual genomes of the studied structures are shown below, while graphs of polynomials are presented in Fig. 3a, b.

The dual genomes of "exactly k of n" structures for n = 7 are vectors:

$$\vec{\eta}_{=1} = (0, 7, -42, 105, -140, 105, -42, 7)^T, \quad \vec{\eta}_{=2} = (0, 0, 21, -105, 210, -210, 105, -21)^T,$$
$$\vec{\eta}_{=3} = (0, 0, 0, 35, -140, 210, -140, 35)^T, \quad \vec{\eta}_{=4} = (0, 0, 0, 0, 35, -105, 105, -35)^T,$$
$$\vec{\eta}_{=5} = (0, 0, 0, 0, 0, 21, -42, 21)^T, \quad \vec{\eta}_{=6} = (0, 0, 0, 0, 0, 0, 7, -7)^T.$$

(a)   **(b)**

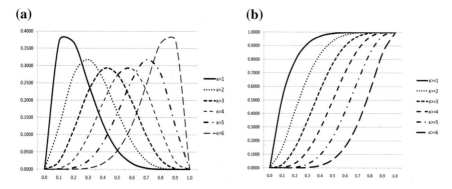

Fig. 3 Graphs of reliability polynomials of structures **a** "exactly k of n", **b** "at least k of n"

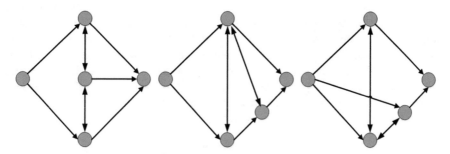

Fig. 4 "Bridged" structures with 7 elements

The dual genomes for "at least k of n" structures for n = 7 are vectors:

$$\vec{\eta}_{\geq 1} = (0, 7, -21, 35, -35, 21, -7, 1)^T, \; \vec{\eta}_{\geq 2} = (0, 0, 21, -70, 105, -84, 35, -6)^T,$$
$$\vec{\eta}_{\geq 3} = (0, 0, 0, 35, -105, 126, -70, 15)^T, \; \vec{\eta}_{\geq 4} = (0, 0, 0, 0, 35, -84, 70, -20)^T,$$
$$\vec{\eta}_{\geq 5} = (0, 0, 0, 0, 0, 21, -35, 15)^T, \; \vec{\eta}_{\geq 6} = (0, 0, 0, 0, 0, 0, 7, -6)^T.$$

Reliability polynomials "exactly k of n" structures are non-monotone, and for "at least k of n" structures—are monotone. Why do these structures belong to structures of the second type? In [17] it is proved that the senior component of the genome (dual genome) of any parallel-series structures is $\eta_n = \pm 1$ and can be greater than 1 ($|\eta_n| > 1$) by modulus only for "bridged" structures. However, the structures homeomorphic to "bridged" ones and consisting of 7 elements can be represented as graphs presented in Fig. 4.

Dual genomes of structures in Fig. 4 represent the vectors:

$$\vec{\eta}_a = (0, 0, 2, 2, -2, -7, 9, -3)^T, \; \vec{\eta}_b = (0, 0, 3, -2, 5, -13, 11, -3)^T,$$
$$\vec{\eta}_c = (0, 0, 2, 1, 0, -9, 10, -3)^T.$$

A senior component represented by the genomes is equal to -3. Senior components of the dual genomes of "exactly k of n" and "at least k of n" structures, except for genome $\vec{\eta}_{\geq 1} = (0, 7, -21, 35, -35, 21, -7, 1)^{T}$, differ significantly from value of -3. Consequently, these structures cannot belong to structures of the first type. However, it should be noted that the structure "at least 1 of n", senior genome component, which is equal to 1, can be represented by a graph depicted in Fig. 5.

The results of the research into the structural reliability and structural survivability of monotone "at least k of 7" structures with use of the original concept of the genome structure are presented in Fig. 6.

Analyzing the presented results, it can be argued that with increasing parameter k, the structural reliability and survivability of these objects decreases monotonically, while the survivability of the "uniform" structures is higher than that of the "non-uniform" ones. This is clearly connected with the monotony property of the considered systems. This should indicate an interesting change in the structural reliability for uniform and non-uniform objects (Fig. 6a). So, if for k = 1, 2, 3, the reliability of "non-uniform" structures becomes higher than the reliability of "uniform" ones, then, for k = 5, 6 "uniform" evaluation becomes higher than "non-uniform" one. The specified property can be observed for all the monotone structures of the first type, which was previously shown in [4]. Therefore, the presented results fail to reveal new features related to reliability and survivability of monotone structures of the second type.

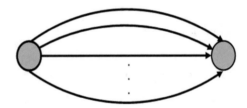

Fig. 5 Structure of "at least 1 of n" type

(a)   **(b)**

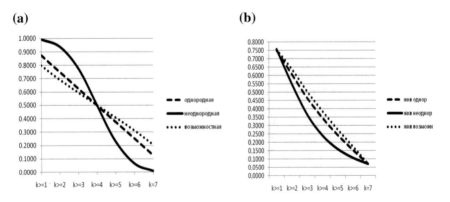

Fig. 6 a Structural reliability of CO of "at least k of 7" type, **b** the structural survivability of CO of "at least k of 7" type

(a)   **(b)**

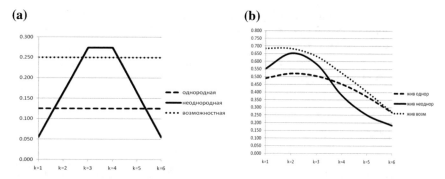

Fig. 7 **a** structural reliability of CO of "exactly k of 7" type, **b** structural survivability of CO of "exactly k of 7" type

Results of the research into the structural reliability and structural survivability of the non-monotone structures "exactly k of 7" are presented in Fig. 7.

After analyzing the presented results, we can indicate some features of "exactly k of n" structures that belong to non-monotone structures of the second type. First, the structural reliability of COs consisting of uniform elements, whether probabilistic or fuzzy-possibility interpretation of failures (reliability) of elements, is constant. The "possibility" evaluation of reliability is higher than "probability" one. Secondly, for non-uniform structures "exactly k of n", function of structural reliability has a pronounced bell-shaped form. For the considered structures with parameters k = 3 and 4 (take the middle position among these structures), reliability is the highest, and the extreme structures have the least structural reliability.

Based on this behavior of structural reliability, one would expect that the structural survivability of CO of type "exactly k of n" for the structures occupying a middle position (k = 3 and 4) will also be the highest. However, it is not the case. The highest structural durability among "exactly k of 7" objects has structure with parameter k = 2 (Fig. 7b). The survivability of the structure with parameter k = 1 is approximately equal to the survivability of the structure with parameter k = 3, and exceeds the survivability of the structure with parameter k = 4. Finally, we should point to the fact that "possibility" assessment of structural survivability is always superior to its "uniform" and "non-uniform" assessments. And a curious change of the structural survivability for uniform and non-uniform objects is observed again.

4 Conclusion

By summarizing the research studies conducted, the following can be concluded. Monotone and non-monotone structures of the second as well as the first type found practical application of structural analysis of CO; especially in the construction of non-monotone structural models in such hazardous industrial sectors as aerospace and space industry, nuclear power industry, oil and gas industry, and others. This situation requires careful consideration of the basic properties of these structures, such as structural reliability and survivability. Therefore, when conducting the research, the revealed features of monotone and non-monotone structures of the second type should be considered.

5 Funding

The research described in this paper is partially supported by the Russian Foundation for Basic Research (grants 15-07-08391, 15-08-08459, 16-07-00779, 16-08-00510, 16-08-01277), grant 074-U01 (ITMO University), project 6.1.1 (Peter the Great St. Petersburg Polytechnic University) supported by Government of Russian Federation, Program STC of Union State "Monitoring-SG" (project 1.4.1-1), The Russian Science Foundation (project 16-19-00199), Department of nanotechnologies and information technologies of the RAS (project 2.11), state research 0073–2014–0009, 0073–2015–0007.

References

1. Ryabinin, I.A., Cherkesov, G.N.: Logiko-veroyatnostnye metody issledovaniya nadeznosti strukturno-slozhnih sistem (Logical probabilistic methods of research of structural complex systems reliability), 264 p. M.: Radio i svjaz (1981). (In Russian)
2. Reinshke, K., Ushakov, I.A.: Application of graph theory for reliability analysis, 208 p. M.: Radio i Sviaz (1988). (In Russian)
3. Ushakov, I.A.: Course on the theory of system reliability, 240 p. M.: Drofa (2008). (In Russian)
4. Pavlov, A.N., Sokolov, B.V.: Structural analysis of disaster-tolerance information system. Trudy SPIIRAN, issue no 8, pp. 128–151. SPb.: SPIIRAN (2009). (In Russian)
5. Osipenko, S.A., Pavlov, A.N.: Research of safety of complex technical objects. News of High Schools. Priborostroenie 53(11), 27–32 (2010). (In Russian)
6. Polenin, V.I., Ryabinin, I.A., Svirin, S.C., Gladkov, I.P.: Application of the general logical-probabilistic method for analyzing the technical, military organizational and functional systems and armed confrontation. In: Mozhayev, A.S. (ed.) Monograph, 416 p. St. Petersburg, NIKA (2011). (In Russian)
7. Pavlov, A.N.: The classification of monotone and nonmonotone information systems based on genome of structure. Trudy SPIIRAN 2(21), pp. 238–248. SPb.: SPIIRAN (2012). (In Russian)
8. Mozhaev, A.S.: Universal graphical-analytical method, algorithm and software module for constructing monotone and nonmonotone logic function performance systems. In: The Third International Scientific School "Modeling and Analysis of Safety and Risk (MASR—2003)", RF, St. Petersburg, August 20–23 (2003). (In Russian)
9. Ryabinin, I.A.: Reliability and Safety of Structural Complex Systems, 248 p. Politechnika, St.Petersburg (2000). (In Russian)
10. Silov, V.B.: Strategic decision-making under fuzzy situation, 228 p. M.: INPRO-RES (1995). (In Russian)

11. Maksimov, V.I.: Structural-targeted analysis of socio-economic conditions. Prob. Manage. (3), 30–38 (2005). (In Russian)
12. Kosko, B.: Fuzzy cognitive maps. Int. J. Man Mach. Stud. **1,** 65–75(1986)
13. Kul'ba, V.V., Mironov, P.B., Nazaretov, V.M.: Stability analysis of socioeconomic systems with the aid of signed digraphs. Autom. Remote Control (7), 130–137 (1993). (In Russian)
14. Podlesnyj, N.I., Rassoha, A.A., Levkov, S.P.: Specific methods of identification, design, and survivability of control systems, 446 p. K.: Vyshcha School (1990). (In Russian)
15. Pavlov, A.N.: Algorithm configuration management structure of complex technical object. Managing the development of large-scale systems. In: Proceedings of the Fifth Conference (October 3–5, 2011, Moscow, Russia), vol. I, pp. 374–377. M .: Institute of Control Sciences V.A. Trapeznikov of Russian Academy of Sciences (2011). (In Russian)
16. Pavlov, A.N., Pavlov, D.A.: Approach for predicting the structural stability of complex objects. T-Commun.—Telecommun. Transp. (6), 65–67 (2013) (In Russian)
17. Pavlov, A.N.: Genome research bipolar network structure. In: IX International Scientific School MA SR-2009 Modeling and Analysis of Safety and Risk in Complex Systems, St. Petersburg, 7–11 July 2009, pp. 429–434 (2009). (In Russian)

The Information Technology
of Multi-model Forecasting of the Regional
Comprehensive Security

Vitaliy Bystrov, Svetlana Malygina and Darya Khaliullina

Abstract The paper focuses on forecasting of the state of the regional comprehensive security by using of multi-model complexes. The informational technology for forecasting of such security based on simulation is proposed. Particular attention is paid the formal description of comprehensive security by using of the theory of matrices and graphs. Computer models designed by using of agent-based modeling and system dynamics method were developed in this study. The results of computer simulation can be considered as analytical information for decision-making in the field of comprehensive security of regional socio-economic and nature-technical systems.

Keywords Agent-based modeling · System dynamics · Regional comprehensive security

1 Introduction

Currently, development level of a state depends on a current condition of its regions which are different from each other on geographical location, climate, level of natural resources endowment, population, infrastructure, etc. Main regional problems are reduced to necessity to identify crisis situations that may endanger national security. Also it is important to study and to analyze these crisis situations because they depend on particularity of region.

V. Bystrov · S. Malygina (✉) · D. Khaliullina
Institute for Informatics and Mathematical Modelling of Technological
Processes of the Kola Science Center Russian Academy of Sciences,
Apatity, Murmansk Region, Russia
e-mail: malygina@iimm.ru

V. Bystrov
e-mail: bystrov@iimm.ru

D. Khaliullina
e-mail: khaliullina@iimm.ru

© Springer International Publishing Switzerland 2016   475
R. Silhavy et al. (eds.), *Automation Control Theory Perspectives*
in Intelligent Systems, Advances in Intelligent Systems and Computing 466,
DOI 10.1007/978-3-319-33389-2_45

Decision making in the field of regional economic development and infrastructure is a complex and time consuming task. This requires the involvement of experts from different fields of knowledge, processing of heterogeneous multidimensional information, implementation of forecasting various scenarios of region development. In connection with the above development of systems of support of decision-making in the sphere of regional complex security is an actual task.

One of the alternatives for building such systems is using of simulation modeling and multi-model complexes. These complexes are used as a means of forecasting possible situation development variants in region according to given conditions. The resulting forecast is analytical information intended for further processing and working out recommendations for decision making.

2 Background

The study of crisis situations one must pay special attention to a notion of regional comprehensive security (RCS). This notion is intimately connected with a notion of sustainable development. In this case we interpret the security system as a state of a complex system when influence of external and internal factors does not worsen the system or not disable its operation and development [1].

One of the primary tasks in the study of RCS is to develop such system of indicators that would reflect existing crises in region objectively and in time. This system should reflect the following aspects: social standard of living, employment, demographics, industrial and scientific-technical potential, fiscal, investment climate, an ecological situation.

The authors proposed an appropriate information technology for studying information support of the problem of RCS forecasting. The information technology is based on the wide possibilities of simulation modeling as a means of scenario forecasting. This technology includes a set of different models, an assessment system of comprehensive security, formal models of situation description, special software. The special software is a set of tools developed in accordance with the architecture of multi-agent systems. This software is designed for managing collaboration of a team of experts in remote mode.

In this report the main attention is paid to the simulated component of the information technology of forecasting of regional comprehensive security.

3 Formal Description of Comprehensive Security

The system of indicators for assessing of RCS was designed in Institute for Informatics and Mathematical Modelling of Technological Processes of the Kola Science Center Russian Academy of Sciences. The system summarizes current indicator systems and forms integral indicators obtained by convolution of some groups of generally accepted security indicators.

1. The economic indicators are: average annual number of employed in the economy; the share of economically active population in total population; transport infrastructure; gross regional product; the share of industries in gross regional product; the consumption of electricity by industries and population; regional productivity; budget revenues per capita; tourism, etc.
2. The social indicators are: population size; natural increase rate; net migration rate; fertility rate; mortality rate; mean annual wages; the average per capita income; life expectancy; aggregate unemployment rate; crime rate; cost of the consumer basket; the degree of depreciation of the housing; the commissioning of residential homes, etc.
3. The bioecological indicators are: background radiation; fuel and energy resources; the reserves of mineral resources; forest resources; change of catch of aquatic biological resources; wastewater discharge into surface water bodies; emission of pollutants into atmospheric air from stationary sources; drinking water quality; air quality, etc.

This system can be represented as a directed graph (Fig. 1) whose nodes are the indicators, and arcs represent the connection between the nodes.

The assessing of RCS is a very difficult task. It is proposed to carry out convolution of the indicators. So the indicators introduced artificially such as an economic attractiveness of the region, an ecological attractiveness of the region and a

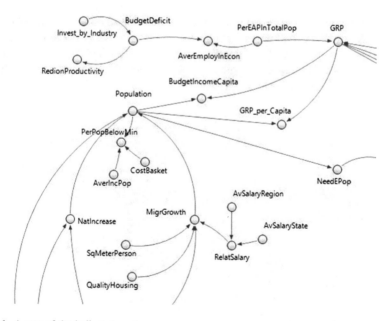

Fig. 1 A part of the indicators system

social attractiveness of the region, are the indicators of the highest level. The remaining indicators are split into levels and groups. The graph theory [2] is used for a formal description of the procedure.

The nodes (v_j) are the indicators of the lower level if their in-degree (the number of incoming arcs) is equal to zero:

$$d^-(v_j) = 0 \qquad (1)$$

The nodes (v_j) are the indicators of the upper level if their out-degree (the number of outgoing arcs) is equal to zero:

$$d^+(v_j) = 0 \qquad (2)$$

The lower level indicators are the parameters for the model, and the upper level indicators are the output of the model. Other indicators can also be divided into sublevels. You can use the following procedure for this: if you remove the nodes of the upper level you derive the upper sublevel, and if you remove the nodes of the lower level you derive the lowest sublevel, etc. (Fig. 2).

The degree of each node can be determined by an incidence matrix $\{a_{ij}\}$, $i = \overline{1,n}, j = \overline{1,m}$, where n is a number of nodes, m is a number of arcs.

$$a_{ij} = \begin{cases} 1, & \text{if the } i\text{-th arc incidenta } j\text{-th node and comes out of it;} \\ -1, & \text{if the } i\text{-th arc incidenta } j\text{-th node and enters it;} \\ 0, & \text{f the } i\text{-th arc not incidenta } j\text{-th node} \end{cases} \qquad (3)$$

Depending on the purpose of modeling it is possible to obtain different outputs. The graph can be divided into subgraphs if the arcs connecting the different blocks (cut-set) will be removed (it is possible to get different blocks such as economic subgraph, social subgraph, ecological subgraph). Formally, this operation can be described by forming the cut-set matrix.

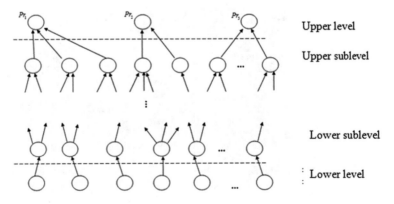

Fig. 2 Distribution of nodes in the graph of the security indicators

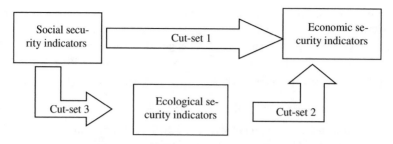

Fig. 3 A schematic depiction of the cut-sets

$$
\cdot q_{ij} = \begin{cases} 1, & \text{if } j\text{-th arc is in } i\text{-th cut-set and its orientation} \\ & \text{corresponds to the orientation of the cut-set;} \\ -1, & \text{if } j\text{-th arc is in } i\text{-th cut-set and its orientation} \\ & \text{does not correspond to the orientation of the cut-set;} \\ 0, & \text{if } j\text{-th arc is not included in } i\text{-th cut-set.} \end{cases} \qquad (4)
$$

In our case we get 3 cut-sets, each of them corresponds to the relations between the blocks discussed earlier (Fig. 3). The direction of the arrows reflects the orientation of the cut-sets.

The matrix of RCS is used as a criterion of assessment of RCS. It is formed with consideration of the offered system of security indicators and includes indicators for each component of security. The matrix allows you to develop formal procedures for comparing different development scenarios of socio-economic and natural-technical systems of region.

This matrix can be formed in two ways:

- One can get the matrix of security specific scenario M_S and in this case the matrix security is formed for each level of indicators.
- All parameters for security assessment are included in a single large matrix of comprehensive security M_{CS}

The general structure of all security matrices has the form:

$$
M_S = \begin{pmatrix} e_1 & b_1 & \cdots & c_1 \\ \vdots & b_2 & \cdots & c_2 \\ e_u & \vdots & \cdots & \vdots \\ \vdots & b_q & \cdots & \vdots \\ 0 & 0 & \cdots & c_l \end{pmatrix} \qquad (5)
$$

The number of columns of the matrix equals the number of top level elements (p). The number of rows equals the maximum number of items associated with the top-level element ($r = \max(u, q, \ldots, l)$). Data from the matrix are the basis for constructing rules. These rules allow us to obtain the values of the elements of the upper level.

All upper sublevel indicators have a range of allowable values [3–6] which defines a critical situation. These ranges give an opportunity to obtain matrices threshold values where matrix $M^- = \left\{ w_{ij}^- \right\}$ is lower range value, and matrix $M^+ = \left\{ w_{ij}^+ \right\}$ is upper range value. The dimension of matrices $(r \times p)$ and their structures correspond to the dimension and structure of the security matrix M_S.

Range of allowable values can be one of three types:

- Type1—the indicator value is not less than a threshold value: $x \geq w^-$.
- Type2—the indicator value is not greater than a threshold value: $x \leq w^+$.
- Type3—the indicator value is enclosed between two threshold values: $w^- \leq x \leq w^+$.

To determine a critical situation it is necessary to carry out normalization of indicators values according to threshold values. For it one can use the following function:

$$g(x, w^-, w^+) = \begin{cases} \frac{w^-}{x}, & \textit{if } \text{there is the first type of range;} \\ \frac{x}{w^+}, & \textit{if } \text{there is the second type of range;} \\ \frac{x - w^-}{w^+ - w^-}, & \textit{if } \text{there is the third type of range.} \end{cases} \qquad (6)$$

The result is a matrix of normalized values of the indicators:

$$K = \{ k_{ij} = g(x_{ij}, w_{ij}^-, w_{ij}^+) \}, i = \overline{1, r}, j = \overline{1, p}. \qquad (7)$$

If at least one of the obtained normalized values of the indicators does not belong to the interval $k_{ij} \notin (0; 1)$ there is a critical situation otherwise there is a non-critical situation.

4 Experiments

Complex of simulation models of RCS is built on the proposed formal model and based on empirical and statistical data. There is its own model for each component of the comprehensive security. The method of system dynamics and agent-based modeling were used to build a model. For example, every industry is represented as an agent with the same type of model structure, but with different indicator values. The complex allows to realize computational experiments and to obtain different results depending on input parameters. The complex of models can be applied as a component of a forecasting system for decision making. For example one can consider a model of manpower needs in the mining industry of Murmansk region.

In this study the term "manpower security of region development" is defined as a set of managerial measures and information technology aimed at management of personnel. The management includes identifying problem areas in the manpower needs of enterprises and working out recommendations to eliminate undesirable effects of manpower policy at all levels of decision making.

The Murmansk region is a region whose economy depends on the development of single-industry towns. For about a third of the population of the region lives in such towns therefore a task of region providing by skilled personnel is a topical issue.

For most industries the reıoasons for the job cuts can be liquidation of enterprises and the adoption of available jobs workers, but for the mining industry in large enterprises, this option can be described by the following equation:

Liquidation business and closing vacancies in the most industries are the reason for job cuts. This process in the mining industry in large enterprises can be described by the following equation:

$$lrm\_kr\_pr = f_1(Ob, dpu), \tag{8}$$

$$lrm\_kr\_pr = 1956 - 0.09 \cdot Ob + 0.043 \cdot dpu, \tag{9}$$

where Ob—turnover of enterprises; dpu—mineral output.

Creation of employment in large enterprises of the mining industry has its own specifics so this dependence is determined as follows:

$$vrm\_kr\_pr = f_2(Ob, dpu), \tag{10}$$

$$lrm\_kr\_pr = 2152 - 0.073 \cdot Ob + 0.092 \cdot dpu, \tag{11}$$

where Ob—turnover of enterprises; dpu—mineral output.

Regression analysis was applied to obtain relations between some indicators.

5 Results

Model verification was carried out with actual data to increase the level of confidence in the results of the simulations. Logical links were checked to confirm accuracy of the logical structure of the model.

Comparison of actual data and simulation results (by the following parameters: vacancies, eliminated workplaces, occupied workplaces) is presented in Fig. 4.

The analysis of these data on mining shows that the simulation generally reproduces the actual results. In the first case, the average error deviation is 7.4 %, in the second case is 7.3 % in the third case is 3.7 %. The verification results give an opportunity to conclude that the behavior of the model is compatible with expert assumptions of the subject area and the model has a correct logical structure. Therefore this model can be used for forecasting and scenario analysis of system behavior of manpower region needs.

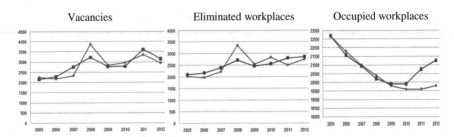

Fig. 4 Comparison of simulation results with actual data mining (■—simulation results, ♦—actual data)

6 Conclusion

The paper presents some results of researches in the field of forecasting of RCS development. It is proposed to use the possibilities of simulation for creating of the information technology for forecasting. The main part of the technology is a multi-model complex. This is built on the offered indicator system of comprehensive security, on the expert knowledge and on data of official statistics. The technology allows us to obtain forecasts for various scenarios of development of regional socio-economic and natural-technical systems that influence on comprehensive security of the region. Recommendations for decision making in the sphere of region development can be based on results of forecasts.

References

1. Zaplatinsky, V.M.: Terminology of science of safety. In: Zbornik prispevkov z medzinarodnej vedeckej konferencie "Bezhecnostna veda a bezpecnostne vzdelanie" (Liptovsky Mikulas: AOS v Liptovskom Mikulasi, 2006) (CD-ROM) (2006)
2. Svami, M., Tkhulasiraman, K.: Grafs, Netwoks and Algorithms, 454p. Mir, Moscow (1984)
3. Novikova, I.V., Krasnikov, N.I.: Indicators of economic security of a region. Tomsk State Univ. J. **330**, 132–138 (2010)
4. Glazyev, S.Yu.: The basis of ensuring the economic security of the country: alternative reformation course. Russ. Econ. J. (1), 8–9 (1997)
5. Voropai, N.I., Klimenko, S.M., Krivorutskiy, L.D. et al.: The essence and main issues of energy security of Russia. In: Proceedings of the Russian Academy of Sciences. Power Engineering Journal, № 3, pp. 38–49 (1996)
6. Kalina, A.V., Savelieva, I.P.: Formation of threshold values of the economic security of Russia and its regions. In: Bulletin of South Ural State University, Series "Economics and Management", vol. 8, № 4, pp. 15–24 (2014)

Dynamic Cognitive Geovisualization for Information Support of Decision-Making in the Regional System of Radiological Monitoring, Control and Forecasting

A.V. Vicentiy, M.G. Shishaev and A.G. Oleynik

Abstract In this paper, we describe the technique of dynamic cognitive geovisualization. Cognitive geovisualization can be used to support user cognitive activity in carrying out tasks of monitoring and forecasting in decision support systems for managing complex natural and technological objects. A feature of cognitive geovisualization is that the construction of geoimages taken into account the peculiarities of perception of the user. Cognitive geoimages improves the efficiency of the visual analysis of large amounts of data, speed and quality of decision-making. We describe the application of dynamic cognitive geovisualization technology in information decision support system for regional radiological monitoring, control and forecasting. To confirm the possibilities of technology we have created a prototype of web geoserver for radiological monitoring, control and forecasting. The main functions of the dynamic cognitive geovisualization technology implemented in this web geoserver as the user tools.

Keywords Cognitive geosualization · Radiological monitoring · Information decision support · Web geoservice

A.V. Vicentiy (✉) · A.G. Oleynik
Institute for Informatics and Mathematical Modelling of Technological Processes
of the Kola Science Center RAS, 24 A, Fersman St., Apatity 184209, Russia
e-mail: alx_2003@mail.ru

A.G. Oleynik
e-mail: oleynik@iimm.ru

A.V. Vicentiy · A.G. Oleynik
Russia and Kola Branch of Petrozavodsk State University,
Kosmonavtov St. 3, Apatity 184209, Russia

M.G. Shishaev
Murmansk Arctic State University, Egorova St. 15,
Murmansk 183038, Russia
e-mail: shishaev@arcticsu.ru

© Springer International Publishing Switzerland 2016
R. Silhavy et al. (eds.), *Automation Control Theory Perspectives
in Intelligent Systems*, Advances in Intelligent Systems and Computing 466,
DOI 10.1007/978-3-319-33389-2_46

1 Introduction

When implementing modern monitoring and forecasting systems for managing complex spatial-organized objects, the operational analysis of large volumes of information problem arises. The problem is compounded by the fact that much of this information is heterogeneous. For example, a regional monitoring system can aggregate and share real-time sensing data, maps, text documents, statistical data, graphic and video data, semantic and other information.

For the analysis of this information requires special tools. These tools should not only provide work with heterogeneous information, but also allows users to analyze different characteristics derived from the processed information. For example, the distribution of various objects in the territories and the time series of different characteristics.

One effective tool for processing large data are special intelligent information technologies that are implemented for specific tasks. Often such intelligent information technology implemented in the framework of object-oriented systems.

To meet the challenges of monitoring and forecasting for managing complex spatially organized natural and technological objects, special decision support systems called intelligent geographic information systems (IGIS) are created. In this paper, under the IGIS understood the geographic information system, which integrates the functions of data processing to assist in forecasting and decision-making. Examples of these features may be data analysis by the selected computational methods, the use of mathematical models of various processes, perform analytical data visualization, use inference engine and other.

But despite the advanced capabilities and powerful tools for intelligent data analysis of modern geographic information systems in the tasks of monitoring and forecasting of potentially dangerous objects and phenomena, can not do without the participation of the person, also known as the decision maker (DMP). Implementation of intelligent information technology in decision support systems ensures the effectiveness of the DPMs.

But even modern information technology can not compete in the field of operational processing of large volumes of heterogeneous geographic information with the human brain. For example, it is one of the reasons for the development of such area of applied computer science, as the Interactive Visual Analysis [1]. The analyzed images are formed in the IGIS on the basis of existing and operational data. In this case, a method of visualization data is critical. Because based on the result of data visualization DMP will perform forecasting and make decision.

1.1 The Problem of Effective Representation of Heterogeneous Data for Monitoring and Forecasting in Decision Support Systems for Managing Complex Spatial-Organized Natural and Technological Objects and Phenomenas

Currently, there are several points of view on what is called data visualization. The main purpose of data visualization, in our opinion, is to reflect the large volumes of data in a convenient form for the visual perception of the DMP. Good visualization is not overloaded with additional elements that are not relevant for making decisions [2]. Thus, data visualization is such a representation of heterogeneous data, which provides the most efficient of their perception, processing and study of the human brain.

The development of the modern understanding of data visualization contributed to the emergence of a number of works on this topic [3, 4]. The result of such work and the development of information technology is that the visual presentation of information and data was used for the study of information and promotion of hypotheses.

Depending on the objectives, emphasis in data visualization can be made on presentation of the data to improve their perception and the analysis and data processing to detect some regularities in them. In modern intelligent analysis systems of heterogeneous data, the data visualization subsystems is one of the most important components. For example, the visualization of raw data is useful for assessing the suitability of the data for analysis. At this stage, often occurs the promotion of hypotheses about the regularities [5] and procedures of the data processing [6, 7].

Visualization of the intermediate processing results can often determine the further direction of analysis. Depending on the results obtained at the intermediate stage of analysis, the analysis target may be adjusted. Visualization of final results provides a convenient way of perceiving the human the data for human brain and serves as a tool for the emergence so called "insight".

A special case of data visualization is a cognitive geovisualization, which allows to combine data about the geographical location of some objects with data from specially organized database. Using the time virtualization allows us to study the complex and distributed in time phenomenon such as landslides, soil erosion, floods, glaciers, etc. The most common geovisualization systems implemented on the basis of some geoservices. Very important is the role of geoservices for monitoring and forecasting of emergency situations or potentially dangerous objects.

In this paper, we describe an approach for building dynamic cognitive geovisualization. We also give a brief description of the developed prototype web geoservice for monitoring and forecasting of radiological situation and decision support in case of potentially dangerous situations (the example of Murmansk region, Russia).

2 Methods

The objective is to describe the method of cognitive visualization of spatial data. The result of cognitive visualization of spatial data is a synthesized geoimage constructed based on the cognitive features of perception of visual information by the user.

The synthesis is carried out as a result of the joint analysis of formal models of the user's request, the visual map stereotypes and user cognitive "settings and preferences" user.

In [8] the problem of increasing cognitive geoinformation is listed as one of the most important problems. Under the influence of this and similar work formed the two main areas of research: (1) research and development of new imaging techniques for spatial information; (2) study of the peculiarities of perception of spatial information to the end user.

At the moment, studies of the perception of the information made significant progress. Even formed a separate interdisciplinary science—cognitive geography. Cognitive geography—the science that studies human perception of space, location and environment. For example, in [9, 10] describes a model of human perception of space under different external circumstances.

As a form of realization of the models we propose to use the ontology. To reduce the cognitive load on the user is prompted to carry out the semantic reducing of geoimage. Thus, the geoimage should more adequately reflect the studied real-world objects, and better meet the objectives of the study data and decision support.

The knowledge accumulated in recent years in areas such as information visualization, scientific visualization, human-computer interaction [11, 12], cartography and others are integrated to develop new strategies for geovisualization.

Many aspects of the visual display large amounts of data, for example, in GIS, in virtual reality systems, in engineering and scientific visualization systems, can be significantly improved, provided greater attention to research in cognitive science (cognitive science, semiotics, psychology of perception, and others.) [13–15]. In this regard, adaptive techniques of cognitive geovisualization for large volumes of complexly organized spatial data must be developed. To do this, should be offered a new model for cartographic stereotype describing and interpretation of user requests in the geovisualization system, develop techniques and algorithmic support of the semantic reduction processes of geovisualization results and create a techniques and technologies to adapt the resulting geoimages for effective visual perception based on a formal representation of knowledge about the cognitive peculiarities of perception users.

Implementation of dynamic cognitive geovisualization technology includes the development of methods and technologies for interactive visualization of spatial data based on cognitive and perceptual characteristics and stereotypes of perception

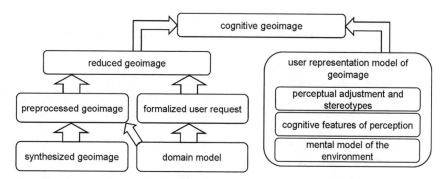

Fig. 1 The scheme of cognitive geovisualization

of visual information to the end user of the system. Generalized basic stages of technology can be summarized as follows (Fig. 1):

1. Development of a formal model of the user's request and descriptions of the visual mapping stereotypes in the geovisualization system to automate the construction of geoimage, which adequate to the task;
2. Development of algorithms of semantic reduction for geovisualization results to reduce the amount of information displayed;
3. Creation of methods and technologies to adapt the resulting geoimage for visual perception based on a formal representation of knowledge about the cognitive user preferences.

To develop models of user request and descriptions of visual stereotypes in the geovisualization system the conceptual modeling and knowledge engineering approaches are used. To ensure effective semantic reduction of synthesized geoimages the method of assessing the quality of cognitive geovisualization are used. The basis of this method is the calculation and assessment of the two formal criteria: (1) informativeness criterion for geoimage; (2) criterion of cognitive load for geovisualization.

Informativeness criterion allows to evaluate the formal loss of geoimage information as a set of graphics primitives. This criterion is used in the first stage reduction. Its value indicates how much of the graphic information can be removed from the geoimage without prejudice to further analysis. The criterion can be applied to the image as a whole and for its individual fragments that relevant in the context of frequent scaling of geoimages.

The criterion of cognitive load allows to evaluate quality of geovisualization in view of features of user representation. The assessment of this criterion gives an idea of how well the visualization of set of graphic primitives was made from a user perspective.

In order that the geoimage has practical value, it should reflect not just the objects some domain, and the specific situation. In addition, for easy user experience situation, it should be presented in accordance with his mental stereotypes. Responding to the situation in this case is associated with cognitive interoperability [16].

One of the key components is a specialized ontology of user representation, describing the visual mapping stereotypes for the different categories of users.

Under cognitive image in this article mean the subjective representation of the state of the monitoring object in a linked set of graphical dynamic components representation of states and rules describing their mutual interference in different situations, that collectively allow the DMP figuratively assess the current situation and elaborate adequate control her actions.

In order to effectively visualize dangerous and potentially dangerous situations for decision support is required to define a language for describing situations and ways of detecting them. Formally, the situation can be described by two ontologies [17]. The first ontology should describe an abstract situation, and the second— particular subject area and classes of situations of this domain. Thus, the domain ontology is built on the base ontology and contains concepts inheriting the concepts of basic ontology. The data on area converted into instances of domain concepts and relations between them and the situation is the notion of a higher level.

The situation S can be represented as a set of all information about the current state and all previous states of the object: $S = \langle 0, R, E, F, Rul \rangle$, where: O—set of physical objects; R—set of areas in space; E—set of events; F—set of predicates, reflecting the relationship between the elements of the sets O, R, E; Rul—set of production rules of the form $(F \Rightarrow G)$ and $(F \Leftrightarrow G)$, that describe the basic mechanisms of logical inference, including restrictions on the value of the predicates F, and also preconditions and consequences for the passage of events E.

Currently, the most common approach in which spatial information can be represented using 4 kinds of objects: the point; line; polygon; raster.

Open Geospatial Consortium in the specification SLD 1.0 describes the possible options for the visualization of these objects. Visualization of objects is described in the file of style in the language xml according to specifications SLD 1.0.

Classes of situations C can be represented as: $C = \langle 0', R', E', F' \rangle$, where: O', R', E'—sets of variables indicating the elements of the sets O, R, E; F'—subset F, each element of which is the predicate, the arguments of which are elements of the sets 0', R', E'.

Then the visualization model situations can be described as follows. The set of classes of situations of different subject areas is divided into a higher level sets. For example, the situation of finding in the way of the radioactive cloud settlement can be attributed to a class of situations in the way of finding a dangerous object of one of the objects of the territory management. Then visualization can be represented as: $V = \langle C', Style \rangle$, where: C'—abstract situation; Style—the display style as a set of style files SLD.

3 Results

The dynamic cognitive geovisualization subsystem is part of the information support of decision-making and monitoring the radiological situation system. At the moment, the system provides a set of radiological situation analysis tools, implemented as a web geoservice.

The information support of decision-making system for regional radiological monitoring, control and forecasting enables users to access the heterogeneous data from different sources, analyze and visualize the data according to their own viewpoint (by setting threshold or selecting the calculation methods) and make decisions. These data includes the radiological monitoring data, spatial data, remote sensing data, thematic maps, statistical data and many other types of data.

Furthermore, the users are able to automatically calculate, analyze and visualize the data on the map in a browser, and share results with other users around the world via web interface.

In the prototype of geoservice users are able to choose their interesting objects and parameters, and visually identify abnormal values of these parameters. Depending on the type of object, users can select the most informative set of data for analysis of the causes and consequences of these deviations.

A feature of the geoservice is the ability to use a variety of spatial data, including images from space satellites (e.g., Landsat [18]). The geoservice has original tools for building composite images from multiple channels of satellite images. These tools can be used in automatic and semi-automatic modes, which facilitates their use by untrained users. Also, users are able to customize the "types of coverage", dynamically generate the most informative in terms of the problem to be solved composites of satellite images channels, and perform other "standard" for geoservices action (determination of coordinates, measurement of distances and areas, etc.).

In the implementation, a prototype system to store various kinds of heterogeneous data from various sources in the database is created. The system receives and stores data different kinds of data, performs the spatiotemporal and thematic data processing and maps the analyzed results onto web map.

The objective of the implementation is to construct a information decision support system for regional radiological monitoring with the possibility of dynamic cognitive geovisualization for sharing and visualization of heterogeneous data via web, which applies for managing complex natural and technological objects.

For the prototype implementation, the database is created by using opensource database PostgreSQL and PostGIS for support geographic objects in a relational database. The other implementation environment is described as below:

- The customer logic implemented by: OpenLayers (JavaScript); HTML 5; CSS 3; GeoExt.
- The server logic is implemented by: GeoServer; PHP; JAVA; GDAL; Apache HTTP Server.
- OS: Mac OS X, Windows Server 2008 R2 Enterprise.

The main sources of data:

- Satellite images of middle and high-resolution;
- Pictures with unmanned aerial vehicles;
- OpenStreetMap data [19];
- Publicly accessible information from web geoservices;
- Problem-oriented database;
- Other sources (external databases, etc.).

To work the client need only a browser and high speed internet access because satellite images and their composites are large (hundreds of megabytes).

The effectiveness of the visualization depends on the understanding of the principles of human thinking, their interpretation of complex data and the possibility of a correct representation of mental models by working with a variety of subject areas.

In the process of practical implementation of the system of information support of decision-making and monitoring the radiological situation it was divided into three major logical unit (Fig. 2): Storage subsystem; Data access and preprocessing subsystem; Information support of radiological measurements, monitoring and decision-making subsystem, the interaction of which is organized by the HTTP and FTP protocols.

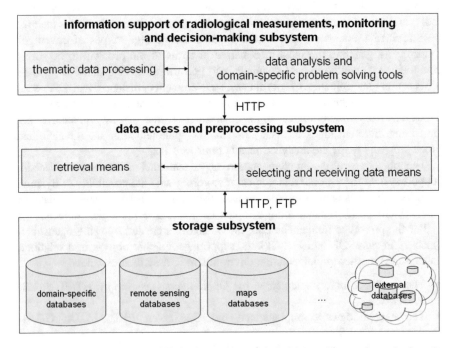

Fig. 2 Basic logic blocks of information support of decision-making and monitoring the radiological situation system

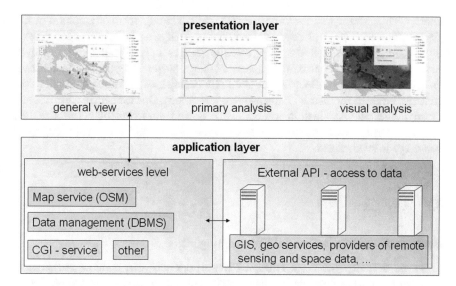

Fig. 3 The architecture of information support of decision-making and monitoring the radiological situation system

Tools for thematic data processing, data analysis tools and tools for object-oriented tasks solutions presented to the user in the form of ready-made web-services with an intuitive interface that allows for more efficient use of them. The implementation unit in charge of search, retrieval and primary processing of data and the physical organization of data storage are hidden from users.

The architecture of the information support of decision-making and monitoring the radiological situation system is represented by two major levels (Fig. 3)—presentation layer that provides the user interface and application layer implementing computing services.

The presentation layer provides all the basic tools available to the user—selection, review, data analysis, visualization, etc. The application layer implements the bulk of the calculations associated with the selection of data from external storage, pre-treatment and the formation of cognitive geoimages with which users interact.

Also, this layer is responsible for the acquisition, thematic processing, and delivery of processed data to the users.

3.1 The General Algorithm of Working with the System

For visual analysis of the results of radiological monitoring the users via an intuitive web interface are able to choose on an interactive map needed points at which the measurements takes place and designate a time interval for which the data of interest. This information is automatically visualized in the form of graphs showing the measurement values for the selected period. Analyzing the extrema are graphs, users define the area on the map, satellite images they would like to analyze additionally, and the approximate dates of taking images and other characteristics. Primary navigation to select the required monitoring points and cartographic substrate is provided by means of opensource technology OpenStreetMap. As a result of cognitive geovisualization system working and attraction of additional data (weather, terrain features, parameters observed objects and others) the cognitive geoimage is synthesized.

3.2 Use Case

Below is an example of using the system (Fig. 4).

1. In the first step the users through the web interface can conduct radiological monitoring. It is possible to obtain additional information about the objects of interest (graphics, history of measurements at various points, etc.), superimpose different time satellite images, browse the various composites and others.

   In this step, the user actively interacts with the databases (domain-specific, remote sensing, maps and others) to obtain all the necessary information. The user can build, compare and analyze various data samples to determine the most interesting area in terms of solved problem.

   When the user defines an "area of interest" which he would like to explore in more detail, he can zoom it and to obtain additional data for this area.

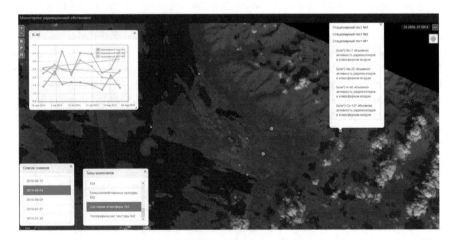

Fig. 4 The user interface of information support of decision-making and monitoring the radiological situation system

In addition, the information decision support system for regional radiological monitoring, control and forecasting provides the user with access to multi-spectral satellite images (and other remote sensing data) of the selected area. And with the help of tools to work with satellite images that are implemented in this system, the user can build a composites of a number of spectra that will best fit problems to be solved by the user. Otherwise, the user can use a system generated sets of composites for solving typical problems.

2. In case the parameters for one or more points abroad of the regulatory limits, check is performed and "hazardous" points are allocated on the map in the operational mode.

   The information decision support system for regional radiological monitoring, control and forecasting is able to detect the exceeding of limit values of one or more parameters in one or more control points in the automatic mode. In such cases the system verifies and identifies dangerous points on the map online. After that, the user can select the points that deserve attention in the first place (the most priority points). He can choose them on the map and view the parameter values and change history for each of them.

3. Further information-analytical system calculates the direction of propagation of the "cloud of pollution", calculates the settlements and other infrastructure facilities that are on the path of "the cloud of pollution", displays them on the screen and calculates the additional parameters, such as the "trajectory of the cloud," "time approach clouds", "the degree of scattering of clouds at the time of approach to human settlements", "radius of danger zone", etc. A set of calculated parameters can be expanded by adding additional software modules for calculating.

The development of such modules and models require a large amount of additional knowledge, such as in fluid dynamics, aerodynamics, the physics of liquids and aerosols, and others. Development of additional modules and models and their inclusion in the information decision support system for regional radiological monitoring, control and forecasting is one of the directions of further development. The development of useful capabilities of the system is also largely due to the filling of additional semantic information. For example, about the population in the cities in the study area, land use maps, cadastre data, etc.

4 Discussion

Implementation of dynamic cognitive geovisualization technology for monitoring, control and forecasting in information decision support systems for managing complex spatially organized natural and technological objects allows operatively generate a mapping interfaces, with high levels of cognition. Formalized in the form of an ontology the domain knowledge provides the ability to automate the analysis of information describing the situation from a variety of sources. Such features of this technology opens up wide possibilities to build on its base comprehensive information support systems for territories management tasks.

To quantify the benefits of using the dynamic cognitive geovisualization technology in information decision support system for regional radiological monitoring, control and forecasting is quite difficult. Quantitative evaluation often highly dependent on the type of solved problem. But, in general, the experimental use of technology showed that the cognitive load on the DMP is reduced by about 10–30 % (depending on a problem) in comparison with the traditional means of information support of decision-making that uses the DMP in their work.

Also, the use of cognitive geovisualization technologies in the project for the development of methodology, modeling tools and information technologies for systemic risk assessment of new exploration of the Arctic will contribute to the integrated use of heterogeneous data, and interdisciplinary knowledge in the formulation of risk-sustainable solutions in the field of safe development of Arctic territories.

5 Conclusion

In this paper, we have presented the dynamic cognitive geovisualization technology that allows to improve the efficiency of the visual analysis of large amounts of geodata, speed and quality of decision-making in decision support systems for managing complex spatially organized natural and technological objects and allows operatively generate a mapping interfaces, with high levels of cognition.

We also have gave a brief description of the developed web geoservice prototype for monitoring, control and forecasting of radiological situation and decision support in case of potentially dangerous situations (the example of Murmansk region, Russia).

Experimental use of the technology in the prototype has demonstrated that the cognitive load on the DMP can be significantly reduced because he uses a synthesized image, constructed considering the cognitive features of visual information perception. The combination of heterogeneous data for monitoring and forecasting and user representation model of geoimage and mapping enables more effective and deep data analysis and decision-making.

Acknowledgement This work is partially supported by RFBR grant № 15-29-06973 "Development of methodology, modeling tools and information technologies for systemic risk assessment of new exploration of the Arctic".

References

1. Oeltze, S., Doleisch, H, Hauser, H., Weber, G.: Interactive visual analysis of scientific data. In: Presentation at IEE VisWeek 2012, Seattle, WA, USA
2. Iliinsky, N., Steele, J.: Designing Data Visualizations. O'Reilly, Sebastopol, CA (2011)
3. Bertin, J., Barbut, M.C.: Sémiologie Graphique. Les diagrammes, les réseaux, les cartes, 431 p. Gauthier-Villars, Paris (1967)
4. Tukey, J.W.: Exploratory Data Analysis, 688 p. Pearson, Reading, Mass (1977)
5. Zenkin, A.A.: Cognitive Computer Graphics. In: Pospelov, D.A. (ed.) 192 p. M.: Nauka (1991). (Зенкин А. А. Когнитивная компьютерная графика/Под ред. Д. А. Поспелова. – М.: Наука., 1991. 192 с.)
6. Pospelov D.A.: Artificial intelligence. Directory. Book 2. Models and methods. M.: Radio and Communications, 304 p. (1990) (Поспелов Д.А. Искусственный интеллект. Справочник. Книга 2. Модели и методы М.: Радио и связь, 1990. 304 с.)
7. Visualization in Scientific Computing. Special Issue, ACM SIGRAPH Computer Graphics, vol. 21, no. 6 (1987)
8. MacEachren, A., Kraak, M.: Research challenges in geovisualization. Cartogr. Geogr. Inf. Sci. **28**(1), 3–12 (2001)
9. Montello, D.R.: A conceptual model of the cognitive processing of environmental distance information. Spatial information theory. In: 9th International Conference, COSIT 2009 Aber Wrac'h, France, September 21–25, 2009 Proceedings, pp. 1–17. doi:10.1007/978-3-642-03832-7_1, Print ISBN 978-3-642-03831-0, Online ISBN 978-3-642-03832-7, Series Volume 5756 Series ISSN 0302-97430
10. Montello, D.R.: The Perception and Cognition of Environmental Distance: Direct Sources of Information. In: Frank, A.U. (ed.) COSIT 1997. LNCS, vol. 1329, pp. 297–311. Springer, Heidelberg (1997). http://dx.doi.org/10.1007/3-540-63623-4_57
11. Cooper, A., Reimann, R.: Dave Cronin About Face 3: The Essentials of Interaction Design, 610 p. Wiley (2007)
12. Hewett, B., Card, C., Gasen, M., Perlman, S., Verplank, W.: ACM SIGCHI Curricula for Human-Computer Interaction. UTILITY doctypa. http://old.sigchi.org/cdg/cdg2.html#2_1
13. Robert, L., Solso, M., MacLin, K., MacLin, O.H.: Cognitive Psychology, 8th edn., p. 592. Allyn & Bacon (September 7, 2007) (2008). ISBN13: 978-0205521081 ISBN10: 0205521088
14. Koffka, K.: Principles of Gestalt Psychology, p. 720. Routledge, NY (1935)
15. Miller, G.: The magical number seven, plus or minus two. Psychol. Rev. **63**, 81–97 (1956)
16. Buddenberg, R.: Toward an Interoperability Reference Model. http://web1.nps.navy.mil/~budden/lecture.notes/interop_RM.htm
17. Matheus, C., Kokar, M. et al.: SAWA: an assistant for higher level fusion and situation awareness. In: SPIE Conference on Multisensor, Multisource Information Fusion: Proceedings of SPIE, vol. 5813, pp. 75–85. Orlando, FL (2005)
18. Data source: Landsat Science. http://landsat.gsfc.nasa.gov/
19. Data source: Openstreetmap. http://www.openstreetmap.org

Remote Sensing for Environmental Monitoring. Complex Modeling

Victor F. Mochalov, Andrei V. Markov, Olga V. Grigorieva,
Denis V. Zhukov, Olga V. Brovkina and Ilya Y. Pimanov

Abstract In this paper the concept of integrated modeling and simulation of the processes of the Complex Technical–Organizational System (CTOS) is presented. Practical directions of the remote sensing for environmental monitoring of the protected area are proposed by the authors. Methodical basis of the integrated modeling and simulation, the process of CTOS operation, and the technology of the remote sensing for environmental monitoring are considered. Results of CTOS remote sensing are shown to adapt models to a changing environment.

Keywords Complex technical organizational system · Structure dynamic control · Parametric and structural adaptation of models · Remote sensing for environmental monitoring · Decision support systems

1 Introduction

Environmental monitoring involves the determination of environmental indicators (first of all, indicators of territorial sustainability under anthropogenic pressure) and also the identification of anthropogenic impact causes and sources. Physical-geographical characteristics and size of the region, its plant species and qualitative composition are the main factors that characterize sustainability of the territory under anthropogenic pressure. Currently, the theory, methods and techniques concerning

V.F. Mochalov (✉) · A.V. Markov · O.V. Grigorieva · D.V. Zhukov
Mozhaisky Aerospace Academy, Saint Petersburg, Russia
e-mail: vicavia@yandex.ru

O.V. Brovkina
Global Change Research Centre Academy of Science of the Czech Republic,
Prague, Czech Republic

I.Y. Pimanov
Saint Petersburg Institute of Informatics and Automation (SPIIRAS),
Russian Academy of Science, Saint Petersburg, Russia

© Springer International Publishing Switzerland 2016
R. Silhavy et al. (eds.), *Automation Control Theory Perspectives
in Intelligent Systems*, Advances in Intelligent Systems and Computing 466,
DOI 10.1007/978-3-319-33389-2_47

the application of mathematical models are widely used. The advanced methodology of environment monitoring based on the Complex Technical-Organizational System (CTOS) model and the application of remote sensing data has been performed [1, 2]. However, such issues as a quality estimation of multi-criteria models, an analysis and classification of applied models, as well as justified selection of task-oriented models are still not well investigated. The importance of the task increases when a research object is described not via a single model, but with a set or a complex of multiple-models including models from different classes or combined models, such as combined analytical–simulation models, logical-algebraic ones, etc. In [1], for example, authors describe the main components of aerospace monitoring system: data processing, archiving, presentation and analysis subsystems, management and performance control unit, and they set the target to maximize its efficiency, improve data handling scheme and develop new methods and tools for their implementation. However, the approach does not rely on complex modeling capabilities of aerospace monitoring system or CTOS including the object of monitoring. This study is aimed to improve the mentioned methodology using CTOS model. The main objective of the present study is to estimate the environmental indicators by choosing an optimal set of satellite data, thematic processing algorithms and technical equipment involved in CTOS functioning.

2 Theoretical Basis

In practice, the processes of CTOS operation are non-stationary and nonlinear. The perturbation impacts initiate the CTOS structure-dynamics and predetermine a sequence of control inputs compensating for the perturbation. In other words, we always come across the CTOS structure-dynamics in practice. There are many possible variants of CTOS structure-dynamics control [3]. In the study we propose the practice of the predetermined modeling where CTOS is a Remote Sensing for Environmental Monitoring.

The modified multiple-model of CTOS is:

$$J_\theta(\vec{x}(t), \vec{u}(t), \vec{\beta}, \vec{\xi}(t), t) \rightarrow \underset{\vec{u}(t) \in \Delta_\theta}{extr}, \tag{1}$$

$$\Delta_\Theta = \left\{ \vec{u}(t) | \vec{x}(t) = \vec{\phi}_\Theta(T_0, \vec{x}(T_0), \vec{x}(t), \vec{u}(t), \vec{\xi}(t), \vec{\beta}_\Theta, t) \right\}, \tag{2}$$

$$\vec{y}(t) = \vec{\psi}_\Theta(\vec{x}(t), \vec{u}(t), \overleftarrow{\xi}(t), \vec{\beta}_\Theta, t), \tag{3}$$

$$\vec{x}(T_0) \in X_0(\vec{\beta}_\theta), \vec{x}(T_f) \in X_f(\vec{\beta}_\theta), \tag{4}$$

$$\vec{u}(t) = ||\vec{u}_{pl}^T(t), \vec{v}^T(\vec{x}(t), t)||;$$

$$\vec{u}_{pl}(t) \in Q_{\theta}(\vec{x}(t), t);$$
$$\vec{v}(\vec{x}(t), t) \in V_{\Theta}(\vec{x}(t), t);$$
$$\vec{\xi}(t) \in \Xi_{\Theta}(x(t), t); \vec{\beta}_{\Theta} \in B; \vec{x}(t) \in X(\vec{\xi}(t), t);$$

$$\vec{\beta}_{\Theta} = ||\vec{\beta}_0^T \vec{w}^T||^T; \vec{w} = ||\vec{w}^{(1)T}, \vec{w}^{(2)T}, \vec{w}^{(3),T}||^T \tag{5}$$

where $\vec{x}(t)$ is a general state vector of the system, $\vec{y}(t)$ is a general vector of output characteristics. Then, $\vec{u}(t)$ and $\vec{v}(\vec{x}(t), t)$ are control vectors, $\vec{u}(t)$ represents CTOS control programs (plans of CTOS functioning), $\vec{v}(\vec{x}(t), t)$ is a vector of control inputs compensating for perturbation impacts $\vec{\xi}(t)$), $\vec{\beta}_{\Theta}$ is a general vector of CTOS parameters. According to [4], these parameters can be divided into the following groups [5]:

- $\vec{w}^{(1)}$ is a vector of parameters being adjusted through the internal adapter. This vector consists of two sub vectors. The first one, $\vec{w}^{(1,n)}$, belongs to the scheduling model, and the second one, $\vec{w}^{(1,p)}$, belongs to the model of control at the phase of plan execution;
- $\vec{w}^{(2)}$ is a vector of parameters being adjusted through the external adapter. This vector consists of the subvector $\vec{w}^{(2,n)}$ belonging to the scheduling model and the subvector $\vec{w}^{(u)}$ including parameters of the simulation model for CTOS functioning under perturbation impacts. In its turn, $\vec{w}^{(u)} = ||\vec{w}^{(2,o)T}, \vec{w}^{(2,b)T}, \vec{w}^{(2,p)T}, ||$, where $\vec{w}^{(2,o)}$ is a vector of parameters characterizing objects in service; $\vec{w}^{(2,b)}$ is a vector of parameters, characterizing the environment; $\vec{w}^{(2,p)}$ belongs to the model of control at the phase of plan execution;
- $\vec{w}^{(3)}$ is a vector of parameters being adjusted within structural adaptation of CTOS SDC models.

The vector of CTOS effectiveness measures is described as:

$$\vec{J}_{\Theta}(\vec{x}(t), \vec{u}(t), \vec{\beta}, \vec{\xi}(t), t) = ||\vec{J}^{(g)T}, \vec{J}^{(0)T}, \vec{J}^{(k)T}, \vec{J}^{(p)T}, \vec{J}^{(n)T}, \vec{J}^{(e)T}, \vec{J}^{(c)T}, \vec{J}^{(v)T}|| \tag{6}$$

Its components are state control effectiveness for motion, interaction operations, channels, resources, flows, operation parameters, structures, and auxiliary operations [4, 6]. The indices "g", "o", "k", "p", "n", "e", "c", "n" correspond to the following models, respectively: models of order progress control ($M_{\langle g,Q \rangle}$); models of operations control ($M_{\langle o,Q \rangle}$); models of technological chains control ($M_{\langle k,Q \rangle}$); models of resources control ($M_{\langle p,Q \rangle}$); models of flows control ($M_{\langle n,Q \rangle}$); models of operations parameters control ($M_{\langle e,Q \rangle}$); models of structures control ($M_{\langle c,Q \rangle}$); models of auxiliary operations control ($M_{\langle n,Q \rangle}$). The transition function $\vec{\phi}_{\Theta}(T_0, \vec{x}(T_0), \vec{x}(t), \vec{u}(t), \vec{\xi}(t), \vec{\beta}_{\Theta}, t)$ and the output function $\vec{\psi}_{\Theta}(\vec{x}(t), \vec{u}(t), \vec{\xi}(t), \vec{\beta}_{\Theta}, t)$ can be defined in analytical or algorithmic form within the proposed simulation system. $Q_{\Theta}(\vec{x}(t), t), V_{\Theta}(\vec{x}(t), t), \Xi_{\Theta}(\vec{x}(t), t)$ are allowable areas for program control, real-time regulation control inputs, perturbation inputs, correspondingly. Expression (4) determines the end conditions for the CTOS state vector $\vec{x}(t)$ at time $t = T_0$ and $t = T_f$ (T_0 is the initial time of the time interval in which the CTOS is being investigated, and T_f is the final time of the interval). The modified multiple-model of CTOS is used in our study.

3 Methods

It is of special importance to provide the required adequacy of the results and control the quality of models. It is obvious that using the model (or multiple-models) $Ob^m_{\langle\rangle}$ in practical studies, we should evaluate adequacy each time. The reasons for inadequacy may be inexact source prerequisites in determining the type and structure of the models, measurement errors in testing and computational errors in processing of sensor data [7]. The use of inadequate models can result in considerable economic loss, emergency situations, and failure to execute tasks posed for a real system.

Based on [7], we consider two classes of modeled systems. By the *first class* we refer to those systems with which it is possible to conduct experiments and obtain the values of some characteristics by measuring. We refer to the *second class* of modeled systems, for which it is impossible to conduct experiments (by the technique presented in Fig. 1) and receive the required characteristics. Large-scale economic and social systems and complex technical systems with function under essential uncertainty of the effect of the external environment are examples of these systems. The human factor plays an important role in these systems (organization structures).

In this figure, we take the following notation: (1) forming the goals of functioning of $Ob^{op}_{\langle\rangle}$; (2) determination of input actions; (3) setting goals of modeling; (4) modeled system (objects $Ob^{op}_{\langle\rangle}$) of the first class; (5) model ($Ob^m_{\langle\theta\rangle}$) of the investigated system $Ob^{op}_{\langle\rangle}$; (6) estimation of the quality of a model (poly-model

Fig. 1 The generalized technique of estimation and control of the quality of models

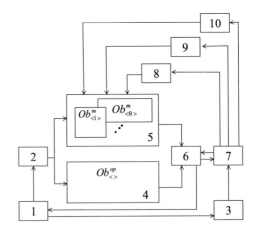

system); (7) controlling the quality of models; (8) controlling the parameters of models; (9) controlling the structures of models; and (10) changing the concept of model description.

All complex technical–organizational systems including complex objects remote sensing processes and systems working in an autonomous mode are examples of systems of the first class. A possible implementation of the proposed technique of estimation and control of models quality for an applied area which is connected with Earth Remote Sensing is demonstrated below.

The ecological monitoring of the environmental conditions is one of the primary ranges of the space imagery application. The monitoring of the pollutions of the three main natural environments (water, land and atmosphere) is provided at local, regional and global levels. The quality of the adaptations of the management decisions is increased as a result of the remote sensing. These adaptations are directed to maintain the ecological safety of the research site and optimize the elimination of disturbances. The executor of the project aims to obtain the required information with high quality and minimum cost. The effectiveness of the project depends on the source data quality (remote sensing data), methodical approach, software and the presentation of the project results. The integrated modeling of the basic technological processes of the remote sensing is performed to synthesize the technical requirements to the hardware-software.

With regard to technical characteristics of the space monitoring and facts influencing these characteristics, it is necessary to choose the capacity of the source data by means of one or some space vehicles and/or the airborne equipment complex, to choose optimal conditions for the survey based on the seasonal and daily variability of the reflectance, radiative characteristics of the landscapes and the mode of equipment operation, to organize the thematic processing of the remote sensing data and the ground measurements using hardware-software, and to present the results of the project in a user-friendly form permitting to make managerial decisions promptly and reasonably (Fig. 2).

The modeling and simulation of the technological processes have some uncertainty and limitations that influence the quality of the optimization task (the vectors $\vec{\beta}_\theta$, $\vec{u}_{pl}(t)$).

The CTOS has several sources of information, such as space imagery in the visible, infrared and super high frequency spectral bands, airborne imagery, unmanned aerial vehicle (UAV) imagery and test ground measurements for verification of the results.

The execution of the space ecological monitoring consists of the next stages.

Firstly, it is the planning and preparation of the tasks (the vectors $\vec{w}^{(1,n)}$ and $\vec{u}_{pl}(t)$). This stage includes the choice of the objects, the list of controllable parameters determining and scheduling of the survey.

Secondly, it is the data acquisition. The stage includes the process of survey and ground-based measurement (the vector $\vec{w}^{(1,p)}$).

The third stage consists of data processing and presentation of the results. Remote sensing data processing and ground measurement, creation of the thematic

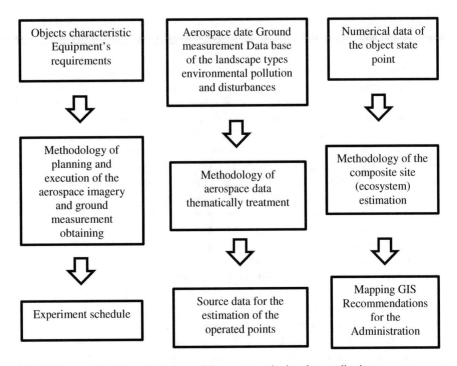

Fig. 2 Framework of the methodology of the space monitoring data application

layer of the digital map, forming the forecast models, calculation of assessments and recommendations are executed at this stage (the vector $\vec{w}^{(u)}$).

The most convenient form for the presentation of the results is the thematic layers of the digital map with the attributive information, database and photo scheme of images.

Moreover, it is possible to estimate the system functioning quality and the choice of the optimal monitoring conditions for the requested image quality. The prediction is based on the optical system characteristics and monitoring conditions. Spatial resolution of the image forms the main predictive parameter and determines an object-background contrast value. The movements of equipment, the Sun height, irradiance of the object, albedo of the site, physical specifications of the atmosphere are taken into account. The modeling and simulation of the particular elements of the space monitoring system and expert evaluations of the system functioning determine the parameters for monitoring system. Future research shall be focused on the development of the generalizing model of the space ecological monitoring for practice issues.

The special software is developed for the automated data processing. The software uses the data from the original library of the spectral landscape features, which are typical for the climatic conditions of the different regions. The technology for the software adaptation is presented in Figs. 1 and 2. The technology allows to expand the number of tasks and to define more forest parameters (Fig. 3).

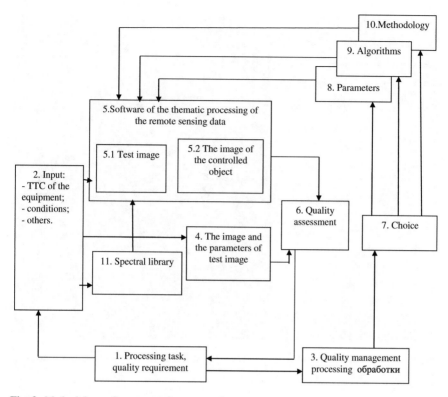

Fig. 3 Methodology of aerospace data processing

The timely adaptation at the parameters level, algorithms and strategies levels is organized for the solution of the thematic tasks. Proposed technology contains the list of new operations and provides the periodic analysis of the quality of data processing in modules of special application.

The quality of software operations is analyzed based on image processing results and field data for the test forest site. This provides the possibility of software adaptation to the specific conditions. In addition, the structural scheme of the model is supplemented by a new original block (7) ensuring the choosing of the parameters for the algorithms or strategies of aerospace data processing.

4 Results

A digital map of ecological zoning nature reserve was created (Fig. 4): zones of protected water (red), vegetation stress (light green), watercourses (blue). The main causes of anthropogenic load are displayed: trash, tree felling, violations of hydrology.

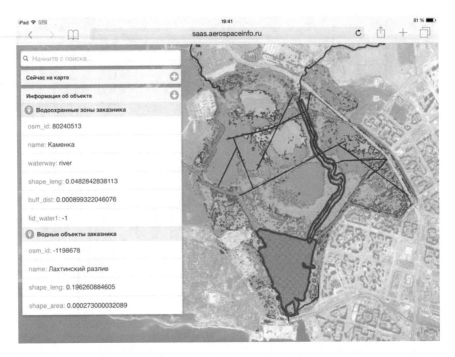

Fig. 4 Fragment of the map of ecological zoning of nature reserve Untolovo (Saint Petersburg, Russia)

The following recommendations have been suggested: equipment for culvert installations, cleaning of drainage ditches, elimination of landfills and further development of the territory, adherence to buffer zones.

Dynamics of the vegetation condition between 2010 and 2015 in percents is represented in Fig. 5.

The total area of the forests and swamps has remained virtually unchanged. The area of deciduous trees has decreased. It can be explained by the natural succession, anthropogenic pressure or changes in hydrogeological conditions.

Complex assessments of nature reserve ecosystems showed that negative environmental changes have increased because of natural and anthropogenic factors. Complex environmental assessment was characterized as satisfactory in 2010. In 2015, the complex environmental assessment was also estimated as satisfactory but close to tense due to the active construction and expansion of the area of anthropogenic load. We can conclude that the environmental situation was tense in 2015.

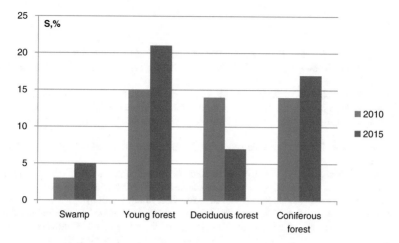

Fig. 5 Changes in vegetation types

5 Discussions

The human ingenuity has developed a new range of sophisticated and powerful techniques for solving environmental problems, such as pollution monitoring, restoration ecology, landscape planning, risk management, and impact assessment. The proposed method is based on an integrated model of remote sensing for environmental monitoring. The method allows carrying out a comprehensive environmental assessment of natural reserves. CTOS complex modeling provides an efficient use of source data, processing algorithms and technical tools. The method of environmental assessment can be adapted to and used in areas of approximately 100 km^2 located under strong human impact. The model can be adapted using various parameters and the accuracy for determination of input parameters.

Further research will focus on an assessment of object-indicators such as soil cover and wetlands, in addition to forest cover, using remote sensing data. This method of complex environmental assessment can be improved by the specific remote sensing data, algorithms for data processing, extension of spectral libraries of landscape elements and by adding units performing the following functions:

- modeling and analysis of dynamics of changing sustainability of the territory under anthropogenic pressure;
- modeling of system reaction to possible anthropogenic pressure;
- modeling of dynamics in environmental change.

All processes shown above will be controlled based on satellite data processing.

References

1. Ohtilev, M.Yu., Sokolov, B.V., Yusupov, R.M.: Intellectual Technologies for Monitoring and Control of Structure-Dynamics of Complex Technical Objects, 410 p. Nauka, Moscow (2006) (in Russian)
2. Ivanov, D., Sokolov, B.: Adaptive Supply Chain Management. Springer, London (2010)
3. Ivanov, D., Sokolov, B., Kaeschel, J.: A multi-structural framework for adaptive supply chain planning and operations with structure dynamics considerations. Eur. J. Oper. Res. **200**(2), 409–420 (2010)
4. Ivanov, D., Sokolov, B.: Dynamic supply chain scheduling. J. Sched. **15**(2) (2012) (Elsevier, London)
5. Ivanov, D., Sokolov, B.: Control and system-theoretic identification of the supply chain dynamics domain for planning, analysis and adaptation of performance under uncertainty. Eur. J. Oper. Res. vol. 224, Issue 2, pp. 313–323. Elsevier, London (2012)
6. Sokolov, B., Zelentsov, V., Yusupov, R., Merkuryev, Y.: Information fusion multiple-models quality definition and estimation. In: Proceedings of the International Confrence on Harbor Maritime and Multimodal Logistics M&S, Vienna, Austria, September, 19–2l, pp. 102–111 (2012)
7. Skurihin, V.I., Zabrodsky, V.A., Kopeychenko, Yu.V.: Adaptive control systems in machine-building industry. M.: Mashinostroenie (1989) (in Russian)

Author Index

© Springer International Publishing Switzerland 2016
R. Silhavy et al. (eds.), *Automation Control Theory Perspectives*
in Intelligent Systems, Advances in Intelligent Systems and Computing 466,
DOI 10.1007/978-3-319-33389-2

Printed in the United States
by Baker & Taylor Publisher Services